MOLECULAR FORCES AND SELF ASSEMBLY

Challenging the cherished notions of colloidal theory, Barry Ninham and Pierandrea Lo Nostro confront the scientific lore of molecular forces and colloidal science in an incisive and thought-provoking manner. The authors explain the development of these classical theories, discussing, amongst other topics, electrostatic forces in electrolytes, specific ion effects and hydrophobic interactions. Throughout the book, they question assumptions, unearth flaws and present new results and ideas. From such analysis, a qualitative and predictive framework for the field emerges; the impact of this is discussed in the latter half of the book, through force behaviour in self assembly. Here, numerous diverse phenomena are explained, from surfactants to biological applications, all richly illustrated with pertinent, intellectually stimulating examples. With mathematics kept to a minimum, historical facts and anecdotes woven through the text, this is a highly engaging and readable treatment for students and researchers in science and engineering.

BARRY W. NINHAM, a pioneer of modern theory describing molecular forces, interactions and self assembly, is currently Professor Emeritus of the Department of Applied Mathematics at the Australian National University (ANU). He has been an active researcher for over 40 years, over which time he has authored or co-authored seven books and more than 400 technical papers. He has received numerous awards, including the Ostwald Award of the German Chemical Society (2005), the Swedish Erlander National Chair of Chemistry (1998), and, in 2008, ANU created the Barry Ninham Chair of Natural Sciences Award to recognize his contributions.

PIERANDREA LO NOSTRO is a Research Fellow and Teacher in the Department of Chemistry at the University of Florence, from where he received his Ph.D. in Chemical Sciences in 1992. His current research interests include macromolecular self assembly, self assembly of biocompatible surfactants specific ion effects (Hofmeister series) and nanomaterials.

Cambridge Molecular Science

As we move further into the twenty-first century, chemistry is positioning itself as the central science. Its subject matter, atoms and the bonds between them, is now central to so many of the life sciences on the one hand, as biological chemistry brings the subject to the atomic level, and to condensed matter and molecular physics on the other. Developments in quantum chemistry and in statistical mechanics have also created a fruitful overlap with mathematics and theoretical physics. Consequently, boundaries between chemistry and other traditional sciences are fading and the term *Molecular Science* now describes this vibrant area of research.

Molecular science has made giant strides in recent years. Bolstered by both instrumental and theoretical developments, it covers the temporal scale down to femtoseconds, a time scale sufficient to define atomic dynamics with precision, and the spatial scale down to a small fraction of an Angstrom. This has led to a very sophisticated level of understanding of the properties of small molecule systems, but there has also been a remarkable series of developments in more complex systems. These include: protein engineering; surfaces and interfaces; polymers; colloids; and biophysical chemistry. This series provides a vehicle for the publication of advanced textbooks and monographs introducing and reviewing these exciting developments.

MOLECULAR FORCES AND SELF ASSEMBLY

In Colloid, Nano Sciences and Biology

BARRY W. NINHAM
Australian National University

and

PIERANDREA LO NOSTRO
University of Florence

CAMBRIDGE
UNIVERSITY PRESS

CAMBRIDGE UNIVERSITY PRESS
Cambridge, New York, Melbourne, Madrid, Cape Town, Singapore,
São Paulo, Delhi, Dubai, Tokyo

Cambridge University Press
The Edinburgh Building, Cambridge CB2 8RU, UK

Published in the United States of America by Cambridge University Press, New York

www.cambridge.org
Information on this title: www.cambridge.org/9780521896009

First published 2010

Printed in the United Kingdom at the University Press, Cambridge

A catalogue record for this publication is available from the British Library

ISBN 978-0-521-89600-9 Hardback

Contents

Preface

This book outlines developments in physical and colloid chemistry over the last decade or two that have changed our understanding of molecular forces and the self assembly of amphiphilic molecules. Within the sciences, a subject or discipline is defined by a freemasonry, the members of which are united by a common lore. The received beliefs of adepts are reaffirmed by recitation and repetition of the lore, and reinforced by the weight of authority embodied in the literature of the discipline. The more venerable the literature, the more conservative is the freemasonry. Challengers to the canon are at first dismissed as heretics or apostates, but if, later, their claims are conceded to have validity, they and their theories are eventually accepted into the fold. The discipline moves on. So, the advocates of a once revolutionary Darwinism, wrongly defined by others by the trite aphorism 'survival of the fittest', retreated into convoluted defences of the dogma, represented by the elegant writings of Stephen J. Gould. Two decades ago the notion that environmental influences such as temperature could affect gene expression invited the ridicule attending to an earlier Lamarkism. Similarly for any questioning of the dogma of the Weissman barrier in immunology. Not any more.

The origins of the discipline of modern physical chemistry can probably be dated most conveniently to Napoleon's scientific expedition to Egypt 200 years ago. Berthelot, one of the expedition's scientists, observed on the receding flood plain of the Nile rocks that were covered with soda lime, sodium carbonate. This was a mystery. Sodium chloride in the Nile waters ought to have stayed in solution and it was calcium carbonate that should have precipitated out. But at the high temperatures of the Egyptian summer to which the rocks were exposed and with desiccation the reaction is reversed (with catastrophic consequences to the Sphinx). So began the notion of temperature (and 'water structure') as controlling variables. Physical and colloid and surface chemistry evolved over the intervening 200 years into a central, enabling discipline of modern science and engineering. Its theories provide an intuition, and its experimental techniques underlie and provide unquestioned support for the work of many other disciplines, from biological to chemical engineering, agricultural and earth sciences.

The interpretation of a measurement depends on a theory. But if the theory is flawed or wrong, so also is the meaning of the measurement. The deficiencies propagate. For a long time adepts of physical chemistry held to a world view that became rigid, dated and lacked predictability. The realization that our theories and intuition were seriously flawed and often misleading occurred gradually over the last 40 years, but the deficiencies became increasingly obvious over the last decade. In that decade a flurry of activity has uncovered defects in the foundations. We have gone some way towards remedying them. A new intuition has been built and is evolving further. From a practical viewpoint, how to exploit this new intuition and put forces to work, the ramifications are large. To say that the developments described in this book are radical is to put the matter too mildly. The background is described in Chapter 1.

We have avoided mathematics as far as possible: on the grounds that those who do not speak the language of mathematics and physics would be none the wiser for its inclusion. For readers who do speak that language it would be superfluous, as the technicalities can easily be tracked down via the literature cited.

There are many good books available that give accounts of overlapping areas in what we might now be entitled to call the 'old' theories of physical chemistry and colloid science. Of the more recent books, those of J. N. Israelachvili [1], D. F. Evans and H. Wennerström [2], V. A. Parsegian [3], R. J. Hunter (*Introduction to Modern Colloid Science*) [4] and O. G. Mouritssen (*Life – As a Matter of Fat*) [5] on lipids are just a few. Some old books, such as Robinson and Stokes [6], Harned and Owen [7] and Friberg et al. [8], still stand. A book by one of us with J. Mahanty [9] developed all the necessary formalism to extend Lifshitz theory and its applications in 1976. Another, by S. T. Hyde et al. [10], is vast in scope and reasonably up to date, on self assembly.

A difficulty is that most of the content of such very good books that inform the discipline give accounts of our knowledge of molecular forces and self assembly that were unquestioned up to 10 years ago. But those theories are now out of date and can be quite misleading. The idea that venerable fundamentals such as pH, pK_as, buffers, ion binding, membrane potentials, ion transport and the classical DLVO theory of colloid stability might have to be revisited, or even scrapped, would have seemed absurd then. Events have moved fast and the long-standing picture of the adepts that included ourselves is certainly no longer valid.

We acknowledge Drew F. Parsons (Australian National University, Canberra) for his invaluable contribution to Chapter 7, where the effects due to dispersion forces are discussed in detail and quantified.

Acknowledgements (B. W. N.)

I owe a debt to R. B. Dingle, who pioneered modern asymptotics, taught me physics and analysis and sent me from the other side of the world from Western Australia to America 50 years ago to study for a PhD in statistical mechanics with Elliott W. Montroll. The scientific world then was a smaller place. Everyone knew everyone, and they knew the genealogy and the history of ideas. It was Elliott who encouraged me to develop this field. Ten years on my friend George H. Weiss allowed me the opportunity of a long sabbatical leave at the National Institutes of Health in Washington. V. Adrian Parsegian and I began there a seminal, tremendously exciting and lasting collaboration on molecular forces in colloid science and biology.

A result of that work was an invitation from my subsequent boss, Sir Ernest Titterton, who pressed the button on the first nuclear bomb explosion, to set up a new Department of 'Applied Mathematics' in the Institute of Advanced Studies at the Australian National University in Canberra. His own mentor, Sir Mark Oliphant, had worked with Rutherford, discovered tritium, was intimately involved with the bomb which he later bitterly regretted and with radar; and was involved in founding the Pugwash conferences. Mark was a founding father of the ANU. When he was compulsorily retired at the grand old age of 65, I gave him haven in the new Department. In return he sometimes gave me advice.

We were not allowed to teach undergraduates or to apply for research grants. 'Applied Mathematics' was defined by me to mean colloid and surface chemistry, and optical sciences. My responsibilities were to do exactly what I pleased. Colloid science was then a poor cousin of physical chemistry; now it struts the stage under a new name, nanotechnology. (Its attractiveness to ANU was in part because of the in-house invention of Synroc for nuclear waste disposal. Success depends on compaction from solution of the colloidal nuclear waste ceramic, a matter of specific ion effects that figure largely in this book.)

From its foundation in 1969, the Department grew and grew. I do not know how because we had no money, but somehow we acquired enough. Oliphant said it was better that way and he was right. The Department became an eclectic international interdisciplinary centre of experimentalists and theoreticians, pure and applied researchers, physicists, chemists, mathematicians, biologists and chemical engineers that still continues after 40 years as successfully as ever. It was a different world and I owe more than I can say to ANU.

Over a period of 30 years or so I was privileged to recruit, to supervise, to mentor, to direct and work with successive waves of astonishingly talented PhD students and young scientists from Australia and many other countries. They came because we had time to think and do science unencumbered. At least 60 of them became senior professors in various countries and disciplines.

I want to express my thanks for seminal interactions with those who introduced me to new fields and taught me things I did not know, especially Stephen Hyde, Kåre Larsson and Sten Andersson, D. John Mitchell and Jacob Israelachvili. I am indebted to them all, as I am to several hundred PhD students, research colleagues and visitors. I hope they will forgive me if I do not list them all.

Besides the ANU, and the NIH, various universities and research institutes gave me respite to work for long periods as a visiting professor without administration and surrounded by students.

In the USA: U. Minnesota; in France: CEA Saclay, the College de France; U. Paris V1; in Sweden: U. Lund, YKI, KTH, Chalmers, and Malmo; in Italy: U. Florence and Cagliari; in Germany: U. Regensberg. My hosts and collaborators who made that luxury possible were: D. Fennell Evans; Tom Zemb and Marie- Paule Pileni; Sten Andersson, Bjorn Lindman, Kåre Larsson and Hakan Wennerstrom, Krister Holmberg and Per Cleasson, Roland Kjellander, Bengt Nordén and Zoltan Blum; Werner Kunz and Maura Monduzzi.

Most importantly Piero Baglioni at the University of Florence gave the two of us the opportunity to work together over the last five years in Florence and we are in his debt.

This book owes much to two earlier books [9,10].

The first contains theory that we shall use, and is mathematically abstruse. Some important parts of it that are relevant are unavailable elsewhere. Unfortunately the book is long out of print and mathematically difficult, so that work still valid 30 years ago is rediscovered in every decade.

The second book sweeps over enormous territory, but the price is out of reach to all but the richest libraries.

Work on complex self assembly and structure in biology and materials sciences and its systemization is moving fast, and we refer the reader to the websites of Sten Andersson and of Stephen T. Hyde to obtain a glimpse of astonishing new worlds of science.

Acknowledgements (P. L. N.)

It is fascinating, during a long walk up in the mountains, to stop, turn back and contemplate not only the scenario in front, but also the narrow and sometimes difficult footpath that we have covered. This book offers me now this opportunity.

Unquestionably my professional experience has the imprint of Enzo Ferroni, my true master at the University of Florence, and the first pioneer in Italy in colloid and interface science. He did not simply contaminate me with his curiosity and passion for physical chemistry, to me the most complete, embracing and appealing discipline among the chemical sciences. His words, proximity and example were the seeds that found a fertile humus in my mind. Briefly, he introduced me to what Maxwell named as 'crossed fertilization' between different aspects (disciplines) of the real world. This attitude has become for me a sort of polar star. And this is why I accepted the challenge to write this book with Barry W. Ninham.

During my PhD I got the chance to work with Sow-Hsin Chen and others at MIT, where my studies on lecithins and fluorinated alkanes began. And since that time I have been focusing my interests on the role of the solvent in self assemblies.

Then the American experience continued at the University of South Florida, where Jack E. Fernandez introduced me to vitamin C amphiphilic derivatives and to their properties.

After my return to Florence, I continued working at the Department of Chemistry with my friend Piero Baglioni. In the following years our group has been growing with excellent younger researchers. Their scientific skills and close friendship support and stimulate my current activity.

My friendship and collaboration with Barry W. Ninham started in 1997. Again, his vibrant passion for unsolved or formerly-solved-but-not-really-so-now issues prompted me to approach the mysterious Hofmeister phenomena. This almost unknown topic – although clearly ubiquitous – goes straight to the core of chemistry, as it reflects the fact that matter is not simply made up of differently coloured ping-pong balls and springs, but behaves in 'specific' ways that can now be understood – at least partially – in terms of intermolecular forces. And that water is not simply a homogeneous and molecularly smooth environment where soluble entities diffuse, but it directly participates at the molecular level in physicochemical phenomena. I learnt from Barry W. Ninham an honest and youthful energy in trying to understand without accepting apparently consolidated, ready-to-use (*ipse dixit*) recipes, not even when they are named after eminent scientists. Any theory is a partial look at a small piece of the world, waiting to be corrected, expanded and fully understood. After all, real science is a well-equipped gym of humility. I recognize that this oxymoron – i.e. accepting, checking and sometimes rejecting the conceptual milestones that our forerunners bequeathed us – is vital to scientists, and opens wide the gate of experimental sciences to the young researchers, giving a glimpse of a much larger horizon.

In the end, this was our wish and we hope that our task will promote debate and curiosity in others.

Front cover illustration

Fresco by Fra' Beato Angelico, fifteenth century, Museum of San Marco, Firenze. The two images on the front cover (courtesy of Piero Baglioni) show the original, damaged by the Arno River floods of 4 November 1966, and the restoration by

Dino Dini. The non-invasive techniques of restoration developed by Professor Enzo Ferroni, at the University of Florence, are a classic exemplar of nanotechnology in action. They have been developed further for preservation of ancient manuscripts. Similar techniques were used by the Mayan civilization. The methods use molecular forces, self-assembled microemulsions, nanoparticle synthesis, water structure and specific ion effects. See P. Baglioni and co-workers, Ref. 11.

31 March 2009, Firenze and Canberra.

References

[1] J. N. Israelachvili, *Intermolecular and Surface Forces: With Applications to Colloidal and Biological Systems*. London: Academic Press (1985–2004).
[2] D. F. Evans and H. Wennerström, *The Colloidal Domain: where physics, chemistry, biology, and technology meet*. New York: VCH (1994).
[3] V. A. Parsegian, *Van Der Waals Forces: a handbook for biologists, chemists, engineers, and physicists*. Cambridge: Cambridge University Press (2006).
[4] R. J. Hunter, *Introduction to Modern Colloid Science*. Oxford: Oxford University Press (1993).
[5] O. G. Mouritsen, *Life – As a Matter of Fat*. Berlin: Springer-Verlag (2005).
[6] R. A. Robinson and R. H. Stokes, *Electrolyte Solutions*. London: Butterworths (1959).
[7] H. S. Harned and B. B. Owen, *The Physical Chemistry of Electrolytic Solutions*. New York: Reinhold (1958).
[8] S. Friberg, K. Larsson and J. Sjoblom, *Food Emulsions*. 4th edn. New York: Marcel Dekker (2004).
[9] J. Mahanty and B. W. Ninham, *Dispersion Forces*. London: Academic Press (1976).
[10] S. T. Hyde, S. Andersson, K. Larson, Z. Blum, T. Landh, S. Lidin and B. W. Ninham, *The Language of Shape. The role of curvature in condensed matter physics chemistry and biology*. Amsterdam: Elsevier (1997).
[11] M. Ambrosi, L. Dei, R. Giorgi, C. Neto and P. Baglioni, *Langmuir* **17** (2001), 4251–4255.

Part I

Molecular forces

1
Reasons for the enquiry

1.1 Molecular forces: some of the background and history of ideas. Why molecular forces?

The matter that concerns us was most clearly articulated nearly a century ago by D'Arcy Thompson in his famous book [1]. He reported the pleas of the early founders of the cell theory, of the then biology, and of the physiologists, that chemists should address the question of molecular forces, then unknown.

We would like to know how it is that molecular forces and the laws of statistical mechanics conspire, with the geometry of molecules and the conformations available to macromolecules, to give rise to the hierarchies of self-assembled equilibrium or dynamic steady states of matter that form cells and dictate biochemical reactivity.

In other words, the game is to link structure and function, the geometry of assemblies of molecules, to the forces that drive self assembly and recognition processes. Any insights ought to allow us to build better, useful connections between the physical and biological sciences. Despite tantalizing hints, that main aim has remained elusive.

D'Arcy Thompson tells us too that of the chemistry of his day and age, Kant said that 'it was wisschenschaft, nicht Wissenschaft; in that the criterion of a true science lay in its reliance on mathematics'. Kant believed that Euclid's geometry was self-evidently that of nature. We now know better. Hyperbolic geometries that we shall come to later better describe the bicontinuous, random honeycombs of nature. In apposition to Kant's position we have it on the authority of Auguste Comte, the founder of the social sciences, that: 'If mathematics should ever hold a prominent place in chemistry, an aberration happily almost impossible, it would occasion a widespread and rapid degeneration of that science'. The quotation is out of context and probably not fair to Comte. But these two opposing views do reflect the ambiguous position of chemistry in science. One view has it that chemistry has to rely on physics and mathematics. The other is that this is a too simplistic, even arrogant expectation.

The word chemistry derives from Al-Khemie, itself supposed to come from a Greek word meaning the black land (renewal of silt from the Nile floods) or Egypt, from which all wisdom came. Some of the magic, mystery and art associated with medieval alchemists still attends some branches of chemistry, the science of molecules, in accordance with Comte's view. But physics, natural philosophy, explores the properties of matter and energy, or the action of different forms of energy on matter generally, and was missing. In the endeavour, and par excellence for physical chemistry, forces and therefore mathematics have to figure centrally.

Newton himself had tried to measure the molecular forces acting between surfaces and reported in Art. 31 of the *Principia* that he failed, because 'surface combinations were owing'. The problems of surface and colloid chemistry were already writ large.

Thomas Young deduced the range of the force between atoms and molecules, this before atoms were agreed to exist. That these forces had to be short-range, unlike gravitational forces, was also known to Newton.

The Reverend Challis of Trinity College, Cambridge, reviewed, in the 1836 meeting of the British Association for Advancement of Science, the then state of play of a dispute between Laplace and Poisson on what we might now call hydration forces at and between surfaces. George Peacock, a professor of mathematics at Cambridge and Young's biographer, furiously reports that Laplace stole Thomas Young's earlier work, on contact angles and surface tension, without attribution. To make it worse, while Young, a writer of great clarity, went to considerable trouble to avoid mathematics, Laplace added an appendix to his Celeste Mechanique on the topic, dressing it up in fancy mathematics. (And incidentally, Laplace got it wrong as he forgot about contact angles formed by a drop of liquid on a solid interface.) Poisson lost out because of a mistake by a factor of two, and to too scrupulous adherence to Ockam's razor by his opponents.

This matter was redressed later in a marvellous article long forgotten on the theory of capillary action by J. Clerk Maxwell in the 1876 edition of the *Encyclopaedia Britannica*, updated by Lord Rayleigh in the eleventh edition. Poisson was right. This apparently absurd dispute between Poisson and Laplace is a conceptual matter, of confusing and central importance. The nature of an interface is with us still, indeed endemic in chemistry. In its simplest form it boils down to this: imagine an ideal, i.e., molecularly smooth surface, like mica. Then, can a liquid that adjoins it be considered to have its bulk density (or with complicated molecules such as water, random molecular orientation) up to an infinitesimal, molecular, distance from the surface (the Laplace approximation)? Or, under the influence of forces between the various constituent molecules, do the liquid molecules exhibit a

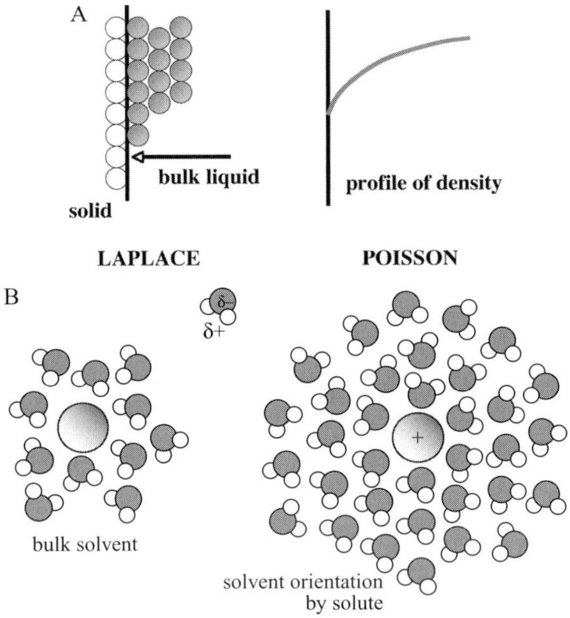

Fig. 1.1. A: Poisson vs. Laplace. B: schematic representation of hydration, or surface induced liquid structure. Around an interacting molecule (such as an ion) the solvent has to be perturbed by the solute molecules.

surface-induced profile of order? That is, does the liquid density gradually change from the bulk value as we approach the surface (the Poisson view)?

Alternatively, and another statement of the problem: consider two inert solute atoms, molecules or ions dissolved in a liquid, approaching each other under the influence of the same forces. In our modelling, can we consider them as hard spheres interacting across the bulk continuum liquid (Fig. 1.1) (the so-called primitive model)? Or must we recognize that the structure of the liquid that surrounds the interacting molecules is itself perturbed by the molecules (the civilized model)? If the liquid is water, we term the result of these indirect forces hydration forces. Of course hydration and hydration forces exist. One has simply to put different salt crystals in water and measure the change in volume of the solution to see bulk manifestation of hydration effects.

Challis invented a new term, mathematical physics, for what we now call colloid and surface chemistry. He suggested first that measurement of molecular forces might best be accomplished by using the then newly possible interferometric techniques.

But it had to wait 150 years for the experiments of Tabor, Winterton and Israelachvili that first measured the molecular forces between surfaces to confirm the quantum mechanical theories that had emerged in the interim [2,3].

It is salutary to reflect on such matters. Because the fact remains that despite all the advances in experimental techniques that are used by biologists, organic chemists and biochemists, all the progress that has been made in the physical chemistry of liquids over the last 50 years since the molecular biology revolution has contributed very little at a conceptual level to the progress of modern molecular biology.

This is genuinely puzzling to physicists and physical chemists who believe, like D'Arcy Thompson, in reductionism. That is, there ought to be useful connections, at least intersections, between the physical and biological sciences beyond the chanting of 'non-equilibrium thermodynamics' or 'chaos' in a mantra that becomes dreary.

Is this so, and why is it so? The problem, this lack of confidence in science, is not peculiar to physical chemistry. It is embodied in the 'science is dead' bleat of the 1990s.

It is embodied in Morris Klein's marvellous book, *Mathematics, the Loss of Certainty*. The mathematicians lost their faith after Godel's theorem proved there was no such thing as absolute proof. There is now not one but many mathematics, little of which is so integrally associated with science as it used to be.

The physicists, with the triumphs of nuclear physics and quantum mechanics and later the solid state, had no doubts about the future and their central role therein. Schrödinger enigmatically speculated somewhat in *What is Life*, and in *Science and Determinism*. Delbruck, a physicist, confronted biology directly with his work on bacteriophages and enthused a generation. Excepting a few hardy biophysicists, the physicists and physical chemists had given up on biology. Some retreated into mumbling on the mystiques attending non-linearity as the source of the New Jerusalem.

1.2 Liquids and computer simulation

A new generation seeks salvation and insights via computer simulations, of liquids, interfaces, membranes and proteins. Perhaps that is as it should be. But nagging doubts remain. The simulation of the structure of a protein in agreement with its X-ray crystal form is a technical triumph. The determination takes up to 30 000 individual effective molecular force parameters, parameters that depend on temperature. Something, apart from water, seems to be missing in this kind of modelling. The structure tells us little of function.

The attractiveness of simulation, a new kind of experimental technique, derives in part from the awkward fact that there is no real molecular theory even of simple liquids. This is because, unlike a solid or gas, a liquid has no ideal molecular reference state. For a solid this is a periodic delta function that describes an array of atoms in a crystal. For a gas a statistical description starts with a completely random distribution of its molecules. This central difficulty for the theory of the liquid state of matter remains.

It is worse for water, the stuff of life. Bernal and Langmuir were probably closer to the mark with their views that water was more like a giant dynamic cooperative entity, as for proteins, rather than a collection of individual molecules. Onsager knew and said the problem was the old one of water. That is indeed a problem.

1.3 Interfaces and colloids

But it is not the only difficulty: we can observe that an interface is a physical reality. A lipid membrane and a biological cell surface exist. In a broader context it is an idealization. The context is in consideration of the totality of real states of matter. Indeed, a macroscopic continuum itself is an idealized notion. Homogeneous gases, liquids and solids that we deal with by thermodynamics never occur other than as theoretical constructs of infinite extent. Soils, clouds, granites and living organisms all are real objects, non-uniform at different levels of organization, even fractal on the entire range of scales, from the atomic to the macroscopic domain.

The usual models of gases or solutions or real crystals are limited in this context. Just as the theory of dislocations in solids allows for some disorder in an ideal array of atoms, so too the virial expansion makes a second-order correction to the ideal gas/solution equation of state. As perturbation theories, these help to account for small deviations from ideality (e.g. the van der Waals equation of state or Debye-Hückel theory of electrolytes). But in passing from the limit of infinite dilution, the mathematical difficulties that confront exact description increase enormously. Additional parameters are invoked to accommodate deviations from limiting laws, and associated uncertainties begin to increase. Theory already breaks down at higher concentrations, even when a gas or a solution still remains homogeneous. It fails and becomes invalid long before the assumption of homogeneity becomes violated by liquid–gas nucleation at supersaturation. Nonhomogeneous and supra- and supermolecular systems can hardly ever be explained with standard theories. It is these systems for which a spectrum of 'unusual' mechanical and optical properties occurs. These states of matter are excluded from consideration in ordinary solid-state physics and fluid mechanics.

The innovative terms 'soft condensed matter' or nanomaterials have come into vogue and have subsumed the old-fashioned term 'colloids'. But colloids remain despite a relabelling. They are springy liquids and sticky solids that never conform with one or the other of the three states of matter. They do not conform to the structures imposed by the universal triad: gas, liquid, solid.

Colloid science attempts to handle the challenge. It has as its aim a theory of all these nonconformist states of matter.

It begins with an unravelling of the problem at its most fundamental level – starting with an interface. Here discontinuity of macroscopic physical properties occurs in its most severe form. The homogeneity assumption here breaks down completely. This seemingly leaves no hope that thermodynamic principles might apply. Equilibrium thermodynamic notions make sense only for uniform continua. Nevertheless, exact thermodynamic results derived for bulk phases do carry over to interfaces. This is because surfaces, although extremely non-uniform, are non-uniform only in a single – normal – direction. In the other two lateral dimensions macroscopic averaging can be done.

Gradients of density and of other thermodynamic properties that occur in a direction perpendicular to a surface are enormous. The associated tangential stress at the surface of a liquid remains at the ultimate limit that a condensed phase is able to sustain. This is the essence of the mechanism by which a liquid terminates at its boundary where it breaks into the vapour. Surface tension is the manifestation of this phase collapse effect. (It is the integral of the profile of this excess pressure that acts in the lateral direction that defines surface tension.)

So interfaces can be investigated by thermodynamics. The theory of capillarity is indeed at the same level of physical rigour that applies to bulk vapours, liquids and solutions. By consideration of different interfaces a great variety of non-uniform chemical compositions that occur in practice can be embraced by a unified physical model.

It is for this reason that thermodynamic quantification of interfaces is such an important issue. By experimenting with interfaces and with their interactions we can extend our exact knowledge further, and then the more complicated systems that we call colloidal dispersions come within reach. The thermodynamic state of a disperse system for which one phase is finely distributed in another is fully defined in terms of interfacial and bulk phase properties.

While a colloidal particle may be compositionally identical to the bulk phases from which it forms, a colloidal dispersion is altogether a quite different entity. It is entirely different in many respects from its parent macrophases. So, for example, emulsions formed from a mixture of transparent Newtonian liquids are opaque and sometimes solid-like. Chemically reduced down to its metal form, gold in a colloidal sol, due to Brownian motion of colloidal particles and double layer repulsion between their surfaces, remains 'dissolved' – suspended in water. The theory of such systems is an extension of molecular-kinetic theory into the world of colloidal dimensions. It is not necessary to give formal mathematical descriptions to conclude that all the amazing diversity of flow and colour patterns seen are brought into existence by interfacial gradients of thermodynamic properties. These occur over molecular distance scales.

1.4 Colloids, polymers and living matter

The word colloid is a term originally associated with dairy processing and similar biotechnologies. The very term coagulation is a notion derived from observations on the clotting of blood.

Colloidal states of matter were first taken to be a specific property of polymers. These giant molecules, far in excess of ordinary molecular sizes, were believed to occur in living matter. Many of these components of bioextracts, of which common examples are soaps or gelatine, have been known since ancient times. They were isolated and identified chemically following progress in biochemistry, often as individual compounds, in the eighteenth and nineteenth centuries. They are not always of very high molecular weight. These molecules nevertheless avoid the formation of ordinary crystals and solutions. They prefer rather to persist in their own micellar and liquid crystalline and amorphous states. A jellyfish with up to 98% water is a challenging example.

It was not until the turn of the twentieth century that it was realized that colloid behaviour is not a unique property of biomaterials. It extends into the realm of inorganic chemistry as well. Swollen clays in soils and gold hydrosols are examples.

1.5 Conceptual locks

There was much progress in colloid and polymer science in the second half of the twentieth century. Rather there appeared to be progress. We shall come to this in detail later.

But the nagging problem of intersections, communication between the physical and biological sciences remained. All the progress in the physical sciences still seemed irrelevant, in this sense: biologists and biochemists certainly use the *tools* developed by physical chemistry and other physical sciences. They use pH meters, buffers, electron microscopes and nuclear magnetic resonance to measure ‘ion binding’ of ions to proteins. They use X-rays to determine protein structure, ultracentrifuges to sort macromolecules. Electrophysiologists use techniques of electrochemistry to measure membrane potentials. But at a conceptual level there is a disjunction.

1.6 The nub of the matter

In seeking an insight into our problem, we take a remark of Stephen J. Gould, who in one of his books on evolution, *Eight Little Piggies*, wrote: ‘I have long maintained that conceptual locks are a far more important barrier to progress in science than factual lacks’. In that aphorism lies a clue to our dilemma. If we can identify those conceptual locks, we might hope to make progress [4].

To make our case we have first to agree on what constitutes a theory, and then to identify the barriers to progress. The word 'definition' has defied definition by the philosophers. We can probably agree, however, that the first stage of science is the naming of things, whence follows awareness of similarities and relations between them. Once those connections can be ordered into a predictive dictionary of events, we have a real theory. The fewer the parameters required to accomplish the ordering, the better is the theory.

For example, the Ptolemaic theory of planetary motion worked quite well. But the Newtonian theory is better, and predicts more. The conceptual lock here, which required more and more parameters as time went on, was evidently the notion that the earth is the centre of the universe. Another example is the modelling of a liquid via computer simulation. With the equations of statistical mechanics only about eight parameters, to specify a molecular potential, are required, to account for the phase diagram of a simple liquid such as argon, given sufficient computer time. But the success of such a computation is tempered a little by the realization that one might do just as well with the same number of parameters, and with molecular potentials of a completely different functional form that have no connection with the actual potentials.

We can agree that thermodynamics works. It must, being a tautology. So instead of doing the statistical mechanics of interacting molecules via simulation, one might just as well measure the boiling or freezing point of argon if one wants those properties.

This trivial example is not so far removed from our theme as first appears. As already remarked, the use of an army of molecular parameters to simulate the folding of proteins is currently popular. Yet from another point of view it is Ptolemy gone mad. Insight is obscured. To simulate what happens when the protein denatures, say, over a very narrow temperature range, one simply changes the parameters. The same used to be true even for such a deceptively simple problem as the calculation of the activity coefficients and osmotic pressure of mixed electrolytes. Change the mix. Then change the parameters. These preoccupations of physical chemists are not theories that will be useful to biologists, or chemical engineers. In this example, the sole point of a theory of activity coefficients or osmotic pressure of simple electrolytes is to see whether the primitive model (continuum solvent approximation) is a valid approximation. It is, and that is useful, for 1:1 alkali halide salts, at low concentrations where the ionic interactions are very long-range. The model fails for interesting ions such as sulfate, nitrate, caesium, phosphate, and for all of them at the concentrations of interest to biology. If we cannot predict the osmotic pressure of a sodium chloride solution in water we are going to be in trouble with a solution of proteins.

The conceptual lock here turns out to be in part due to the too-simple approximation that the solvent, water, is a continuum. But it is due too to the fact that all the classical models of electrolyte chemistry depend on electrostatic models of ionic interactions. These ignore all the quantum mechanical forces acting between ions that are just as important, and determine the very specificity that we are looking for![1]

1.7 Molecular forces in self assembly

If we come to self-assembling amphoteric molecules like the phospholipids of cell membranes we enter into more difficult ground. Here is a statement of our problem that sums it up: 'Despite enormous progress in understanding the genetics and biochemistry of molecular synthesis we still have primitive ideas of how linearly synthesized molecules form the multimolecular aggregates that are cellular structures. We assume that the physical forces acting between aggregates of molecules and between individual molecules should explain many of their associative properties; but available physical methods have been inadequate for measuring these forces in solids or liquids.' These few succinct opening sentences from an old review by V. A. Parsegian stand [5]. They embrace and define the whole grey area bridging chemistry, physics and biology which is our concern. They imply a formidable injunction. For while it is axiomatic to the physicist or chemist that structural changes in any system should be reduced to a consideration of forces or free energies which cause those changes, the burden of proof lies with the proponent. The axioms of physics do not always receive so ready an acceptance from biologists whose whole thinking in the past has been centred on the role of geometry to the almost complete exclusion of forces and entropy. The burden of proof becomes especially great if one considers the increasing sophistication of those few successful theoretical advances in our understanding of condensed matter. To be convincing, and to have any hope whatever of reducing to some semblance of order the vast complexity of those intricate multimolecular structures that are the subject of biology, any successful theories must have as a minimal requirement extreme simplicity, to make them accessible to the biologist who has enough concerns of his own not to be dragged into the subtleties of modern physics.

[1] An extreme example of the absurdities this path to theory can lead to is the idea that if one put all the nuclear particles and electrons together and wrote down their potentials of interaction and solved the equations one would come up with a universe that is us. It works, say for the excitation spectra of complicated molecules, but in the process all notions of chirality are lost. The Born–Oppenheimer approximation that fixes the positions of the nuclei does retain shape, but at the expense of a loss of computational accuracy.

There is merit in the view that forces and entropy are important. There is merit in the view that geometry (of molecules) is a determining factor in self assembly. And there were, at least 30 years ago, few attempts at modelling self-assembly problems that embrace both views. Then theories of self assembly with a minimal number of parameters began to develop. They seemed to be on the right track.

A little later, to try to understand where we were several decades ago, we could say this: until a few years ago the possibility that all observations on association colloids could ultimately be handled by a single theoretical framework seemed remote. It became less so following attempts to extend the ideas of Tanford and others on dilute micellar aggregates to larger surfactant associations such as cylindrical micelles, vesicles and bilayers. The main point of departure lay in quantifying the part played by molecular geometry (packing) in determining allowed structures. It was an old idea that had been allowed to lie fallow. And it worked. Theory does appear to be on the right track. While there are gaps, parts of the jigsaw puzzle have been filled in more or less satisfactorily for dilute surfactant solutions. Certainly many of the physical properties of micelles and vesicles such as size and shape, critical micelle concentration and polydispersity appear to be accessible without a detailed knowledge of the complex intermolecular forces involved.

Our purpose here is twofold. (1) To attempt to define better and to explore some of the basic assumptions which underlie ideas currently extant. (2) To see how these ideas might be extended to include multicomponent systems (microemulsions).

From a pragmatic point of view, one main aim of such studies in the subject must surely be: to elucidate the phase diagrams of water–surfactant (and cosurfactant)–hydrocarbon mixtures; in particular to identify which structures form, when and why; and as a corollary: how to maximize solubilization of oil in water, or water in oil, with a minimum surfactant (cosurfactant) concentration.

This aim is ambitious, and the problem of such complexity is that, to paraphrase and borrow a remark made by Stillinger: 'It is essential to maintain a respectable balance between the sterile intricacy of formal theory and the seductive simplicity of poetic "explanation" ' [6].

Before beginning our study it may be useful to expand this dictum. In attempting to make a theory there are two extreme approaches. A fundamental treatment using statistical mechanics which takes into account complex surfactant molecule interactions in water is possible in principle. However, even the hydrophobic interaction between two simple molecules in water is still a matter of dispute. Further, the simplest prototype for aggregation, the problem of nucleation (and consequent phase transition) in a van der Waals gas, is an open subject. Moreover, the high road via statistical mechanics is necessarily so complicated that physical insight tends to be wholly obscured.

At the other end of the spectrum one can avoid detailed models as far as possible and search for a unified picture for micelle and bilayer formation. But thermodynamics is tautological and can only go so far. At a certain point some details of molecular interaction must be invoked. Our guiding principle in attempting to steer a middle road between these two extremes will be that these details must be minimal. Otherwise with too many (unknown) parameters theory tends to become an exercise in curve-fitting, a numerical game that loses predictive capacity and credibility.

An immediate consequence is that language must be used with some care. The problem is here doubly compounded and confounded by the fact that words such as micelle, amphiphile, hydrophobic, hydrophilic, lipophilic and aggregate are either anthropomorphic in origin (disguising their complexity) or intuitive, ill-defined, and are so familiar that we tend not to question their meaning. This question of definitions will plague us throughout and is unavoidable. A result is that our essay has elements of schizophrenia, with necessary appeal to formal statistical mechanics relegated to an appendix (even the definition of a micelle is non trivial) interspersed with (hopefully) an occasional insight gleaned from intuition and simple models.

Well then, there typically we all were. The confidence expressed by this new wave of physicists becoming chemists, or chemists becoming physicists and enamoured of computers, was evident, not just in self assembly, but in measurement of forces, in polymers, surface phenomena generally, and much else that ought to have a bearing on model biological systems. And indeed there was much progress. But this progress had little to do with biology.

1.8 Classical theories

The physical chemist, thinking about biological cells, or reading the books of his biochemist or molecular biology colleagues, makes what to them is an absurd abstraction. There are surfaces, in membranes made up of lipid–protein mixtures. There are 'surfaces' with polymers, DNA, RNA, and proteins, and of the glycocalyxes of cells and bacteria. There are salt solutions. There are interactions between all these entities that result in self-assembly processes. There are phase separations in sol–gel transitions in cell division and cell motility. There is adhesion to substrates and cell recognition that occurs between proteins, and between cell surfaces. There is catalysis. Enzymes evidently involve cooperative interactions. The words hydrophobic, hydrophilic, water, gas, crystals, solubility and so on belong to the common language used by both. And there is one difficulty. The words are undefined. That is a problem.

But if we accept the abstractions, it seems that the results of the colloid and surface and polymer chemist or the soft condensed matter physicist, working on model systems, ought to be applicable somehow to the complex systems.

The subject of colloid and surface science used to be an arcane art. It began to take quantitative form as a central area of modern science only about 50 years ago. There were several key developments. One was the theory of colloid stability of Deryaguin, Landau, Verwey and Overbeek which has underpinned the subject ever since and still holds centre stage. The other was the theories of Langmuir and Onsager which dealt with phase transitions in suspensions of clay platelets, tobacco mosaic viruses and spherical objects like latex spheres. Another was in the understanding of self assembly of surfactants and microemulsions, and their microstructures. But these advances again do not seem to have connected to biology at all. Why? Because, as Stephen J. Gould told us, of the conceptual locks.

1.9 How we lost the farm

We can identify several places where what ought to be the enabling discipline of a physical, colloid and electrochemistry programme lost its way.

1.9.1 Hofmeister effects

Theories of forces between surfaces did not include the all-important specific ion (Hofmeister) effects. These 'specific ion effects' were discovered over a century ago and ignored. They are as important in the scheme of things as Mendel's work was to genetics. And ignored. The classical theories of physical chemistry and colloid science were seriously flawed or incorrect. That statement is true for molecular forces between colloidal particles, for electrolytes, membrane and zeta potentials, interfacial tensions, even for matters as fundamental as pH and buffers.

We are now beginning to understand why, and how to improve the situation.

1.9.2 Hydrophobic interactions

Direct measurements of molecular forces between surfaces appeared to have confirmed the classical theory of lyophobic colloid stability and its extensions due to Verwey, Overbeek, Deryaguin and Landau. But as time has gone on practically all measurements interpreted in terms of these ideas invoke extra fitting parameters such as effective charge of interacting surfaces. These vary from surface to surface and electrolyte to electrolyte in a bewildering manner. Even allowing such fitting parameters, much more sophisticated theory (Lifshitz and its extensions) that called in the full apparatus of quantum electrodynamics still failed in many cases, and new forces had to be called in. These are variously called hydration, secondary hydration, and that bugbear of the 1990s, long-range 'hydrophobic' attraction.

1.9.3 Dissolved gas and water structure

The role of dissolved gas and other solutes in interactions, liquid structure and free radical production has been completely ignored. This is important if we are to ever understand chemical reactivity. It is an area that had hardly been recognized and was virtually untouched. Dissolved gas at atmospheric pressure in water is about $2{\cdot}10^{-3}$ M, and about 10 times as much in oil. Dissolved gas is intimately coupled to the range of the mysterious long-range 'hydrophobic' interactions. It changes interactions by factors of up to 100! The microstructure of water with dissolved gas and electrolyte, and it does depend too on electrolyte type, is a subject about which we know nothing much. Work on optical cavitation, sono-luminescence and related phenomena is beginning to reveal more of the extraordinary complexity of water. This, the nature of water, is an essential key that remains to be unlocked.

1.9.4 The cubic phases and bicontinuous states of matter

There is yet another conceptual lock. Until recently theories of self assembly, of surfactants, lipids, microemulsions, polymers and mixtures thereof were constrained by an intellectual mind set that limited thinking to a particular set of shapes. These are those provided by Euclidean geometry-points, spheres, cylinders and planes. For example, for surfactants and phospholipids, we tended to think in terms of monomers, micelles, hexagonal phases, vesicles, lamellar and reverse phases. It turns out that hyperbolic geometries, everywhere bicontinuous, random or regular with zero (cubic phases) or constant average curvature, are the rule in Nature. The same holds for their two-dimensional analogues, the mesh phases, which provide a richer framework in which to think of biomembranes and their action than the older Danielli–Davson bilayer model. That model relegates the lipids to an inconsequential nonspecific supporting role for the proteins and DNA.

1.10 The way ahead

The theme of this book is that the reasons for the present disjunction in connectedness are becoming clearer. Over the last decade, these matters have all taken a positive turn. The flaws in theory have become starkly apparent. Hofmeister effects can now be taken into account at a first predictive level.

This, together with the role of dissolved gas, has resulted in new theoretical insights that as we shall see do have application and intersection with biosciences, and even real chemical engineering! A paradigm shift is in progress.

In parallel with these realizations, there have been experimental and theoretical advances, in understanding the limitations of primitive models, in oscillatory or depletion, hydration and double layer forces. There has been a reconciliation of older ideas on water structure, 'kosmotropic and chaotropic', 'hard and soft' ions.

'Hydrophobic' forces, the bugbear of the 1990s, and cooperative cavitation begin to make sense, and with dissolved gas begin to provide insights into how some enzymes might harness weak physical intermolecular interactions to provide chemical energy and reactivity.

'Polywater' and gels begin to make sense.

In summary, after a long time, we have a field in ferment, and a new beginning.

References

[1] D. W. Thompson, *On Growth and Form*. Cambridge: Cambridge University Press (1917).

[2] D. Tabor and R. H. S. Winterton, *Nature* **219** (1968), 1120–1121.

[3] J. N. Israelachvili and D. Tabor, *Proc. R. Soc. London, Ser. A* **331** (1972), 19–38.

[4] S. J. Gould, *Eight Little Piggies. Reflections in natural history*. New York: W. W. Norton (1993).

[5] V. A. Parsegian, *Annu. Rev. Biophys. Bioeng.* **2** (1973), 221–255.

[6] F. H. Stillinger, *J. Solution Chem.* **2** (1973), 141–158.

2

Different approaches to, and different kinds of, molecular forces

2.1 Molecular potentials

2.1.1 General remarks

In trying to tease out the complex nature of molecular forces in solution, in colloidal particle interactions and self assembly, and between macromolecules there are several approaches. The first we can call a 'bottom-up' approach. Here we can start with the force between two molecules. Then we can try to estimate the interactions between two colloidal particles by adding up the forces between the individual molecules of each particle.

There are several kinds of potentials that provide the building blocks for such an approach. These formulae and their derivation can be found in all standard texts, e.g. Refs. 1–4.

If the molecules are charged ions then at large distances there is an electrostatic interaction between them with a potential (repulsive for like charges, +, or attractive for unlike charges, −):

$$V(r) = \pm\frac{|ze|^2}{4\pi\varepsilon_0\varepsilon_r r} \tag{2.1}$$

where r is the distance between them, ze the charge, ε_0 the permittivity in vacuum ($8.854 \cdot 10^{-12}$ $C^2 \cdot m^{-2} \cdot J^{-1}$) and ε_r the static dielectric constant of the solvent.

The Coulomb forces are long-range, are invariably screened by the cooperative effects of neighbouring ions and are not additive. They have to be treated by statistical mechanics (cf. Chapter 3). The potential of mean force between two ions in an electrolyte is then, to first approximation:

$$V(r)_{\text{eff}} \propto V(r)e^{-\kappa_D r} \tag{2.2}$$

where κ_D is the inverse of the Debye screening length given by $\kappa_D = \left(\frac{\varepsilon_0\varepsilon_r k_B T}{e^2\rho}\right)^{-1/2}(\nu_1 z_1^2 + \nu_2 z_2^2)^{1/2}$, ρ is the number density of ions per unit volume,

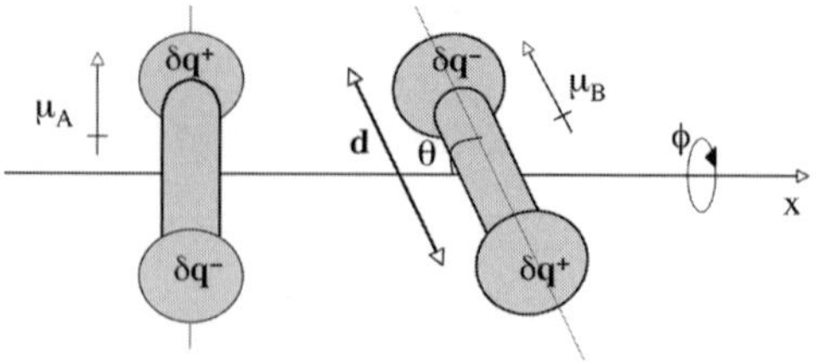

Fig. 2.1. For molecules with permanent dipole moments like water $\mu = \delta q \cdot d$; the electrostatic potential depends on orientation.

T the absolute temperature, ν_i is the stoichiometric coefficient for each ion in the electrolyte and k_B Boltzmann's constant.

If they have a permanent dipole moment μ_A and μ_B, the potential is angle dependent (Fig. 2.1).

A thermal average over all angles of rotation of this potential gives the so-called Keesom (or orientation) potential

$$\langle V_{\mathrm{orient}}(r)\rangle = -\frac{2}{3k_B T}\left(\frac{\mu_A \mu_B}{4\pi\varepsilon_0\varepsilon_r}\right)^2 r^{-6} \tag{2.3}$$

Another force between charged molecules is called the induction force. If the ions are spherical the charge on one induces a static polarization (charge separation) in another and vice versa. The electrostatic potential due to this charge fluctuation effect due to the kinetic motion of the ions in a solvent is:

$$V_{\mathrm{ind}}(r) = -\frac{[\alpha_A(0) + \alpha_B(0)]e^2}{2(4\pi\varepsilon_0\varepsilon_r)^2} r^{-4} \exp(-2\kappa_D r) \tag{2.4}$$

Again, like the Coulomb force, this potential is screened with a factor of $\exp(-2\kappa_D r)$ due to the effects of surrounding ions.

Next, a molecule with permanent dipole μ will induce an instantaneous dipolar response in a neighbouring molecule with static polarizability $\alpha(0)$ and give a potential called the Debye potential:

$$\langle V_{\mathrm{ind}}(r)\rangle = -\frac{\alpha_A(0)\mu_B^2 + \alpha_B(0)\mu_A^2}{(4\pi\varepsilon_0\varepsilon_r)^2} r^{-6} \tag{2.5}$$

When the interacting molecules are uncharged a different kind of force operates. Since all molecules are polarizable, the instantaneous fluctuating field due to their electrons gives rise to an instantaneous dipole moment in one. It sets up a field that polarizes the other. The mutual interaction gives rise to a quantum mechanical potential variously termed a van der Waals, London or dispersion potential of the form (in a vacuum):

$$V_{\text{disp}}(r) = -\frac{3\alpha_{\text{A}}\alpha_{\text{B}} I_{\text{A}} I_{\text{B}}}{2(4\pi\varepsilon_0)^2(I_{\text{A}} + I_{\text{B}})} r^{-6} \tag{2.6}$$

where α is the polarizability and I the first ionization potential, for each intervening molecule. (The term 'dispersion' forces came about because it was realized that the original quantum mechanical perturbation theory result could be expressed in terms of dispersion '*f*-values' or oscillator strengths of the atoms. It is sometimes retained now to describe all quantum mechanical and electrostatic fluctuation forces. The distinction between the different kinds of forces in condensed media will be seen later to be artificial. We shall see later that this standard approximation is not very good quantitatively and the actual dispersion forces are much more subtle.)

Beyond this there are higher-order, shorter-range contributions due to quadrupoles, octupoles and so on.[1]

These contribute more and more at smaller and smaller distances.

These classical potentials of interaction are a consequence of the fact that at large distances a molecule, to first approximation, can be viewed as a point particle. Any instantaneous or fixed electronic charge distribution of the molecule, whatever its shape, can be expanded at large distances r in an infinite series sum of inverse powers of distance, r^{-n}. The first power is a monopole, the second a dipole and so on. Each power has a coefficient which is a spherical harmonic with a weight that depends on shape. These forms break down at close distances where this mathematical expansion becomes a poor approximation. It is usual to approximate the short-range interaction typically by an ad hoc form, say an infinitely repulsive hard core, or a softer Lennard-Jones potential, $V(r)_{\text{repulsive}} \approx r^{-12}$.

The net van der Waals potential between two neutral spherical molecules then would look like that in Fig. 2.2.

[1] In fact one of the earliest theories in 1920, pre-quantum mechanics, to account for the van der Waals equation of state for gases evolved from the observation that the van der Waals constant '*a*' was connected to the refractive index of the gas. This is determined by the polarizability of its constituent molecules. Debye proposed a theory that the polarizability is the cause of molecular forces. In his treatment he assumed each molecule had a permanent quadrupole and that the electric field of this quadrupole polarizes a neighbouring molecule. The force between them is due to the quadrupole and induced dipole moment. The potential varies as r^{-8} and the constant of proportionality depends on the polarizability which determines the induced dipole moment. Debye obtained reasonable agreement with experiment. Keesom extended this further, noting that the existence of a quadrupole moment entails orientation effects in the interaction energy of two molecules [1,5]. Although only of historical interest, it is worth noting the recent work of Kjellander and coworkers, who show that an r^{-8} potential exists in primitive model electrolytes due to a coupling between Coulomb and dispersion interactions. Formally this could dominate screened Coulomb interactions [6,7].

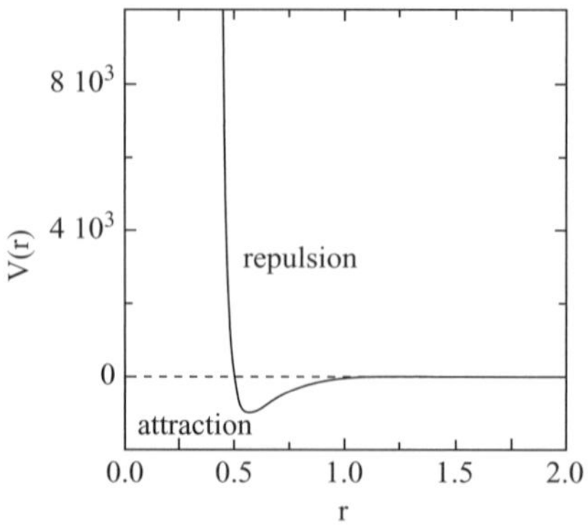

Fig. 2.2. Lennard-Jones potential $V(r) = Br^{-12} - Ar^{-6}$. At short distance the interaction becomes repulsive, unless the atoms form a chemical bond. It is customary for computational convenience to approximate the short-range potential as a simple hard sphere or softer, e.g. ad hoc Lennard-Jones potential. It is only recently that more soundly based analytic forms for short-range potentials have been calculated [8]. There are further, sometimes very long-range forces, between uncharged, especially linear molecules due to the presence of electrolytes [1]. These will occur later. With this catalogue of individual molecular forces to start with, we can try to build up to the thermodynamics of a system of many interacting molecules via statistical mechanics. We could write down a so-called partition function for a solution of molecules that form a liquid or a gas. In practice this is possible only in dilute solution. Or else we could use Monte Carlo and molecular dynamics computer techniques to evaluate the partition function and thermodynamic functions of the system.

2.1.2 Pairwise summation of two body forces

Suppose we wanted the van der Waals force between two solid colloidal particles across a vacuum. We can idealize and visualize them as molecularly smooth planar media, made up of molecules at a constant density. Then we can add up the forces between all the molecules that comprise the colloidal particles across the gap [1–4].

We can correct such results, roughly, for the presence of an intervening liquid.

The van der Waals potential energy of interaction between two such half planes a distance L apart per unit area (Fig. 2.3) is:

$$E(L) = -\frac{A}{12\pi L^2} \tag{2.7}$$

A is the Hamaker constant, after Hamaker, who first did the sum. The constant A is given by $\pi^2 \rho_1 \rho_2 \Lambda_{12}$, where ρ is density of the material, and Λ_{12} is:

$$\Lambda_{12} = \frac{3\hbar}{2}\alpha_1(0)\alpha_2(0)\frac{\omega_1\omega_2}{\omega_1 + \omega_2} \tag{2.8}$$

where $\hbar$ is Dirac's constant (or reduced Planck constant, $h/2\pi$), $\alpha(0)$ is the static polarizability and ω is the frequency of principal absorption [1–4]. The factor

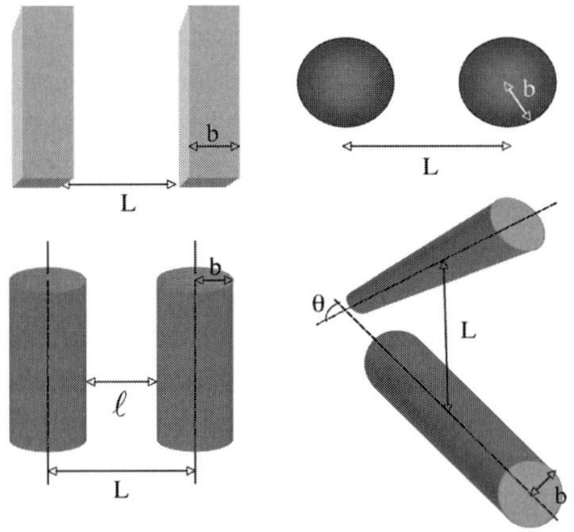

Fig. 2.3. Different idealized geometries that mimic the colloidal situation. A complete formulary can be found in Refs. 1–3.

12π was introduced by Hamaker so that the sphere interaction behaved as in Equation (2.9).

For spheres:

$$E(L) = -\frac{A}{6}\left[\frac{2b^2}{L^2 - 4b^2} + \frac{2b^2}{L^2} + \ln\left(1 - \frac{4b^2}{L^2}\right)\right] \tag{2.9}$$

For cylinders at small separation ($\ell = L - 2b \ll b$):

$$E(\ell) = -\frac{A}{24b}\left(\frac{b}{\ell}\right)^{3/2}\left[1 - \frac{\ell}{b} + \frac{1}{\sqrt{2\pi}}\ln\left(\frac{\ell}{b}\right) + \cdots\right] \tag{2.10}$$

For cylinders at an angle θ:

$$E(L) = -\frac{A\pi}{2}\left(\frac{b}{L}\right)^4\left[1 + \frac{5b^2}{L^2} + 21.875\frac{b^4}{L^4} + \cdots\right] \tag{2.11}$$

Cylinders at an angle θ experience a strong torque that lines them up.

The bottom-up approach, starting with individual molecular interactions, seems simple and straightforward. Pairwise additivity of molecular forces is the basis of statistical mechanical theories of liquids and gases and of modern molecular dynamics and Monte Carlo simulation techniques. These go further by allowing kinetic energy to the interacting particles and so include entropy.

2.1.3 Interfacial energies

If we wanted some idea of the interfacial tension of a liquid or of a solid we could go further. Consider the energy of interaction per unit area of our molecularly smooth slabs on ‘contact’. Contact is an atomic separation distance. So if we simply do

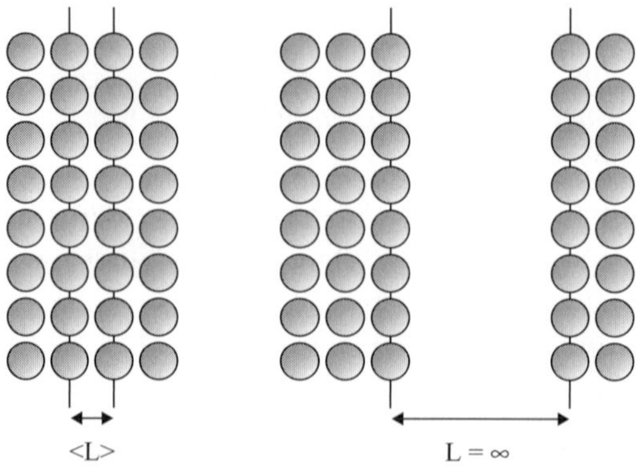

Fig. 2.4. Crude estimates of van der Waals interfacial energy. On contact the interaction energy is given by Equation (2.7). At infinite separation this is just twice the interfacial energy of a single surface. The agreement works for simple liquids such as argon and alkanes, but for nothing else.

the thought experiment of pulling the surfaces apart, the interaction energy should be just twice the surface energy (see Fig. 2.4) [1,3]. Thus for hydrocarbon liquids, we could take the refractive index of the liquid, $n^2 = 1 + 4\pi\rho\alpha$, where ρ is the number density of CH_2 groups, α the static polarizability of CH_2 groups at the measured adsorption frequency or an average thereof.

The numbers come out about right – for alkanes around 40 mN/m.

But the formula follows from what is really a dimensional argument. Thomas Young deduced the same result from the strength of materials as long ago as 1805. Newton knew that the potential of interaction between molecules had to be $\sim r^{-6}$. The numerical agreement is illusory for reasons we will explain later.

2.1.4 Problems with pairwise additivity: cooperativity of permanent dipolar interactions

The bottom-up approach assumes that the molecular forces are pairwise additive. This is generally not so. That is clear for electrostatic, Coulomb forces discussed in the next chapter that operate between ions in an electrolyte. They are inherently collective, long-range many-body forces. The matter presents considerable technical problems, especially for simulation methods, that remain unsolved.

For permanent dipolar forces such as exist in water, we have the same problem. They are not additive! The Keesom force above is a thermal angle average force between two freely rotating molecules with permanent dipoles. It varies with temperature as $\approx (k_B T)^{-1}$. But in condensed media the interaction, say of dipolar proteins in water, is proportional to $k_B T$ [1,2,9]. These temperature-dependent many-body forces, we shall see, generally constitute from 30 to 70% of the interaction energies in biological (roughly like oil-in-water) systems. They contribute about 70% of the total interfacial tension of water [1,3]. So we can expect that pairwise summation methods will fail in condensed media when the molecules have permanent dipoles.

The relative magnitude of these molecular forces in vacuum or corrected for an intervening medium, and their size compared with the thermal energy of a molecule ($k_B T$), are easily estimated. They can be found e.g. in Refs. 1–4. We do not dwell on these because the comparisons are misleading. The situation in condensed media, such as oil or biomembranes or proteins in water with which we are concerned, turns out to be very different, due to cooperativity of molecular interactions, temperature effects and others. In the domain of colloid science for which the classical theory of DLVO was devised the focus was on interactions between inorganic particles in water, for which pairwise dispersion forces dominated. Hamaker constants were an order of magnitude larger than for biological materials. But even for inorganic materials in water, pairwise summation gives estimates defective sometimes by an order of magnitude.

2.1.5 Many-body dispersion forces

Even for these shortrange van der Waals, dispersion forces, non-additive three- and many-body dispersion interactions [1] are significant. Liquid argon is about the simplest liquid one can imagine. It is composed of essentially hard spheres interacting via van der Waals interactions.

Even for this simplest imaginable liquid, three-body potential contributions to the simulation of the interfacial tension of liquid argon are about 13% of the total.

A more subtle reflection of many-body interactions is this. The binding energy (energy of formation) of a molecular crystal has two parts. One is due to the short-range repulsive interactions between the molecules. The other is due to dispersion interactions. They crystallize in a face-centred cubic (fcc) structure. But the hexagonal close-packed structure (hcp) has the same number of nearest neighbours, and the same distances between neighbours.

Pairwise additivity gives identical energies. It was necessary to include all many-body multipolar interactions to decide the issue (see Ref. 1, Chapter 4; and Ref. 7).

2.2 Liquid structure at solid interfaces: many kinds of forces

In estimating the forces between molecularly smooth solid surfaces such as mica or calcium fluoride, in a vacuum, it seems reasonable to allow molecular smoothness. What that means can be appreciated by an analogy of Israelachvili, who first measured molecular forces directly. He remarks that the situation is analogous to taking two spheres as big as the moon, and moving them 20 cm together by stages starting from several metres apart; each time measuring the forces between the moons. On this scale the surfaces made up of 'molecules' have the roughness of close-packed footballs.

(It is almost impossible to avoid contamination of high-energy surfaces like mica, even formed by cleaving the crystal under vacuum. The surface energy drops due to adsorbed molecules such as oxygen by an order of magnitude even if extreme care is exercised.)

But if liquids are involved, or at liquid–gas surfaces, a more complicated story pops up.

This goes back to the Young–Laplace, Poisson argument that we have already come across in Chapter 1.

The pairwise summation formulae of Section 2.1 for the forces between ideal colloidal particles separated by a liquid assume that the liquid has bulk properties (density, average molecular dipolar orientation) right up to a molecular distance from the solid surface. As we have mentioned in Chapter 1, the validity of that assumption was Poisson's concern. It will remain ours.

We outline a number of different idealized but real situations that do occur involving liquid structure at interfaces, and how it affects molecular forces between colloidal particles (see Fig. 2.5). (For discussion of experiments on measurements of molecular forces see Refs. 1, 3, and references in Chapters 4–8.)

2.2.1 'Young–Laplace': continuum liquid approximation

The standard approximations for attractive forces always assume bulk liquid up to an interface.

2.2.2 Forces due to molecular granularity, oscillatory and exponential

Molecular structure (granularity) of a simple liquid influences forces to give rise to oscillations (Chapter 4). These forces are smoothed and decay exponentially at 'rough' surfaces with a range of about a molecular spacing in the liquid. The surface does not perturb the bulk liquid. These forces merge into continuum approximation (power law) van der Waals forces after around 4–6 molecular layers.

2.2.3 The 'Poisson' case: surface-induced liquid order

If the liquid molecules are anisometric, e.g. suppose that they have a large dipole, or are needle-like, like a nematic liquid crystal, they will be oriented by steric effects at

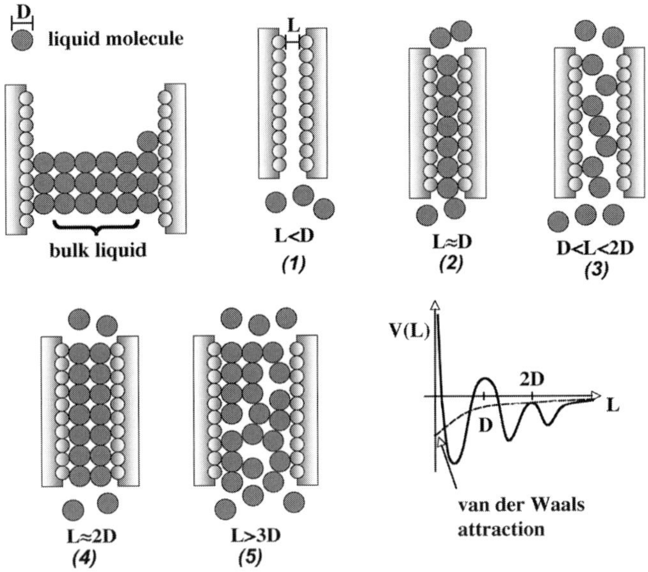

Fig. 2.5. Illustration of different situations involving liquid structure between colloidal and solute particles, and plot of the interaction as a function of the interlayer distance L. (1) The liquid molecules are too big to squeeze into the gap, and interaction is across a vacuum. (2) The liquid molecules can adsorb into the gap and form an anisotropic single layer. The interaction is screened and becomes weaker. (3) The liquid molecules adsorb but their density is lower than in (2). The gap is partially filled and more vacuum-like. The interaction increases again. (4) The layer of liquid squeezes in. The gap is full, filled with two layers, more like bulk liquid. The interaction is more screened than in (3), and it decreases. (5) The intervening liquid is almost an isotropic bulk layer and behaves like a continuum. The interaction goes over to a low-power attraction. Adapted with permission from Ref. 10. Copyright 1980 American Chemical Society.

the surface. They will also be oriented by the van der Waals and electrostatic forces between molecule and surface, which are also highly anisotropic. The surface-induced order is cooperative, and decays exponentially. The overlap of the profiles of order from each surface gives rise to a force due to this surface-induced liquid structure (see Fig. 2.6).

At a single surface such as an oil–air or oil–water interface this average orientation effect occurs. The oil molecules line up and orient increasingly with increasing chain length. to give an anisotropic dielectric profile. At an interface involving water the cooperative dipolar (and hydrogen bonding) water molecule interactions ensure that the water interface is always anisotropic in both its density profile and dielectric properties and this effect gives the major contribution to the interfacial tension.

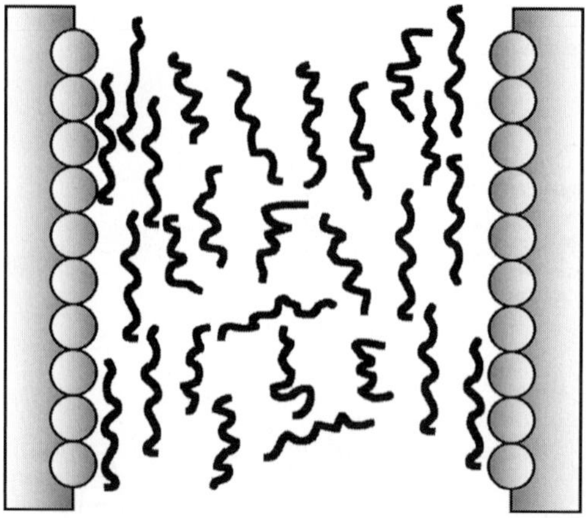

Fig. 2.6. Even with simple alkanes, steric effects and van der Waals forces, anisotropic due to their more or less cylinder-like nature, cause a surface-induced ordering. These effects tend to line them up at a surface and disordering due to opposing entropic effects is reflected in a distance-dependent order profile. The competition between them gives rise to strong temperature-dependent forces.

2.2.4 Hydration forces

If the surfaces are made up of anisotropic molecules, typically the phospholipids of biomembranes, or other surfaces formed from self-assembled surfactant aggregates (Part II of this book), the head groups have a very strong affinity for water and are hydrated. The strongly bound surface hydration layers impose their order on vicinal water layers (see Fig. 2.7). The hydration forces have a range that again decays exponentially with decay length about that of water molecule thickness. These vary enormously in specificity as the 'bound water' of hydration depends on stereochemistry of the headgroup and of their dipolar flexibility. (Phosphatidylcholine has a large headgroup and dipole moment with little freedom to fluctuate. Phosphatidylethanolamine has a small headgoup that fluctuates more and gives much reduced hydration forces.) These in-plane fluctuation processes are cooperative (see Chapter 5).

2.2.5 Double-layer electrostatic forces

These are at the heart of colloid science and electrochemistry and, opposed to the van der Waals forces, form the basis of the classical theory of colloid stability.

The surfaces become charged on contact with water if they have dissociable surface molecular groups. We can consider at first level that the water is a continuum, with the 'liquid' now being the electrolyte ions. The charged surface sets up a potential that attracts counterions and ions of the same charge. It repels those of different charge (see Fig. 2.8).

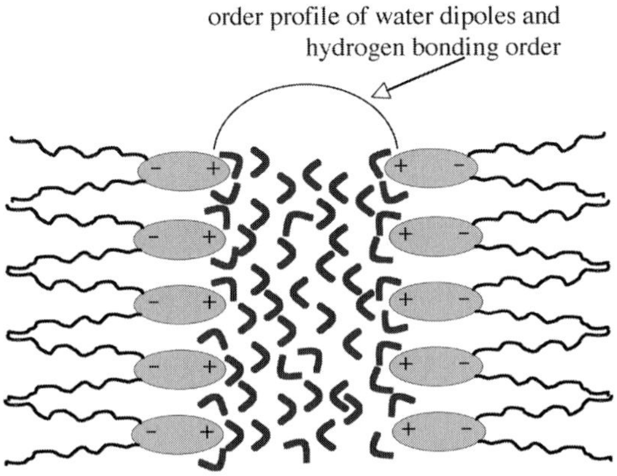

Fig. 2.7. The strongly hydrated phospholipid headgroups at the surface of a monolayer impose their order (hydrogen bonding, dipolar orientation) on the neighbouring water molecules. This is randomized by the effect of the large headgroup motions, which depend on their headgroup motion restriction due to size and dipole moments. For example, the large phosphatidylcholine headgroup has steric limitations on its freedom to rotate, while the smaller phosphatidylethanolamine headgroup can oscillate about its dipolar axis. Hydration forces decay more rapidly in that case.

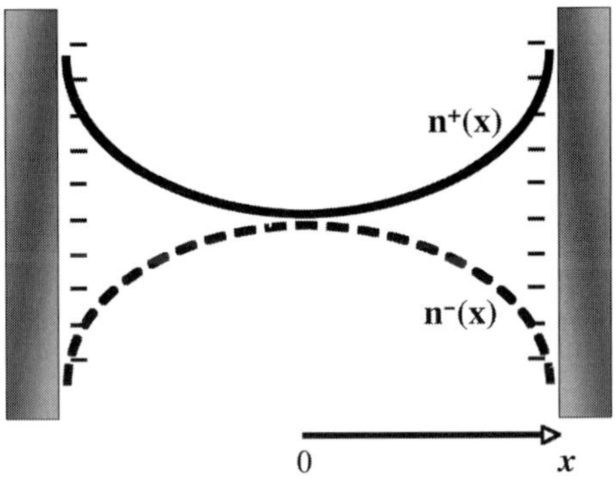

Fig. 2.8. Charged interfaces of colloidal particles attract cations, and repel anions. The profiles of cation (full line) and anion (dotted line) are par excellence an example of surface-induced liquid structure. The 'liquid' is here the electrolyte, the water a passive continuum. A self-consistent potential is set up by the cations and anions. The difference between their potential and that of the surrounding electrolyte gives an osmotic force between the particles (Chapter 3).

The overlap of profiles of ionic density gives rise to an osmotic repulsive double-layer force between like surfaces. The force decays exponentially at large distance with decay length given by the Debye length (Chapter 3).

The real picture is more complicated: the surfaces can 'feel' or recognize the influence of their neighbours, and dissociate, more or less, to regulate surface pH and charge and the interaction.

With charged amphoteric surfaces, almost universal with biosurfaces, the electrolyte strength regulates the forces.

In fact as we shall see the dispersion forces and electrostatic forces cannot be so glibly disentangled from each other and from hydration forces.

2.2.6 Secondary hydration forces

A different sort of hydration is seen in cases associated with adsorption of counterions onto molecularly smooth mica with double-layer force measurements between mica in an electrolyte.

At close distances of approach the decay length of the double-layer force changes from the Debye length. It goes over to a stronger short-range hydration force. Fitted to an exponential form, the decay lengths of these forces were successively 1, 2 and 3 nm for 1:1, 2:1 and 3:1 electrolytes, respectively. They are not explainable by classical electrolyte theory. They were called 'secondary hydration forces' by Pashley, who first measured them [11,12].

With hydronium ions instead of cations there are no such forces, a matter attributed to the crystal structure of mica which allows the H^+ ion to be adsorbed into the lattice. Their explanation reflects oscillatory forces imposed on the exponentially decaying double layer and depends on hydrated ion size of the cations, and on missing dispersion forces acting on ions (Chapter 7). These forces exist and can be exploited in many applications to control flocculation.

2.2.7 Hydrophobic forces

Another kind of force comes about with water and other hydrogen-bonded liquids. The surfaces are hydrophobic – water-hating, which generally means a high or at least non-zero contact angle (see Chapter 8).

Surface-induced fluctuations in water molecule orientation lead to lowering of the density between the interacting surfaces. The fluctuations are due to cooperative water dipole orientations.

This lowering of water in the gap leads to lower pressure than the outside pressure, and this gives rise to an attractive force of the same (Hamaker-like) form as a van der Waals force, but an order of magnitude or more larger. The term 'hydrophobic forces', being anthropomorphic in origin, embraces large a multitude of confusions and sins.

Other kinds of long-range 'hydrophobic' forces are due to charge-regulated two-dimensional phase transitions of mobile adsorbed ionic surfactants or polymers.

2.2.8 Capillary forces

The oldest and strongest and forgotten forces are those due to capillary cavitation. When the contact angle is greater than 90° the fluctuations become critical, and a bridging cavity forms between the two surfaces (see Chapter 8).

2.2.9 Effects of dissolved gas

With sparingly soluble hydrophobic solutes such as methane, or dissolved atmospheric gas, the water fluctuations are extended by the solute particles. (This complicated situation is explored in Chapter 8. It is probably the most important omission in the entire body of knowledge in physical chemistry.) The 'hydrophobic' forces then become extremely long-range, significant at ten and more nanometres, and orders of magnitude larger than any van der Waals force.

Additional dissolved salt enhances or opposes such effects (see Chapter 8). This last, the effects of dissolved atmospheric gas on forces, rather puts the cat amongst the pigeons. It can change predicted net forces by orders of magnitude.

2.2.10 Hofmeister effects

Another matter that will be a central concern is that of ion specificity. Changes in salt type at the same concentration can induce dramatic changes in forces. These, known as Hofmeister effects, and ignored for 150 years, are ubiquitous in chemistry and in biochemistry.

It is only recently that we have come close to understanding specific ion effects and their interplay with dissolved gas. The ramifications, from pH to buffers and surface potentials, are immense (see Chapters 6–8).

We will be dealing with all of these forces and more in what follows. It will be evident that approaching the problem of interactions via bottom-up, two-body force interactions is going to be limited. The separation of forces is artificial (see Chapters 5–8). All manifestations of forces depend ultimately on short-range quantum mechanical (dispersion, dipolar) and electrostatic forces. The question is how they do so, and do so cooperatively.

The idealization of a colloid particle as a molecularly smooth plane is clearly just that, an idealization. It is realized in practice only for materials like cleaved mica.

For a macromolecule such as a protein the validity of its approximation as a model smooth surface depends on the situation. It is sometimes useful, but clearly requires a judicious eye and healthy scepticism.

2.3 Liquid structure at other interfaces and around solutes

At a liquid–gas interface it is much more obvious that there must be some profile at least in density as the liquid merges into the vapour phase.

But the problem of even *defining* an interface quantitatively is very difficult (see Chapter 4 Section 3 for more detailed discussion). At the critical point of a liquid the interface is infinitely wide. We have seen in Section 2.1 above that taking the Laplace assumption of molecular sharpness we could arrive at what seemed a reasonable estimate of the interfacial tension. If we look a little closer this is not so. If we take the measured handbook data for interfacial tensions of alkane liquids as a function of temperature we can decompose it numerically into entropy and enthalpy. The enthalpy has a density dependence with alkane chain length that shows the chains are lined up more as needles rather than at a uniform density. There is a profile of alkane molecule order at the interface reflected in their average orientations.

In fact there are two profiles, as the atmospheric gas, at 1 mol/L, drops to a density of about $5 \cdot 10^{-2}$ M over the first few water molecules of oil. For water the bulk density of dissolved gas is about $5 \cdot 10^{-3}$ M at 25 °C. So there are significant quantities of gas molecules at such an interface, as indeed for solid hydrophobic surfaces. This affects propagation of surface-induced liquid structure and forces between surfaces, liquid or solid.

We shall see astonishing manifestations of such effects.

No simulations or theories of interfacial tensions include effects of dissolved gas. Nor do experiments on liquids or colloids remove dissolved gas. The effects of dissolved gas on colloidal particle interactions can be very large (Chapter 8). So theories that ignore this reality probably deserve to be taken with at least some caution.

2.3.1 Other solid–liquid surfaces

Silica–water surfaces, and others that form soft gels, are too complex even to describe here. The description occupies several large volumes. We simply note the existence of cement as a problem.

2.3.2 Hydrogen bonding

The problem of quantifying the interface between a surface and an adjoining liquid runs through and besets any thinking about water. It is the same problem for a solute molecule that perturbs the surrounding water 'structure'. In the bottom-up approach from which most intuition has been drawn, we think of separate contributions from the forces between molecules: electrostatic, dipolar, van der Waals, induced dipole–dipole and others. In condensed matter that division is too simple. The first two are long-range, cooperative and involve many molecules, but still physical forces.

The classical concept of a 'hydrogen bond' sits somewhere between that of a 'physical' and a 'chemical' bond. It is derived from a quantum calculation that involves just two water molecules. The energy of binding involves a dipolar interaction between an electronegative atom and a hydrogen atom bonded to another more electronegative atom (such as oxygen, nitrogen, chloride, fluoride, and so on). In (dynamic) water and in solute–water interactions the interactions are all shared between many water molecules. Hydrogen bonding energies are variously estimated experimentally, in different situations, to range from 5 to 30 kJ/mol, i.e. they range from weak 'physical' to strong 'chemical' bonds. The distinction between the two kinds of interactions disappears. The notion of a molecule itself disappears. This uncertainty reflects our lack of knowledge of water and of its cooperativity. Langmuir thought of water itself as a giant molecule. For simulation studies an effective potential can be used to mimic the properties of bulk water. But this best potential still cannot explain the density maximum at 4 °C and why ice floats on water, problems of the cooperativity of hydrogen bonds in water. That problem, which exercised the disputes between Galileo and his colleagues, has only recently been partially resolved. It requires a radically different approach to the problem [13].

(A larger problem for simulation studies is that molecular density functional theory from which the effective water molecule potential is usually derived treats dispersion forces wrongly or not at all. A recent density functional theory 'first principles' simulation of water at constant pressure gets the density wrong – too low – by 20%, which indicates that the simulation approach is still very much a work in progress.)

2.3.3 A remark on dissolved gas and effective potentials

A great deal of work on theories of liquids, interfacial tensions and interactions between colloidal particles involving water (or oil, or proteins) uses computer simulations based on statistical mechanics. They usually use quantum mechanical, angle averaged, effective interaction potentials of mean force. These are extracted via statistical mechanical methods. These predict and appear to account for the bulk and surface properties, measured statistical mechanical pair distribution functions measured by, for example, neutron scattering. However, the experimental data are invariably obtained on water or other liquids that contain dissolved atmospheric gas. As already remarked (cf. Chapter 8), on removal of dissolved gas, water has quite different properties. 'Hydrophobic' interactions disappear! The tensile strength of water as measured by pressures applied to effect cavitation is several hundreds of times less than that of defect-free, small solute- and gas-free water. The water potentials are not those of real gas-free water. Theories and simulations based on such water 'potentials' are derived from fitting to experimental water that contains gas. These are then applied to explain the properties of pure water and interactions therein. This circumstance can mask and confuses the real mechanisms that drive interactions. For hydrophobic interactions it is precisely the defects in the liquid structure due to the dissolved gas molecules that extend their range.

Simulations of the effects of salts on the interfacial tension at an air–water interface are another example of such complications due to the real world. The experimental data are obtained from water in contact with the atmosphere, at 1 M. But the theoretical modelling ignores dissolved gas, which changes from 1 M to $5 \cdot 10^{-2}$ M across the interface. Any claimed agreement between theory and experiment may not be entirely meaningless, but certainly has to be viewed circumspectly.

2.3.4 Solvent structure around solutes: molecular size

Much the same problems outlined schematically in Fig. 2.2 for interfaces bedevil the problem of solutes in water.

The question of how to define an interface is replaced by: what do we mean by the radius of an atom or an ion? We remark on assumptions and limits of validity of the idea of molecular size.

2.3.5 'Size' in van der Waals and ionic interactions

The classical derivation of the r^{-6} potential for interatomic interactions follows from quantum perturbation theory applied to two atoms or molecules treated as *point* molecules. Each atom is in its ground state, with the energy levels obtained from Schrödinger's equation. The perturbation is the instantaneous electrostatic field between the electrons of each atom which gives rise to a simultaneous mutual polarization. The time average of the instantaneous induced dipole moments gives a change in the energy of the combined system. This depends on their distance apart. Point atom or ion models are valid at far enough distances. But how far is 'far' and how small is 'small'? The approximate interaction for point molecules becomes infinite (diverges) when the distance is zero.

At close distances, when the size of the atoms due to their spread-out electron clouds become significant, shorter-range contributions due to induced dipole-quadrupole ($\sim 1/r^{-8}$) and higher-order interactions become progressively more significant. (Mathematically, as already remarked, this is because the expansion of the electric potential due to a distribution of charges into a series of spherical harmonics breaks down at distances of the order of the size of the (e.g. spherical) atom. The interacting atoms 'see' each other as spheres, not points.) One way that this is handled in dealing with the statistical mechanics of liquids is to use the approximate form $V(r) \sim r^{-6}$ up to a 'hard core' cut-off distance. This is taken as the sum of the 'radii' of the interacting atoms. Alternatively it is customary to deal with the short-range effects via a phenomenological short-range softer potential (Lennard-Jones potential) of the form $V(r) \sim Br^{-12} - Ar^{-6}$. The first substantial simulation via molecular forces of the phase diagram of argon [14] used such a potential, with the parameters A and B taken from the binding energy of solid argon. The measured X-ray spacing gives the 'contact' distance, assuming the given potential. The process leads to a van der Waals parameter A wrong by a factor of 2, as compared with a first-principle quantum mechanical calculation. This is a theoretical artefact. It arises because of the too crude approximation to the short-range interactions.

2.3.6 More accurate descriptions: intermediate distances

Even if we confine ourselves to the induced dipole–induced dipole point molecule approximation, the r^{-6} potential is a zero-temperature approximation. The r^{-6} potential is the leading term of (technically) a non-uniform asymptotic expansion. It is valid only at small, but not too small, distances. The true interaction potential valid at any temperature behaves as r^{-6} modulated by a factor that depends on the specific molecular frequencies of an intervening liquid and decays exponentially with distance. The change in form is significant around 2–4 nm, in the region of interest for colloid interactions. At larger distance still it reverts to the inverse 6th power behaviour again and becomes proportional to temperature (Chapter 5).

The point is that the choice of a simple form for a potential to model liquids or interactions between particles that give apparently good agreement with averaged, integrated thermodynamic quantities in no way is a confirmation of that form. It is that same problem that occurs in numerical analysis – a polynomial fit to the exponential curve $y = e^{-x}$ may give an excellent numerical approximation to the function over a finite range of values of x. But this does not mean that the function is a polynomial in x. At the risk of labouring the point, exactly this problem is endemic in the entire field.

2.3.7 'Small' distances

When the point dipole approximation breaks down at atomic distances, extensions that capture the essence of what goes on have been given only recently [8].

The inclusion of finite size of atoms and ions into the theory of interactions between atoms and ions will be the focus of much that follows. With this generalization of theory, interaction energies do not diverge on contact [1,15].

2.3.8 Water structure and ion size

The entire gamut of situations in Fig. 2.5 applies equally to ions in water. In the absence of a better theory of water, 'water structure' and 'hydration' can rarely be captured, except in terms of such cartoons. But the cartoons do convey and imply a language that is useful, sometimes quantified by neutron scattering experiments of formidable elegance.

The work of Collins [16–18] in elucidating a phenomenology of hydration, embodied in the terms chaotropic and kosmotropic ions, is insightful for Hofmeister problems that we shall come to later.

But the entire intuition here and throughout our subject rests on electrostatic forces alone. We shall see that this underlying credo is incorrect. This deficiency is one that that we will come to grips with.

References

[1] J. Mahanty and B. W. Ninham, *Dispersion Forces*. London: Academic Press (1976).
[2] V. A. Parsegian, *Van der Waals Forces: a handbook for biologists, chemists, engineers, and physicists*. Cambridge: Cambridge University Press (2006).
[3] J. N. Israelachvili, *Intermolecular and Surface Forces: with applications to colloidal and biological systems*. London: Academic Press (1985–2004).
[4] D. F. Evans and H. Wennerström, *The Colloidal Domain: where physics, chemistry, biology, and technology meet*. New York: VCH (1994).
[5] H. Margenau and N. R. Kestner, *Theory of Intermolecular Forces*. 2nd edn. Oxford: Pergamon (1971).
[6] R. Kjellander and B. Forsberg, *J. Phys. A: Math. Gen.* **38** (2005), 5405–5424.
[7] R. Kjellander, *J. Phys. A: Math. Gen.* **39** (2006), 4631–4641.
[8] K. Cahill and V. A. Parsegian, *J. Chem. Phys.* **121** (2004), 10839–10842.
[9] V. A. Parsegian and B. W. Ninham, *Nature* **224** (1969), 1197–1198.
[10] B. W. Ninham, *J. Phys. Chem.* **84** (1980), 1423–1430.
[11] R. M. Pashley and J. N. Israelachvili, *J. Coll. Interface Sci.***97** (1984), 446–455.
[12] J. N. Israelachvili and R. M. Pashley, *J. Coll. Interface Sci.* **98** (1984), 500–514.
[13] S. Andersson and B. W. Ninham, *Solid State Sci.* **5** (2003), 683–693.
[14] J. A. Barker, D. Henderson and W. R. Smith, *Mol. Phys.* **17** (1960), 579–592.
[15] D. F. Parsons and B. W. Ninham, *J. Phys. Chem. A* **113** (2009), 1141–1150.
[16] K. D. Collins, *Methods* **34** (2004), 300–311.
[17] K. D. Collins, G. W. Neilson and J. E. Enderby, *Biophys. Chem.* **128** (2007), 95–104.
[18] X. Chen, T. Yang, S. Kataoka and P. S. Cremer, *J. Am. Chem. Soc.* **129** (2007), 12272–12279.

3

Electrostatic forces in electrolytes in outline

3.1 The assumptions of classical theories

The classical theory of electrolyte solutions starts with the assumption that ionic interactions in solution or between colloid particles are dictated by Coulomb, electrostatic forces alone. An ion is considered to first approximation to have a charge distribution confined to a hard sphere of a given radius. In the 'primitive' model the ions are immersed in water (or another solvent) within which they interact by electrostatic interactions. The solvent is treated as a passive dielectric continuum. The radius of an ion is not always just its crystallographic radius. It is an effective radius that includes one or two water layers of 'hydration'. What occurs in the theory for the free energy of interactions involves the sum of hydrated ion radii besides the Coulomb force. The hydrated ion size is derived as a fitting parameter from comparison of theory with experiments.

For interactions between ions the long-range Coulomb interactions dominate for very dilute solutions below about $5{\cdot}10^{-2}$ M, and ion size is irrelevant. At higher electrolyte concentrations, around and above about 10^{-1} M, the cooperative electrostatic interactions that dominate at low concentrations become progressively less important. Shorter-range forces subsumed in the ionic 'radii' begin to come into play. When short-range interactions between the ions become significant the molecular structure of the water in the immediate neighbourhood of the ions (hydration) becomes a dominant feature. Hydration and local hydrogen bonding are words that attempt to describe this ion-specific, local water structure induced by the ions. The continuum solvent approximation then starts to break down (Fig. 3.1). The same is true for the interactions of ions with interfaces and proteins and colloid particle interactions.

So a better description of electrolytes is called for. Where the theory breaks down, usually always at any realistic concentration, it has been customary to handle matters by introducing extra adjustable parameters. The 'civilized' model

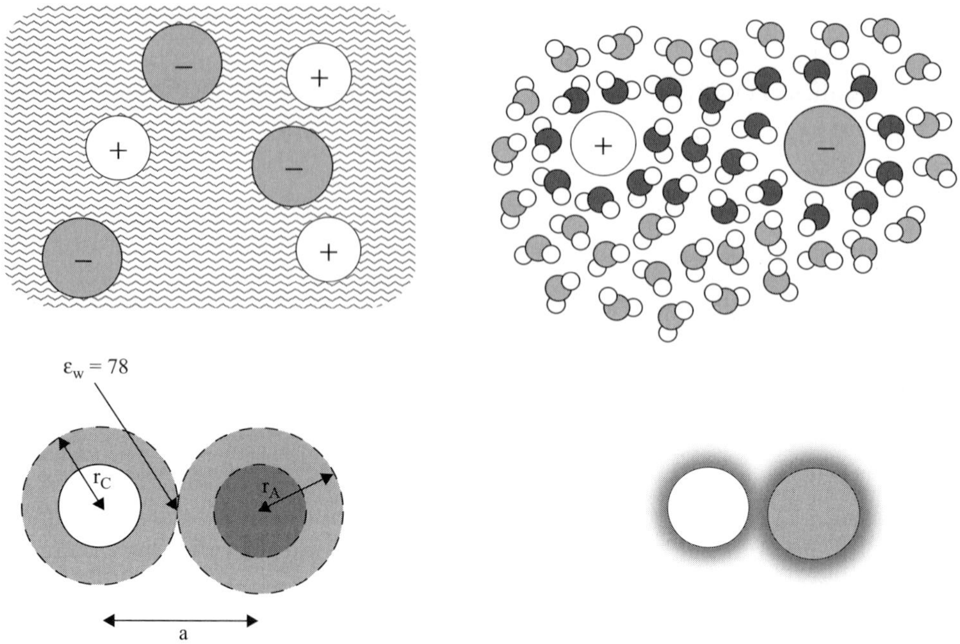

Fig. 3.1. Schematic picture of ions in water on which classical ideas are based. Left: primitive model. Upper left: bare ions in a continuum. The intervening water has a dielectric constant of 78 at room temperature. Lower left: hydrated ions in the primitive model. The 'hydrated' cations and anions ions have radii r_C and r_A. The sum of hydrated radii $(r_C + r_A) = a$ is the distance that parameterizes the short-range interactions between ions. Right: 'civilized model'. Lower right: the hydration is diffuse. Upper right: the short-range forces are due to reorientation of the hydrated water molecules due to the overlapping profiles.

drops the continuum solvent approximation and treats the water as a molecular liquid also. That extension leads to the idea of 'hard' and 'soft' radii for hydration. The short-range interactions are here due to overlap of the hydration shells around each ion.

The civilized model still has inadequacies in that it completely ignores all the quantum mechanical (dispersion) forces operating between ions and between ions and water molecules that give rise to the most interesting ion-specific effects!

The statistical mechanics of strong electrolytes, in solution and at interfaces, whether in the primitive model approximation or including solvent model interactions, analytical or via simulation, is the subject of a vast literature.

Other parameters are binding constants for adsorption–dissociation equilibrium of ions and solution and at charged surfaces, or at the amine or carboxylate groups of proteins, and association constants for weak electrolytes.

It is only recently that progress that allows some inclusion of quantum forces that underlie the field has become possible via quantum chemistry. These developments will be outlined in Chapter 7. Quantum, i.e. Fermi–Dirac and Bose–Einstein, statistical mechanics that includes spin, as opposed to classical statistical mechanics, was a field of much excitement 50 years ago and led to modern theories of low-temperature condensed-matter physics. Applied to electrolytes it rapidly became intractable. The only treatment of an electron gas that includes quantum statistical mechanics analytically is that in Ref. 1.

To first approximation, the cooperative electrostatic interactions between ions in a solvent of dielectric constant ε decay exponentially with an effective potential of interaction of the form:

$$V(r) \approx -\frac{|z_1 z_2| q^2}{4\pi\varepsilon_0\varepsilon_r r} \exp(-\kappa_{\mathrm{D}} r) \tag{3.1}$$

The long range ($\sim 1/r$) Coulomb interaction between two ions is screened by its surrounding neighbours. The main parameter that characterizes the electrostatic interactions is the Debye length $\lambda_{\mathrm{D}} = \kappa_{\mathrm{D}}^{-1}$. It measures the range of electrostatic interactions between ions and the (double layer) forces between colloid particles. ε is the dielectric constant of water (about $78\varepsilon_0$ at room temperature), and q is the unit charge. For an electrolyte $C_{\nu_1}^{z_1} A_{\nu_2}^{z_2}$, where C and A are the cation and the anion with charge $z_1 q$ and $z_2 q$, at concentration c, for example for $(K^+)_2(SO_4^{2-})$, $z_1 = +1$, $z_2 = -2$, $\nu_1 = 2$ and $\nu_2 = 1$, the Debye length is given by:

$$\begin{aligned} \lambda_{\mathrm{D}}^{-1} = \kappa_{\mathrm{D}} &= \left[\left(\frac{\varepsilon_r k_{\mathrm{B}} T}{4\pi e^2 \rho} \right)^{-1/2} \left(\nu_1 z_1^2 + \nu_2 z_2^2 \right)^{1/2} \right]_{\text{CGS units}} \\ &= \left[\frac{c^{1/2}}{0.304} \left(\frac{\nu_1 z_1^2 + \nu_2 z_2^2}{2} \right)^{1/2} \right]_{\text{SI units}} \end{aligned} \tag{3.2}$$

In this expression, ρ is the number density of ions per unit volume, T absolute temperature, and k_{B} Boltzmann's constant. In the right-hand formula if the concentration c is measured in moles per litre, distances are in nanometres. The results and those below are derived in all textbooks on physical chemistry and colloid science [2–8].

The range of the electrostatic forces is then typically: $\kappa_{\mathrm{D}}^{-1} = 10$ nm (at 10^{-3} M), 3 nm (at 10^{-2} M), 0.8 nm (at 0.15 M) and 0.3 nm (at 1 M). Biological concentrations of most interest are greater than or about physiological concentrations of 0.15 M. The diameter of a water molecule is about 0.3 nm. So we can see that a 0.8 nm Debye length relevant to a salt concentration of 0.15 M corresponds to the distance of two ions separated by at most two water molecules. A continuum solvent approximation is at best then dubious.

In fact the primitive model breaks down beyond about $5{\cdot}10^{-2}$ M salt concentrations. This makes such a theory apparently not so interesting for any system of real interest.

Nonetheless the primitive model theory of electrostatic forces with a continuum solvent model, extrapolated to higher concentrations via adjustable parameters supposed to take account of the 'hard core' size of the ions and changes in water structure induced by them, is of interest. This is because it provides the classical theoretical framework that underlies the interpretation of measurements of all fundamental quantities such as pH, pK_as, buffers, activities, hydration, ion binding, membrane potentials, the zeta potentials of colloid science and for electrochemistry generally. That the classical theory of electrolytes is inadequate is well recognized. There is a paradox here, because the same theory informs our entire intuition. More importantly it provides the basis for the interpretation of measurements of all the above fundamental quantities. This is generally unquestioned. However, if the theory is flawed, then the meaning of the measurements is called into question.

Because of this, it is useful to recall the results of the classical theory. We omit details of derivations and refer to standard books on physical chemistry.

3.2 The electrostatic self energy of an ion and the Debye–Hückel theory

3.2.1 Born self energy

The electrostatic self energy of an ion arises out of the interactions of the (spread out) ionic charge of the ion with itself in the presence of other ions [2,3]. If the charge e_ν is assumed to reside on the surface of a sphere of radius a:

$$E_\nu = \frac{e_\nu^2}{4\varepsilon_r \kappa_D a^2}\left(1 - e^{-2\kappa_D a}\right) \tag{3.3}$$

If the electrolyte is dilute so that $\kappa_D a \ll 1$, we have:

$$E_\nu = \frac{e_\nu^2}{2\varepsilon_r a} - \frac{\kappa_D e_\nu^2}{2\varepsilon_r}\left[1 + O\left(\kappa_D a\right)^2\right] \tag{3.4}$$

The first term gives the Born self energy of an ion immersed in a dielectric medium. The second term gives the correction to the self energy due to the electrolyte.

This can be used to provide some estimate of the free energy of transfer of an ion from water to another medium, say oil or the interior of a membrane of a different dielectric constant.

It can be used to get some estimates of solubility by comparing the free energies of a solid ionic crystal with the free energies of transfer of ions to the solute. It can be used to get some intuition on hydration reflected in measured partial molal volumes, entropies and enthalpies of solution.

For the state of affairs on these important issues in classical electrolyte theory and electrochemistry see the review papers Refs. 9–12. A major problem in basing analyses on electrostatic forces alone embedded in this literature is, as we shall see, that fitting parameters, e.g. temperature-dependent 'ionic effective radii', have to be invoked. That seems reasonable. 'Hydration' of ions in water can be expected to depend on temperature, as water loses its hydrogen bonding nature with increasing temperature.

Even within the confines of an electrostatic primitive (continuum solvent) model there is a missing and significant temperature-dependent ion-specific electrostatic 'induction' contribution, due to an ion charge polarizability effect [2]:

$$\Delta F_{\mathrm{SO}} = -\frac{2k_{\mathrm{B}}T}{\sqrt{\pi}a}\kappa_{\mathrm{D}}^{2}\alpha(0) = -\frac{8}{\sqrt{\pi}}\frac{\alpha(0)}{a\varepsilon_{\mathrm{r}}}\sum_{\nu} n_{\nu}e_{\nu}^{2} \tag{3.5}$$

This contribution shows up as a 10% correction to the measured hydrophobic free energies of transfer of the hydrocarbon tails of ionic surfactants in micellization [13]. Ion-induced polarization of ions leads to some very long-range forces in particular circumstances, as will be discussed in Chapter 6.

Larger contributions still that include specific ion effects on self energies are due to quantum mechanical dispersion forces missing from the theory. These give ion-specific additional contributions of 20–50% as indicated in Refs. 14 and 15. Quantitative analysis in the light of later developments is discussed in Chapter 7. So we leave the self energy problem for the moment and consider now interactions in ionic solutions.

3.2.2 *Digression on the Debye–Hückel interaction energy and activity coefficients*

In the presence of other ions in solution there is a change in the self energy of an ion due to interactions with the others. The classical theory is as follows. If we return to Equation (3.4) and sum over all the ions we get:

$$E = -\sum_{\nu}\frac{\kappa_{\mathrm{D}}n_{\nu}e_{\nu}^{2}}{2\varepsilon_{\mathrm{r}}} \tag{3.6}$$

The extra interaction Helmholtz free energy then follows from thermodynamics that gives rise to the Debye–Hückel correction to the perfect gas (Raoult's) law:

$$F(\mathrm{int}) = T\int_{T}^{\infty}\left(\frac{E}{T^{2}}\right)dT = -\frac{2}{3}\left(\frac{\pi}{k_{\mathrm{B}}T}\right)^{1/2}\left[\sum_{\nu}\frac{e_{\nu}^{2}n_{\nu}}{\varepsilon_{\mathrm{r}}}\right]^{3/2} \tag{3.7}$$

The extension of Debye–Hückel theory in the case $\kappa_{\mathrm{D}}a \approx 1$ gives [8]:

$$E_\nu = \frac{e_\nu^2 \kappa_{\mathrm{D}}}{2\varepsilon_{\mathrm{r}}(1 + \kappa_{\mathrm{D}}a)} \tag{3.8}$$

With an adjustable value of the parameter a for each ion pair, such expressions appear to give good agreement with the observed activity coefficients of electrolytes even though they are obtained on the basis of the unjustified assumptions that the dielectric constant ε_{r} retains its bulk value right up to the 'surface' of the ion and that the solution is ideal. There is still no satisfactory general theory of electrolytes. In a model that allows the ionic charge to be smeared out the energy of an ion would be [2]:

$$E_\nu = \frac{e_\nu^2}{4a^2 \varepsilon_{\mathrm{r}} \kappa_{\mathrm{D}}} \left(1 - 2\kappa_{\mathrm{D}}a - e^{-2\kappa_{\mathrm{D}}a}\right) \tag{3.9}$$

which is practically indistinguishable from the above. The difference comes about because the usual treatment imposes the condition $\nabla^2 \phi = 0$ 'inside' the ion, where ϕ is the electrostatic potential.

In the primitive model approximation besides these purely electrostatic contributions to the free energy of interaction there are additional repulsive hard ('hydrated core') interactions. Together these interaction free energies lead to measured mean molar activity coefficients $\gamma_\pm$ that characterize electrolyte solutions [3–8,16].

If we approach the problem from thermodynamics, interactions show up in activities (a) and activity coefficients (γ) in the chemical potential (μ) equation. These measure the deviations of the electrolyte solution from ideality. They are defined by:

$$\mu_i = \mu_i^0 + RT \ln a_i = \mu_i^0 + RT \ln(\gamma_i c_i) \tag{3.10}$$

The addition of a strong electrolyte, for example KCl, to water produces significant changes in the solution properties of the mixture. These changes depend on the concentration of the ions and their interactions with water. Experimental parameters such as the density, viscosity, pH, solvating power for organic molecules and gases, conductivity, refractive index, surface tension, activity and osmotic coefficient, viscosity, etc. all depend on the specific kind of salt, especially when its concentration is greater than about 10 mM. For lower concentrations the effects of different salts are much less specific in their nature, and depend only on their concentration.

Figure 3.2 shows the variation of $\mathrm{Log}\gamma_\pm$ as a function of the salt concentration (in molal units) for different electrolytes up to 1.0 m. For concentrations higher than 0.1 m the curves flatten and then rise more or less linearly, or may continue to fall,

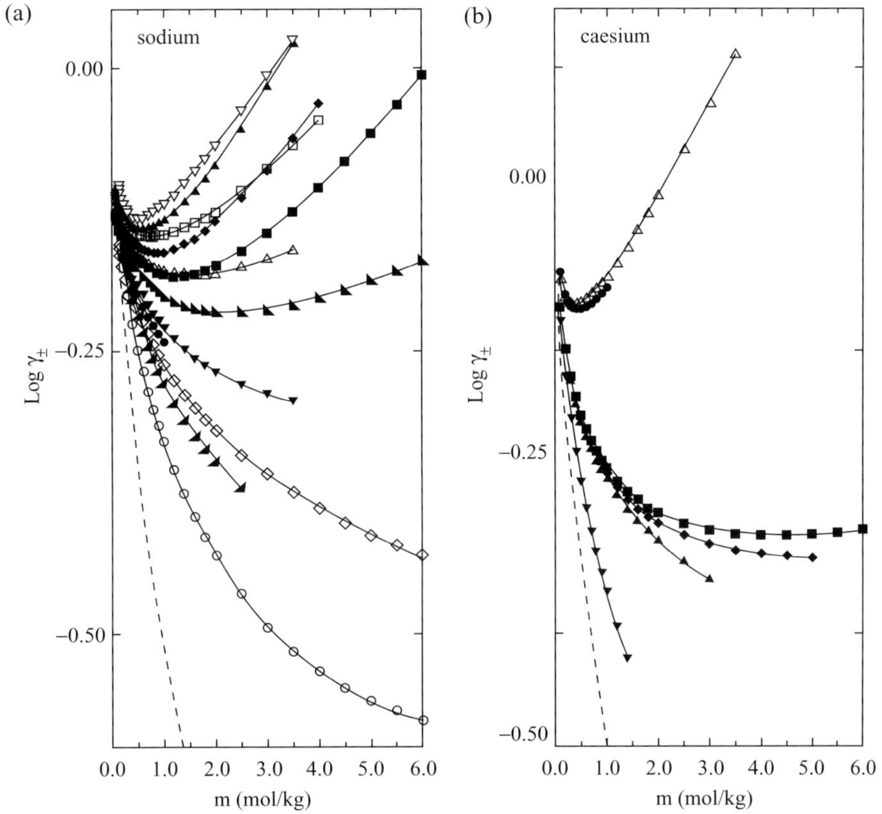

Fig. 3.2. Mean ionic activity coefficient for different sodium (a) and caesium (b) electrolytes in aqueous solution between 0 and 6 m, plotted as Log$\gamma_\pm$ versus m, at 25 °C. The lines are the fitting curves obtained with Equation (3.11). (a) Sodium salts: NaF (●); NaCl (■); NaBr (♦); NaI (▲); $NaClO_3$ (▼); $NaClO_4$ (◣); $NaBrO_3$ (◢); $NaNO_3$ (◇); NaHCOO (△); NaOAc (▽); NaSCN (□); NaH_2PO_4 (○); Debye–Hückel theory for 2:1 electrolyte (□); Li_2SO_4 (♦). (b) Caesium salts: CsOH (●); CsCl (■); CsBr (♦); CsI (▲); $CsNO_3$ (▼); CsOAc (△). In both plots, the dashed line indicates the prediction of the Debye–Hückel theory for a 1:1 electrolyte. Data taken from Ref. 8.

depending on the specific solute ion pair. In this region the effects of short-range interactions become more important and eventually dominate [8].

The deviation of the salt solution from ideality is expressed as the chemical potential change with respect to the standard state (Equation (3.10)). In the standard state (infinite dilution) the solution obeys Henry's law and the activity coefficient γ is taken as unity.

In a salt solution cations and anions are always together, unless some sort of work is applied (such as electrical work). This means that we cannot separate $\Delta\mu_+$ from $\Delta\mu_-$ and therefore $\Delta\mu_{\text{tot}} = \Delta\mu_+ + \Delta\mu_- = RT(\ln a_+ + \ln a_-) = RT(\ln c_+c_- +$

$\ln \gamma_+\gamma_-$). To continue, we need to consider an average activity coefficient given by $\gamma_\pm = (\gamma_+\gamma_-)^{1/2}$. This is valid for a 1:1 electrolyte (such as CsBr). For a generic salt M_pX_q (such as $MgCl_2$ or $Al_2(SO_4)_3$) then the average $\gamma_\pm$ is given by $(\gamma_+^p\gamma_-^q)^{1/(p+q)}$. This assumption means that the cations do not affect the behaviour of anions (in terms of deviation from the ideal state), and that their effect is equally distributed all over the solution. In doing so we already neglect the specific interactions among ions. The approximation is true at very low concentrations, but it cannot hold for real solutions (see for example later the formation of ion pairs).

The expression that relates the activity coefficient $\gamma_\pm$ to the ionic strength I of the salt solution is obtained as:

$$\mathrm{Log}\gamma_\pm = -\frac{A|z_+z_-|\sqrt{I}}{1+Ba\sqrt{I}} + bI \approx -A|z_+z_-|\sqrt{I} + (A|z_+z_-|Ba + b)I \quad (3.11)$$

which shows that the activity coefficient data can be fitted with an electrostatic term ($\sqrt{I}$) and a non-electrostatic term that depends on the specific salt. The second term vanishes at low concentrations, and the Debye–Hückel limiting law model applies. Electrostatics dominates the behaviour in the solution.

For moderately concentrated salt solutions the second term dominates, electrostatic interactions are screened, and salt specificity shows up. In Fig. 3.2 we show the experimental values for aqueous solution of caesium and sodium salts. The lines represent the fitting curves to the data according to Equation (3.11).

This equation, known for a long time [8], lies at the core of all inferences on the properties of electrolytes. It can be modified to include a more refined treatment of the electrostatic interactions and those due to 'hard core' interactions [16]. However, the inferences from comparison of theory and experiment remain the same.

A careful perusal of the $\mathrm{Log}\gamma_\pm$ versus molality m for a large set of different electrolytes in water at 298 K shows the following points:

1 In dilute solutions, $\gamma_\pm$ decreases with m. Often (but not always) $\gamma_\pm$ reaches a minimum value, and then increases and may reach a very large value.
2 In general, activity coefficients that reach very high values are associated with greatly hydrated ions.
3 Activity coefficients that have moderate values are interpreted in terms of the formation of ion pairs (that essentially remove free ions from the solution).
4 Very low values of $\gamma_\pm$ result from complex ion formation (zinc and cadmium salts).
5 The chloride class shows that all curves reach a minimum after which $\mathrm{Log}\gamma_\pm$ increases again: these minima shift to larger concentration values from H^+ to Cs^+.
6 For chlorides, bromides, iodides, nitrates, chlorates and perchlorates, the order in $\gamma_\pm$ is: Cs < Rb < K < Na < Li. It is reversed for hydroxides, formates and acetates.

7 Keeping the cation constant, the order for simple halide ions is Cl < Br < I for lithium, sodium and potassium, but it is inverted for rubidium and caesium.
8 The potassium salts with nitrate, chlorate and perchlorate show low values of $\gamma_\pm$, probably due to the formation of ion pairs. Instead, perchlorates of bivalent cations show very high values of activity coefficients.

The peculiar behaviour of $CdCl_2$ aqueous solutions is well known: while strong electrolytes present minima in their Log$\gamma_\pm$ vs. m plots, in the case of cadmium chloride the activity coefficient keeps decreasing almost to the saturation limit [17]. This behaviour was ascribed to the formation of ion pairs, and of different complexes between Cd^{2+} and Cl^- ions ($CdCl^+$, $CdCl_2$, $CdCl_3^-$ and $CdCl_4^{2-}$). Similar trends have been recorded for $CdBr_2$ and CdI_2; and for zinc salts, but with the order reversed.

3.2.3 Osmotic coefficients

The molal osmotic coefficient ϕ is defined through:

$$\ln a_\mathrm{w} = -\frac{\nu m M_\mathrm{w}}{1000}\phi \tag{3.12}$$

where a_w and M_w are the activity and molecular mass of the solvent (water), ν is the number of ions produced by one mole of electrolyte (2 for NaCl, assuming that the salt is a 'strong' electrolyte, i.e. it dissociates completely), and m the molality of the solution. Using the Gibbs–Duhem equation it is straightforward to derive the integrated form [8]:

$$\ln \gamma_\pm = (\phi - 1) + \int_0^m (\phi - 1) d \ln m \tag{3.13}$$

that relates $\gamma_\pm$ to ϕ.

Figure 3.3 shows the osmotic coefficient for some common salts in aqueous solution at 25 °C as a function of molality.

Activity coefficients or osmotic pressures are thermodynamic entities that involve integrals of ion–ion pair distribution functions of statistical mechanics, or equivalently potentials of mean force. As integral quantities they are not too sensitive to adjustable parameters.

Initially the decrease in the curves is explained by the Debye–Hückel theory which gives rise to a net attraction. Around 0.1 M, however, the curves turn up. This reflects the onset of repulsive short-range forces, and is modelled by an additional hard-core interaction. The parameter that occurs is the sum of the ionic radii of cation and anion (anion–anion and cation–cation are assumed not to come

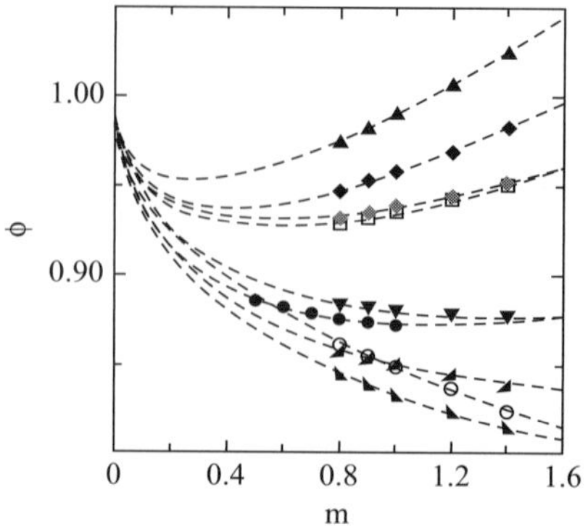

Fig. 3.3. Osmotic coefficient (ϕ) as a function of salt concentration (m, in molal units) for: NaF (•), NaCl (□), NaBr (♦), NaI (▲), $NaClO_3$ (▼), $NaBrO_3$ (◣), $NaNO_3$ (◢), NaOAc (♦) and NaSCN (○). Values taken from Ref. 8. The lines are the fitting curves for the relationship $\phi = 1 + A\sqrt{m} + Bm$.

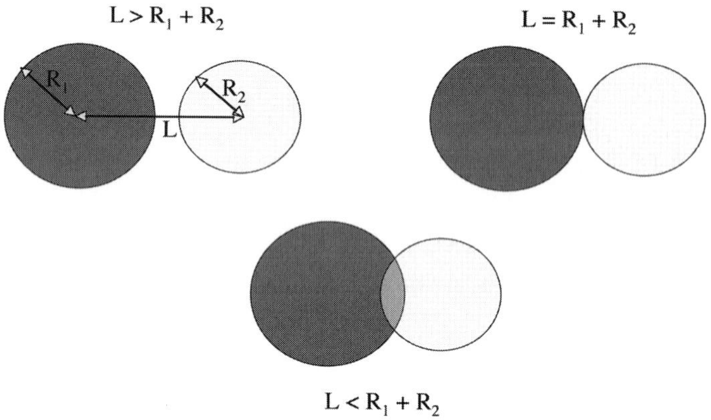

Fig. 3.4. Upper left: 'hydrated' ions have effective hard cores greater than the sum of bare ion radii. Upper right: unhydrated ions can contact each other. Lower: sometimes to fit the data the cations and anions have to interpenetrate. This is impossible and the smaller fitted radius required reflects compensation for a missing attractive force.

close to each other due to electrostatic repulsion; at 1 M the screening length is 3 Å and electrostatics is irrelevant). In fact except for (presumed) weakly hydrated ion pairs such as Na^+Cl^-, the fitting parameter does not coincide with this distance. A fitting parameter greater than the sum of ionic radii might be accommodated by invoking hydration, which provides a larger radius, or interpenetration of such shells (see Fig. 3.4).

The theory 'works' for some alkali halide pairs in the sense that the radii deduced are consistent and additive, e.g., a(LiCl) – a(LiBr) = a(NaCl) – a(NaBr), etc. [16].

Such an explanation seems to capture part of but cannot be the whole story. For some simple ion pairs such as Cs^+Cl^- or Cs^+Br^- that are not expected to be hydrated, the data can only be accommodated by using a 'hard core' that is less than the sum of the two crystallographic radii. That makes sense only if there is another missing attractive force operating between the ions, and this is indeed so, as we shall see. The same is true for many other electrolytes such as nitrates and sulfates, the one highly nonspherical and the other spherical.

The same modified Debye–Hückel theory is used to 'predict' changes in the pH of a buffer with added salt [18,19].

But the hard-core radii that have to be invoked vary from buffer to buffer and salt to salt and are often clearly unphysical. A 'radius' of 3 nm for a simple buffer anion is not meaningful.

Extensive sources of data with parameters that 'predict' properties of mixed electrolyte properties at high concentrations are those available from US Geological Survey publications. The prediction of such data is essential to earth sciences and to industry. The absence of a predictive theory of specific heats is a major safety issue. To compare with theories we note also that the experimental measurements are done at constant pressure (Gibbs free energy, Lewis Randall picture). The theories of free energy of the electrolyte solution are done at constant density (the Helmholtz free energy ensemble of statistical mechanics, McMillan Mayer picture) [16].

A comparison of theory and experiment therefore requires a conversion from constant pressure to constant density, a conversion not usually done.

3.2.4 Ion pairs

When the attraction between two ions of opposite charge is greater than the thermal agitation that keeps them apart, they can form a new entity, stable enough to remain intact in spite of the collisions with the other molecules. The scheme for ion pair formation is: $M^+ + X^- \rightleftharpoons MX$. Ion pairs do not have charges, but they do have dipole moments; therefore for example they do not participate in electrical conductivity, but they do modify the thermodynamics of salt solutions because $2N$ charged species are substituted by N dipolar entities. If the original salt is asymmetric, the situation is complicated by the formation of still ionic but new species, for example $CaCl^+$ from $CaCl_2$. According to Bjerrum, the criterion that decides whether two ions form a pair is when their mutual distance becomes less than a particular value d. This is chosen somewhat arbitrarily as the distance at which the mutual potential energy of the two ions equals $2k_BT$, i.e., $|z_1z_2|e^2/\varepsilon_r d = 2k_BT$. For a 1:1 electrolyte in water at 298 K, $d \approx 3.57$ Å. If two ions of different charge approach closer than d, then they form an undissociated ion pair.

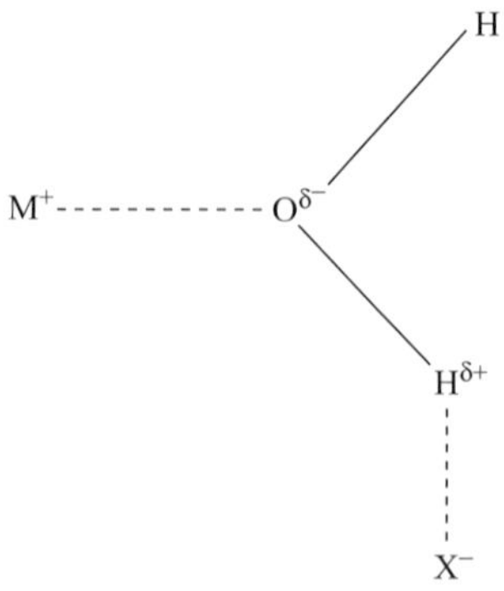

Fig. 3.5. Interaction between water and ions.

Therefore the Debye–Hückel model applies only to those ions at a distance greater than *d*.

Solvents with smaller dielectric constant will favour the formation of ion pairs [12]. In a variety of cases the formation of ion pairs in water solution is reported. Very interestingly, this occurs for several nitrates that exhibit strong anomalies in the activity coefficient trend. And more interestingly, this behaviour was attributed to the planar structure of this anion, which can allow closer approaches than would be so if they are modelled inappropriately, as effective spheres [8]. Usually polyatomic anions (ClO_3^-, ClO_4^-, BrO_3^-, $H_2PO_4^-$) and large scarcely solvated cations (Rb^+ or Cs^+) are considered to form ion pairs. Robinson and Stokes concluded that if Cs^+ is not solvated, it promotes polarization effects in oxoacid anions [8].

A particularly intriguing effect is that shown by hydroxides. In fact while chlorides, bromides, iodides, nitrates, etc. show a trend for $\gamma_\pm$ of $Li^+ > Na^+ > K^+ > Rb^+ > Cs^+$, consistent with the decreasing hydration from lithium to caesium, the reverse order is found for $\mathrm{Log}\gamma_\pm$ with hydroxides. To account for this behaviour, Robinson and Harned proposed a mechanism known as 'localized hydrolysis'. The mechanism involves hydration water molecules and anions that can operate as proton acceptors (formate, acetate, phosphates) as depicted in Fig. 3.5.

Apparently a similar effect could also take place with halides. The picture shows that a cation M^+ polarizes a nearby water molecule, which then is sandwiched between the positive ion and the corresponding counterion X^- that interacts directly with the hydrogen of the same water molecule.

These anomalies indicate that it is the specific nature (i.e. the chemistry) of the ions that determines their behaviour in solutions that are not dilute. The onset of specific behaviour derives from different contributions, depending on the structure of the interacting ions, and their delicate balance results in the experimentally observed deviation from the theoretical expectation.

3.2.5 Other physicochemical parameters

The dependence of the activity coefficient on the salt concentration for moderate values of the ionic strength (see Equation (3.15)) is found also in other parameters, such as the viscosity and the solubility of organic substances in water. This observation may seem trivial, as the activity coefficient reflects the local specific interactions that involve the ions in solution. However, it brings about a significant insight to the discussion of the interplay between electrostatic and, so far in our development, as yet unquantified hydration and non-electrostatic forces.

The viscosities of aqueous solutions of electrolytes show significant specific salt effects, with a strong dependence on the nature and concentration of the ions [10]. Some ions increase the viscosity, while others make water more fluid. This different behaviour is the basis for a classification of salts into kosmotropic and chaotropic species [20–22].

Kosmotropes are supposed to 'bind' water molecules more strongly and more closely. (Examples are phosphate, carbonate, sulfate, fluoride, magnesium and aluminium.) Chaotropes establish weak interactions with water and actually prefer to adsorb at interfaces, so escaping from solvent molecules (iodide, thiocyanate, perchlorate, caesium).

The empirical expression:

$$\frac{\eta}{\eta_0} = 1 + A\sqrt{c} + Bc \tag{3.14}$$

relates the viscosity of the solution (η) with respect to the viscosity of pure water at the same temperature (η_0) to the concentration c of the solute. According to Jiang and Sandler [23], the coefficient A reflects the ion–ion (electrostatic) interactions and the ionic mobilities, while B depends on (non-electrostatic) solute–solvent interactions. When B is positive, $\eta > \eta_0$, and the salt is classified as 'structure-making' or kosmotropic (ordering). Instead when $B < 0$, $\eta < \eta_0$, and the electrolyte is a 'structure breaker' or chaotropic (disordering). Usually Equation (3.14) holds well for $c < 0.1$ M, and for more concentrated salt solutions further terms are needed (for example an extra term Dc^2 is added) [20].

Similarly, the solubility of a polar non-electrolyte (such as benzene or a protein) in salt solutions follows an equation that defines the Setschenow constant:

$$\mathrm{Log}\frac{S_\mathrm{w}}{S_\mathrm{s}} = k_\mathrm{s}c_\mathrm{s} \tag{3.15}$$

where S_w, S_s, k_s and c_s are the solubility of the polar solute in pure water and in the salt aqueous solution, the Setschenow constant and the electrolyte concentration, respectively [24–26]. The constant k_s is related to the ion induction term (see Equation (3.5) and Chapter 7 in Ref. 2).

Long and McDevit had already in 1951 pointed out that a square root term ought also to appear in Equation (3.15) [27] but at that time there was not sufficient experimental evidence to support this expectation. Later von Hippel and Schleich [28] showed that the solubility of proteins in water did indeed satisfy the equation:

$$\mathrm{Log}\frac{S_{\mathrm{w}}}{S_{\mathrm{s}}} = -k_i\sqrt{c} + k_{\mathrm{o}}c_{\mathrm{s}} \tag{3.16}$$

where k_i and k_o (both positive) are the salting-in and salting-out constants for each electrolyte. First – at low concentration, where the term in $\sqrt{c}$ dominates – the protein solubility increases (salting-in). Then, at higher salt concentrations (when the term in c dominates) the protein solubility decreases (salting-out) [25].

More recently it has been shown, for example in the case of the critical micelle concentration (CMC) of some surfactants [29], in the optical rotation of amino acids [30,31] and for the growth of some bacterial strains in the presence of salts [32], that these quantities are well fitted by equations that comprise a $c^{1/2}$ and a c term.

The conventional interpretation of these data that are fitted to an equation like (3.11) is as follows:

- It establishes that both electrostatic and non-electrostatic forces are at work. The electrostatic term is proportional to $c^{1/2}$ and therefore dominates at low concentrations, in the very same way that it operates within the Debye–Hückel model.
- While the electrostatic contribution is constant for all electrolytes, on the other hand, salt specificity is introduced by the linear term in c. This dominates at higher concentrations of electrolyte, and depends on the nature of that specificity.
- The equation introduces a still semi-empirical but more general way to classify and capture specific salt effects.

However, there are several complicating factors in the electrostatic interactions in mixed electrolytes (proteins plus salt) that mean such an interpretation is much too simple. This will be discussed below.

It is significant that inferences of hydration 'radii' from a variety of different experimental techniques give as many fitted radii as there are techniques, indicating that something is missing.

3.2.6 Further quantification of ion specificity

Usually salt effects are reported in the literature as a more or less significant deviation of one or more properties due to the addition of an electrolyte to a particular system under study. The experimental parameters that can be monitored are: nature of and/or concentration of the electrolyte, concentration of the other co-solutes, temperature, pressure, pH, etc.

The experimental results with the different salts are then compared to the salt-free case, and ordered in terms of specific salt-induced changes. This results in the compilation of the 'Hofmeister series' that was obtained for the first time by F. Hofmeister in the 1870s, for the salt-induced precipitation of egg albumin from water dispersions. He also studied the same effects in colloidal suspensions of ferric oxide and sodium oleate. Typical concentrations that induce precipitation are around 0.3–1 M. At such a salt concentration the Debye length is 0.3 nm and electrostatics can have no direct influence on matters [33].

With fixed cation the effectiveness of anions in precipitation follows the sequence:

$$SO_4^{2-} > HPO_4^{2-} > CH_3COO^- > Cl^- > NO_3^- > Br^- > ClO_3^- > I^- > ClO_4^- > SCN^-$$

That is, for example, sodium sulfate will cause precipitation of the protein at a lower concentration than sodium iodide. With fixed anion the effectiveness of cations follows the sequence:

$$NH_4^+ > K^+ > Na^+ > Li^+ > Mg^{2+} > Ca^{2+} > C(NH_2)_3^+$$

However, the order is not always the same; there are reorderings along the series, depending on the specific case. This reflects in the main specific counterion interactions with specific surface groups of the particular protein (depending on its isoelectric point and pH). In some cases there is a total inversion of the series.

From the beginning the explanation of reversal of the Hofmeister series with change of pH above and below the isoelectric point has posed a major challenge, only recently coming to be understood [34–37]. Similar sequences are followed by activity coefficients (cf. Figs. 3.2 and 7.1).

However, if sulfate has a certain effect, perchlorate will have the opposite one. This is the conceptual basis of the Hofmeister series. Historically, and due to the water structure model, ions have been categorized into two main classes: kosmotropes and chaotropes. Apparently, both classes perturb the structure of the water hydrogen bonding. The former strongly interact with solvating molecules and modify the structure of water by increasing its ordering, while the latter do not firmly 'bind' solvent molecules and therefore decrease the pre-ordered structure of water. Although this model is still debated, and more recent results corroborate some criticism [38,39], it is useful albeit confusing to keep the two terms 'kosmotrope' and 'chaotrope'.

It is interesting to observe that kosmotropic and chaotropic ions have specific roles in nature. Although the number of chaotropes is larger than that of kosmotropes (especially for anions), most of the inorganic and biological world is based on a few dominating kosmotropic species: Na^+, Ca^{2+}, Mg^{2+}, SO_4^{2-}, SiO_4^{4-}, CO_3^{2-}, phosphate (as PO_4^{3-}, HPO_4^{2-} and $H_2PO_4^-$). Bones, teeth, shells and rocks represent the most important examples. Apparently air pollution depends also on the kind of inorganic particulate that is present in the atmosphere [40].

In order to quantify the effect of different salts, and to investigate the mechanism involved in specific salt effects, various experimental parameters can be compared.

Some of these are simply empirical quantities such as the lyotropic number or the Setschenow constant (see Chapter 7). Other parameters are the electromagnetic frequency-dependent polarizability of the ion in solution (α), its free energy or entropy of hydration (ΔG_{hydr} or ΔS_{hydr}), the surface tension molar increment ($\sigma = \partial\Delta\gamma/\partial c$) and the partial molal volume. (It will emerge that these are directly related to the quantum mechanical dispersion forces of Chapter 2, and reflect the specific nature of each ion. The effects of anions are usually larger than those of cations, and this, we shall see, seems related to the excess of electrons in anions and through that to dispersion forces missing from the conventional theories.)

A wide variety of studies have correlated the specific salt effects to one or more of these parameters that show up in Hofmeister effects.

At this point in our outline of the classical theory we remark that an entire intuition on ions in solution has been built up, simple conceptually and apparently sensible, from the idea of electrostatic forces, augmented by 'hydration', ion-induced 'water structure' itself a consequence of charged dipole and 'hydrogen bond' interactions. The theory lacks predictability as the parameters that are used to characterize specificity are sometimes obviously just fitting parameters that have no particular direct interpretation. Something is missing. These ideas have been extrapolated further and further beyond their range of admissibility into the biological and biochemical milieu. Taken too far they wreak havoc, as we shall see.

The Hofmeister effects show up universally, as we shall see in later sections [21,22].

3.2.7 Interfacial energies due to electrolytes

Another illustrative example of specific salt effects is the variation of the surface tension at the air–water interface due to the addition of an electrolyte [41]. At 25 °C water has a surface tension of $\gamma = 72$ mN/m. If a salt is added, the interfacial tension can increase or decrease with concentration depending on the particular salt. If the salt concentration is greater than 0.1 M, $\Delta\gamma$ is simply proportional to the electrolyte concentration c. Figure 3.6 illustrates the variation of $\Delta\gamma$ with the salt concentration for a wide range of different electrolytes.

The original theory is due to Onsager and Samaris [43] and is based on the repulsive electrostatic image potential that an ion experiences near the air–water

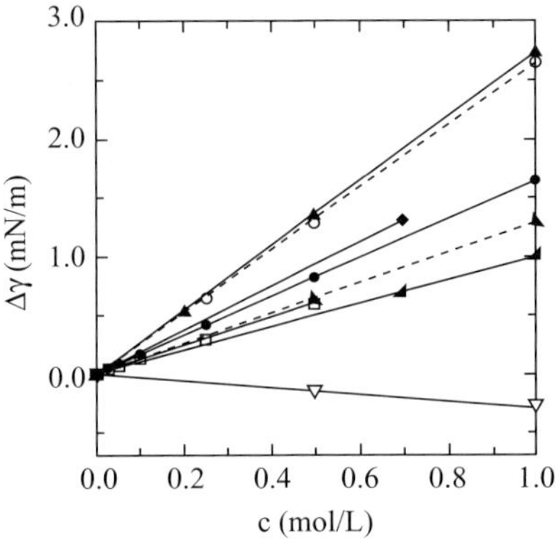

Fig. 3.6. Surface tension change ($\Delta\gamma$) versus electrolyte concentration c (molar units) for different aqueous salt solutions at 20 °C. Most simple electrolytes exhibit a linear relationship. HCl (▽), NaI (◢), $NaNO_3$ (□), NaBr (◣), NaCl (•), NaOH (♦), Na_2CO_3 (○) and Na_2SO_4 (▲). Adapted with permission from Ref. 42. Copyright 2007 American Chemical Society.

interface, due to the change in dielectric constant across the interface. We omit details, summarize the results, and refer to Refs. 41 and 43. We shall need for later reference an analytic expression for the change in interfacial tension at the limit of extremely low concentrations. It is [41]:

$$\Delta\gamma = -\sum_i c_i v_i^2[\ln(2\kappa_{\mathrm{D}}a_i) + O(\kappa_{\mathrm{D}}a_i) + \cdots] \tag{3.17}$$

This is truly a limiting law analogous to the Debye–Hückel theory. The validity of the expansion is restricted to $2\kappa_D a \ll 1$, where a is the size of the ion. In practice this (theoretical) logarithmic form is limited to concentrations much lower than 1 mM, and the occurrence of an ion size 'a' in the formula is an artefact of linearization of the image charge distributions. At higher realistic concentrations, only one length scale is involved in reality, the Bjerrum length $e^2/(\varepsilon_r k_B T)$. But if this were the whole story, there would be no ion specificity at all, and all 1:1 salts should in principle produce the same $\Delta\gamma$. Besides, an electrostatic origin for the experimental results is ruled out because it would *always* give an increase in interfacial tension with salt addition. So some essential forces that drive the phenomenon are missing.

3.3 A first appearance of dispersion forces

This problem remained unanswered for a very long time. Stairs [44] and others dating back to Onsager argued cogently for the inclusion of image charge-induced dipole interactions. It is an important question. While one can argue away the necessity for non-electrostatic forces in the electrolyte problem by invoking electrostatically induced 'hydration', it is not possible here.

It can be shown that dispersion forces dominate the electrostatic image forces and indeed electrostatic interactions between ions above 0.1 molar concentrations where ion specificity begins to appear strongly [41,45]. These ionic dispersion forces depend on ionic polarizabilities and their frequency dependence, i.e. on the electronic configuration of the ionic species. Their interactions with an interface can be attractive or repulsive depending on the nature of the ion and they can account for ion specificity [45].

3.3.1 Simplified sketch of technical details and digression

We give here a brief schematic account of how this comes about. Ions in an aqueous solution experience different interactions near the air–water interface. The image forces are repulsive. This leads to a depletion of ions at the interface, and to a decrease in interfacial tension. If x is the distance of the ion from the interface, $U_{\pm}(x)$ is the external potential, and ϕ is the electrostatic potential due to the ions, then the accumulation of the ions at the interface – expressed as number of ions per unit area excluded from the surface – can be written as:

$$\Gamma_{\pm}(c) = c \int_0^{\infty} \exp\left[\frac{U_{\pm}(x) \pm e\phi}{k_B T} - 1\right] dx \tag{3.18}$$

The ratio $\Gamma_{\pm}/c$ is essentially the length over which the ions are excluded. The Gibbs adsorption equation provides the value for $\Delta\gamma$ as:

$$\Delta\gamma(c) = -k_B T \int_0^{c} \left[\frac{\Gamma^{+}(c) + \Gamma^{-}(c)}{c}\right] dc \tag{3.19}$$

Onsager and Samaris assumed that the potential experienced by the ions is simply the repulsive electrostatic image potential $U_{\text{im}}(x)$ calculated as:

$$U_{\text{im}}(x) \approx \frac{e^2 \Delta}{4\varepsilon x} \exp\left(-2\kappa_D x\right) \tag{3.20}$$

where $\Delta = \frac{\varepsilon_w - \varepsilon_{\text{air}}}{\varepsilon_w + \varepsilon_{\text{air}}} \approx 1$, and ε_w and ε_{air} are the dielectric constants for water and air. κ_D is the inverse Debye length, and c is the molar concentration of the 1:1 salt. At 300 K, $\kappa_D = 0.329\sqrt{c}$ Å^{-1}. $U_{\text{im}}(x)$ does not depend on the charge of the ion, and since there is no separation of charges normal to the interface, the electrostatic potential ϕ is zero. This is an oversimplified model that does not explain the experimental findings. Onsager and Samaris had already noted that other forces must be involved.

When ions approach the air–water interface, they displace some water molecules. If α_w and α_i are the polarizability of water and of the ion, then the dispersion potential $U_{disp}(x)$ can be shown to be schematically of the form:

$$U_{disp}(x) = \frac{(n_w^2 - n_{air}^2)\alpha^*(0)\hbar\omega_i}{8x^3} = Bx^{-3} \tag{3.21}$$

where n_w and n_{air} represent the refractive index for water and air, $\alpha^*(0)$ is the static excess polarizability of the ion in solution, and $\hbar\omega_i$ is a typical ionization potential of the ion. $\alpha^*(0)$ is the difference in polarizability of an ion compared to an equivalent volume of water and can be deduced from the partial molar volume of the bulk aqueous solution. Considering that ω_i is in the IR–UV range of frequency, e.g. 10^{15} rad/s, $\alpha^*(0) \approx 2$ Å^3 and $n_w = 1.33$, a realistic estimate for B is $2{\cdot}10^{-50}$ J·m^3.

If the cation and the anion experience exactly the same dispersion potential $U_{disp}(x)$, so that there is zero charge density at any point in the solution, then there is no effect due to electrostatics, and the surface excess Γ of each species is simply proportional to the concentration c. At high concentration, the image potential is screened, in fact for $c = 1$ M, $\kappa_D \approx 3$ Å^{-1}, and dispersion forces dominate. If B is positive, then $\Delta\gamma$ is given by:

$$\Delta\gamma(c) = -2k_BTc\int_0^\infty \left[\exp\left(-\frac{B}{k_BTx^3}\right) - 1\right]dx = 2.7k_BTc\left(\frac{B}{k_BT}\right)^{1/3} \tag{3.22}$$

Usually $\partial\Delta\gamma/\partial c$ spans between $5{\cdot}10^{-4}$ and $1.6{\cdot}10^{-3}$ J·m^3·L/mol, which means that B must be between $1.7{\cdot}10^{-51}$ and $5{\cdot}10^{-50}$ J·m^3. Equation (3.22) correctly accounts for the linear dependence in c, as observed experimentally.

Electrostatics come back into play if anions and cations do not experience the same forces, and therefore a charge separation occurs. In this case ϕ must satisfy the Poisson–Boltzmann differential equation:

$$\nabla^2\phi = -\frac{4\pi e}{\varepsilon}\sum_\nu n_\nu \exp\left(-\frac{e_\nu\phi}{k_BT}\right) \approx \frac{4\pi\phi}{\varepsilon k_BT}\sum_\nu e_\nu^2 n_\nu = \kappa_D^2\phi \tag{3.23}$$

where the concentrations of the ionic species are expressed as:

$$c_\pm = c_0 \exp\left(\mp\frac{e\phi + U_{im} + B_\pm x^{-3}}{k_BT}\right) \tag{3.24}$$

The equation needs to be solved numerically, which is non-trivial. A mathematical consequence of overall charge neutrality is that the surface excesses for cations and

Table 3.1. *Values of the surface tension increment (in $\mu N{\cdot}m^2/mol$) for selected sodium electrolytes. Adapted with permission from Ref. 42. Copyright 2007 American Chemical Society*

Electrolyte	$\partial\Delta\gamma/\partial c$
NaCl	1.76
NaBr	1.71
Na_2SO_4	2.99
$NaNO_3$	1.35
$NaClO_4$	0.62
NaOAc	1.41
$NaClO_3$	0.72
NaOH	2.03

anions must be equal, although the two species have a different spatial distribution. For example, cations may have a negative excess polarizability and certainly have a smaller polarizability than anions. In that case they could be positively adsorbed and accumulate closer to the air–water interface.

Technical problems in the experimental measurements of γ at different salt concentrations make it difficult to determine an accurate value for $\partial\Delta\gamma/\partial c$. The purity of the electrolytes and the evaporation of the solvent may affect the final results, as well as the presence of a considerable amount of atmospheric gas at the interface.

Table 3.1 lists the experimental values for $\partial\Delta\gamma/\partial c$ as reported in the literature for different salts.

What emerges from such analyses [41,45] is that:

1. The inclusion of dispersion forces is essential.
2. Their inclusion gives results of the right order of magnitude.
3. It can also accommodate both increased and decreased interfacial tension with different electrolytes.
4. It can account for phenomena such as positive adsorption of KBr at the oil–water interface and negative adsorption at the air–water interface.
5. Many other ion-specific Hofmeister phenomena begin to fall into place, such as salt elution from Sephadex columns and activities of electrolytes.

We have treated ions and the air–water interface in the primitive model, that is, assumed that the profile of water at the interface behaves like continuum, bulk water of given dielectric properties. Clearly the interface has a profile of water structure spread over two or three water molecule diameters that also includes atmospheric gas diminishing from 1 M in the gas to 1 mM in the bulk water over the same distance. But despite the extreme simplifications, the numbers do come out about right. How they are affected by such complications, i.e. reality, is not known, and will be discussed further in Chapter 7.

3.4 Electrostatic forces at and between charged interfaces

3.4.1 The electrostatic double layer

The double-layer force between two colloidal particles is at the core of the entire subject. The literature dating back a century or more is vast. Good accounts of the theory and references to the literature can be found in the standard books. One way or another, directly or indirectly, it is embedded into the interpretation of nearly all measurements and characterization in colloid and electrochemistry. This includes pH, pK_as, buffers, zeta and streaming potentials, electrophoresis and separation technologies for macromolecules and ionic polymers such as DNA. It is the backbone of the DLVO theory of interactions between colloidal particles to be discussed in Chapter 4. Because of this centrality of the double layer, we cannot avoid a discussion of the classical theory to fix the background for later developments. It will turn out that the theory is deeply flawed. Worse than that, it is incredibly boring.

The classical theory idealizes two interacting colloid particles '2' separated by an electrolyte medium '1' as bounded by planar interfaces that have ionizable groups. The surface species dissociate in an electrolyte, leaving the surfaces charged. This sets up an inhomogeneous profile of cationic and anionic density. The ionic distribution follows from the equation:

$$div\mathbf{E} = div\nabla\phi(x) = 4\pi\frac{\rho(x)}{\varepsilon_r} \tag{3.25}$$

Here $\mathbf{E}$ is the electric field, $\phi(x)$ the potential at distance x from the interface, ρ the charge density. The boundary condition is that $(\varepsilon_1 E_1 = \varepsilon_2 E_2) = 4\pi\sigma$ where σ is the surface charge and ε_1 and ε_2 are the static dielectric constants of the particle and of water.

For a simple 1:1 electrolyte like NaCl, if the surface is negatively charged the charge density profile can then be described by the Boltzmann distribution:

$$\rho(x)_+ = \rho_0 \exp\left(+\frac{e\phi}{k_B T}\right); \qquad \rho(x)_- = \rho_0 \exp\left(-\frac{e\phi}{k_B T}\right) \tag{3.26}$$

where e is the unit charge and ρ_0 the salt concentration far from the surface.

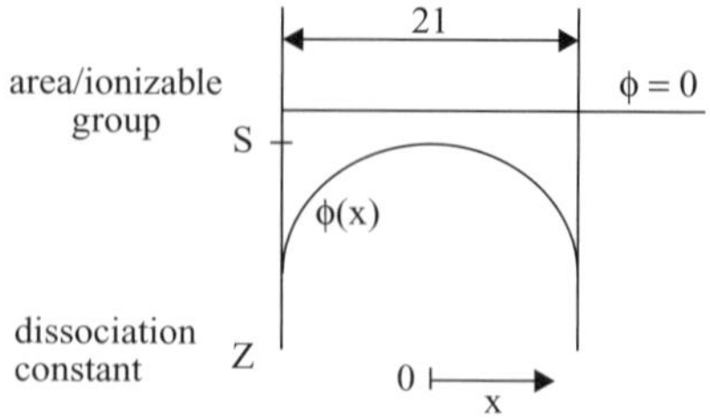

Fig. 3.7. Two planar charged surfaces (vertical lines) separated by an electrolyte solution. The boundary conditions taken to solve for the potential, and hence anion and cation distributions between the surfaces, are usually taken to be constant surface charge or constant surface potential. The difference between the potential ϕ at the mid-plane and in the bulk reservoir $\phi = 0$ produces an osmotic repulsive force due to the differences in ionic concentrations between and outside the interacting particles. The plates are separated by a distance $2l$. In general the charge comes about because the surfaces have ionizable groups of surface area S, each group having a dissociation constant Z. In that case the surface potential or charge is not constant. It is regulated due to the influence of the other surface. The surface pH then changes, as well as the ionic distribution near the plates with separation [50]. Adapted from Ref. 50 with permission from Elsevier.

Subject to the boundary conditions assumed, which could be constant surface charge or constant surface potential, the required self-consistent potential follows from the Poisson–Boltzmann equation:

$$div\nabla\phi(x) = \frac{4\pi}{\varepsilon_{\mathrm{r}}}\left[\rho(x)_{+} + 4\pi\rho(x)_{-}\right] = 8\pi\frac{\rho_0}{\varepsilon_{\mathrm{r}}}\sinh\left(\frac{e\phi}{k_{\mathrm{B}}T}\right) \tag{3.27}$$

For two interacting surfaces an additional boundary condition is required, that the derivative of the potential at the mid-plane is zero (see Fig. 3.7).

Once the problem is solved a standard, but not trivial, textbook argument gives the repulsive force between the plates per unit area as Equation (3.28).

$$F(l) = k_{\mathrm{B}}T\left[\rho\left(\tfrac{1}{2}\right)_{+} + \rho\left(\tfrac{1}{2}\right)_{-} - 2\rho_0\right] \tag{3.28}$$

The difference between the ionic densities at the mid-plane 1/2 and that in the bulk reservoir surrounding the particles gives rise to an osmotic pressure that gives the force. (For oppositely charged particles the force is attractive.)

The calculations can be extended to a multi-component mixture of ions. It can be modified to include shape and curvature. The substantial result is that the force per unit area has the form (when $\kappa_{\mathrm{D}}l > 1$):

$$F(l) = f(\sigma)e^{-\kappa_{\mathrm{D}}l} \tag{3.29}$$

The prefactor depends on the assumption of constant surface charge or surface potential and can be bounded.

At closer separations the formulae are very complicated and increase more rapidly at smaller distances, at which the assumptions of the model start to break down.

3.4.2 Charge regulation

In a more general and more realistic situation the assumption of constant charge or constant surface potential has to be modified. For example, one can imagine a situation with given bulk pH, and a mixture of sodium chloride at 0.15 M with a small admixture of 2.5% of the electrolyte being calcium ions, as in physiological solutions. The negatively charged surfaces of biomembranes contain ionizable groups, typically amines and carboxylates, that ionize depending on pH. Surface pH, and consequently degree of ionization of the interacting surfaces, changes with separation. The differences between bulk and surface pH are considerable. Further, even with such a small amount of divalent calcium the concentration of calcium at the surfaces is comparable to that of the monovalent sodium counterion. Because of the Boltzmann factor in Equation (3.26), the divalent calcium is attracted more strongly to the surfaces than the monovalent sodium and weakens the interaction. Charge regulation, a kind of recognition of one surface by another, occurs in all double-layer problems and is usually ignored. (The computations are not trivial.) A complete analytical treatment of the problem for mixtures of divalent and monovalent electrolyte is shown in Ref. 50.

As for bulk electrolytes and interfacial tensions, the theory could only be expected to work when electrostatic forces dominate others. As for electrolyte solutions too, a myriad complications and fitting parameters can be called into play. The Stern layer takes account of finite ion size. Inner and outer planes near the surfaces that describe the diameters of adsorbed hydrated ions of the electrolyte can be assigned different dielectric constants. With so many additional parameters such generalizations of theory lose all predictability.

One firm inference that can be made concerns mixed electrolytes containing multivalent ions. The Boltzmann factor that determines distributions of ions near charged surfaces depends exponentially strongly on ionic valence as well as the signs of the ionic charges. Because of this factor that enhances adsorption, an extremely small concentration of multivalent ion in the bulk electrolyte compared with that of a univalent ion such as sodium can have a very large concentration at a charged surface. This explains why it is that trace elements of multivalent ions such as chromium, selenium, vanadium, zinc, copper, etc. are so important in agriculture with poor soils lacking such elements.

3.5 Mixed electrolytes and pixie dust

The presence of impurities and multivalent ions affects the range of electrostatic interactions in real systems (Debye length).

In the real world a variety of phenomena occur that often make no sense in terms of conventional intuition based on the theories of interactions in colloid systems containing electrolytes that we have described. The Debye length is the principal measure of the range of electrostatic, double-layer forces operating in colloidal particle suspensions. In many applications the theories simply do not work. This is so, both in biological and in many chemical engineering applications. Some examples are given below.

While the breakdown is often due to the approximations already made in the primitive model, it sometimes resides in more complicated electrostatic phenomena than those captured by the Debye–Hückel theory (which involves, technically, linearization of a Poisson–Boltzmann distribution). These non-linear electrostatic effects are encompassed by more sophisticated statistical mechanical treatments such as the hypernetted chain approximation (HNC).

3.6 The Debye length in multivalent electrolytes

The range of Coulomb interactions in simple electrolytes and in colloidal particle (double layer) interactions is measured by the Debye length of the solution as in Equation (3.2) above. But the theory begins to break down for simple electrolytes containing multivalent ions. This is so, often at even small concentrations of multivalent species where hard core, specific ion, water structure, hydration, hydrophobic and other interactions ought to be irrelevant. Long-range electrostatic interactions should dominate.

The reason that the intuition is misleading and wrong is this: the range of screened electrostatic interactions built into Debye–Hückel theory that describes interactions emerges theoretically from a linearization of the Boltzmann factor. This describes the distribution of the ionic cloud around a charged ion, colloidal particle, surface or macromolecule. Around a highly charged ion this distribution is highly asymmetric. Linearization of the distribution is not possible. The distributions of charged ions with charge zq near a surface S, for example, is given by:

$$n_S(r) = n \exp\left[-\frac{zq\phi(r)}{k_B T}\right] \tag{3.30}$$

where $\phi(r)$ is the electrostatic potential and n the bulk concentration.

It turns out that in an asymmetric electrolyte the actual screening length λ is not the Debye length. The ordinary Debye length above is just the leading term in

Table 3.2. *Correction term to the classical Debye screening length for simple asymmetric electrolytes. B is the ratio of κ/κ_D*

Electrolyte	B (1 mM)	B (10 mM)	B (100 mM)
1:1	1.00	1.00	1.00
2:1	1.02	1.06	1.18
3:1	1.10	1.31	2.00

a so-called asymptotic expansion in the concentration. The statistical mechanical theory that leads to such results is very complicated. The result for the screening length for a $C_{\nu_1}^{z_1} A_{\nu_2}^{z_2}$ electrolyte is:

$$\lambda^{-1} = \kappa = \kappa_D \left[1 + \frac{7 \ln 3}{24\sqrt{2}} (c^{1/2}) \frac{(\nu_1 z_1^3 + \nu_2 z_2^3)^2}{(\nu_1 z_1^2 + \nu_2 z_2^2)^{3/2}} + O(c^{3/2} \ln c) \right] \tag{3.31}$$

Here $\lambda_D = \kappa_D^{-1}$ is the ordinary Debye length and c the salt concentration [51,52].

Technically, the validity of this expansion is limited by the condition that when the second term is about half that of the leading term the expansion will have broken down. When the analytic expansion breaks down and is no longer useful for computation, a different form then describes the nature of the electrostatic interactions. (In a symmetric electrolyte, e.g. 1:1 or 2:2 electrolyte, the second term in the full expansion cancels out identically and the simple limiting law Debye formula has a wider range of validity than one expects.) Extra correction terms in the analytic formula impose the additional condition $\kappa_D a \leq 1/2$, where a is the radius of the largest ion or colloidal particle in the mixture.

Table 3.2 shows this non-linear valence-dependent effect for simple 2:1 and 3:1 salts. Use of this correct form for the screening length does yield correct activity coefficients for simple asymmetric electrolytes [53]. At first sight this non-linear valence effect does not seem too serious when compared with other approximations involved. But small amounts of highly charged proteins or polymers in an electrolyte change matters considerably. Direct measurement of forces between mica in a solution of protein cytochrome c (a 12:1 or 8:1 electrolyte) shows how dramatic these effects can be (Table 3.3). Even at *micromolar* concentrations, the Debye length is wrong by a factor of 2 [54,55]. (The forces between proteins or between colloid particles are then 10-times smaller than expected, even at micromolar concentrations.)

These measurements confirm that the theory can be used accurately for prediction. The cited paper [54] gives a complete formula for the Debye length in an

Table 3.3. *Correction term to the classical Debye screening length for simple asymmetric electrolytes. Comparison of experimental decay lengths for double layer forces in cytochrome c solutions. This is a 12:1 electrolyte. Columns indicate: cytochrome c concentration (μM), Debye length from usual formula (κ_D^{-1}, in nm), actual Debye length predicted (λ, in nm) and measured screening length (nm)*

Bulk concentration (μM)	κ_D^{-1} (nm)	λ (nm)	Experimental decay length (nm)
0.4	54.4	44.7	45.6
0.87	36.9	28.0	26.0
2.2	23.2	15.4	16.0
3.6	18.1	11.0	11.2
4.1	17.0	10.0	10.5

arbitrary mixture of any number of components useful for adaptation for any mixed electrolyte system. It is [55]:

$$\frac{\kappa}{\kappa_D} = 1 + \frac{\sqrt{\pi}\ln 3}{4}\left(\frac{q^2}{\varepsilon k_B T}\right)^{3/2} \times \frac{\left(\sum_i \rho_i z_i^4\right)\left(\sum_i \rho_i z_i^2\right) - \sum_{i=1}^{n-1}\sum_{j=i+1}^{n} \rho_i \rho_j z_i^2 z_j^2 (z_j - z_i)^2}{\left(\sum_i \rho_i z_i^2\right)^{3/2}} \tag{3.32}$$

The effects can be very large for even trace amounts of highly charged species present in a background salt, as can be seen by direct substitution in the formula. The effects come about because of the highly non-linear nature of the problem.

3.7 Salting in and salting out

The Debye length in a whole mixture of protein and salt is the crucial determinant of the free energy of the solution. On addition of salt (or protein) the counterion can sometimes apparently ‘unbind’. But that counterintuitive result is only counter-intuitive if we have the idea that the Debye length is that appropriate to the bathing salt solution, not that of the whole mixed electrolyte that contains the proteins also. When the correct Debye length is used the observations are consistent with thermodynamics. (With ion binding to proteins and polyelectrolytes for which the counterion is, say, sodium, addition of further NaCl can cause apparent ‘unbinding’ of the sodium ion. This is apparently absurd. It violates thermodynamics. The law of mass action would demand that more, not less, sodium should be bound on addition of NaCl.)

Collins has explained the specific salt effect in protein solubilization as due to the specific ion adsorption at the protein surface and in terms of competition for water molecules between the protein and the kosmotropic ions [21]. This is an additional effect beyond the more fundamental result above and its consequences.

3.8 Applications to colloidal suspensions

The analytical formula for the true Debye length is subject to an additional restriction that limits its range of validity. The further condition ($\kappa_D a \leq 1/2$) must be satisfied. Here a is the radius of the largest ionic species or macroion in the solution. (A typical protein has a diameter around 3 nm, as for a typical micelle formed from surfactants. Theories will start to go pear-shaped certainly above 10 mM electrolyte that contains such beasts.) When the expansion breaks down we have to move to a different approximation scheme.

If we consider now a dilute colloidal suspension, the stability of the suspension depends on the screening length of electrostatic interactions across the background electrolyte solution. All standard theories used to characterize and interpret the surface charge or potential of the particles depend on the Debye length. This is so for direct force, zeta potential or biomembrane potential measurements, and even for pH determination, or of pK_as in biomembranes. If the Debye length, a fundamental parameter used to interpret the measurement, is incorrect, the potential inferred is not meaningful.

If the background mixed electrolyte satisfies the conditions of Equation (3.32), we can modify the standard theory to arrive at a correct, or at least better, interpretation of the measurement.

But suppose that the electrolyte contains a small concentration of multiply charged macroions. These might be proteins, polyelectrolytes or self-assembled micelles made up from charged surfactants. If the additional constraint $\kappa_D a \leq 1/2$ is violated also, the formula that provides our intuition breaks down. There are two effects. In such conditions the charge on the macroion is not its bare charge. It behaves as a 'dressed' macroion, as a macroion together with its cloud of bound counterions and co-ions. The free electrolyte ions and the dressed macroions are then the appropriate effective entities as far as the colloidal particles are concerned.

(This leads to a non-intuitive colloid–colloid effective pair potential which varies with the volume fraction of the colloidal particles present in the system. A striking consequence is that under some conditions, correlations in the spatial distribution of particles can persist over four orders of magnitude of the volume fraction. These correlations may not be a monotonically increasing function of the particle charge [56].)

If the formula still does not satisfy the conditions for applicability, a new kind of effective force which is oscillatory in nature due to the granularity of the liquid between the particles takes over and can act to stabilize the dispersion [57].

Some of the best experimental measurements of double-layer forces are discussed in Refs. 58–62. Such effects will be dealt with in Chapter 4.

Awareness of these complications is useful in unravelling the often magic or 'pixie dust' effects of trace amounts of impurities in real systems.

The complications occur nearly always in practical situations and for biosystems. For example, attempts can be made to understand bioadhesion of bacteria to substrates by varying (low) background salt concentration. The forces between the bacterial membrane and the substrate change with concentration. But standard theories of colloid chemistry for the relevant forces, double-layer interactions or zeta potentials show little correlation with experiment. The reasons are in part due to the fact that a bacterium approaching a surface experiences a change in local physicochemical environment. This induces changes in metabolism that results in extrusion of charged proteins as part of the process of building a compatible adhesion substrate. These extruded charged proteins are then part of the electrolyte separating substrate and membrane. In turn these proteins, extruded in response to the change in physicochemical conditions, affect the range of electrostatic forces operating, at least partially, in ways that can appear to be counterintuitive.

The same occurs in major industrial processes such as the separation (beneficiation) of minerals by froth flotation. In drilling muds used for oil exploration, very small amounts of polymeric additives such as phosphonates, as low as a millionth molar or less, have a dramatic effect on stability and viscosity of the colloidal clay suspension essential to operation of the drill.

There are many reasons besides electrostatic effects for such apparent breakdowns in classical theories that we shall explore in what follows.

Such effects, lumped together under the generic title 'pixie dust' or 'magic', are very familiar in chemical engineering and formulation industries, in 'nanotechnology', biochemistry and biotechnology. They are also known under the name Murphy's law.

References

[1] M. Diesendorf and B. W. Ninham, *J. Math. Phys.* **9** (1968), 745–752.
[2] J. Mahanty and B. W. Ninham, *Dispersion Forces*. London: Academic Press (1976).
[3] J. Israelachvili, *Intermolecular and Surfaces Forces*. New York: Academic Press (1991).
[4] D. F. Evans and H. Wennerström, *The Colloidal Domain*. New York: Wiley-VCH (1999).

[5] A. W. Adamson, *Physical Chemistry of Surfaces*. 2nd edn. New York: Interscience (1967).
[6] H. S. Harned and B. B. Owen, *The Physical Chemistry of Electrolytic Solutions*. New York: Reinhold (1958).
[7] R. H. Hunter, *Foundations of Colloid Science*. New York: Oxford University Press (2000).
[8] R. A. Robinson and R. H. Stokes, *Electrolyte Solutions*. London: Butterworths (1959).
[9] C. Kalidas, G. Hefter and Y. Marcus, *Chem. Rev.* **100** (2000), 819–852.
[10] G. Hefter, Y. Marcus and W. E. Waghorne, *Chem. Rev.* **102** (2002), 2773–2836.
[11] Y. Marcus and G. Hefter, *Chem. Rev.* **104** (2004), 3405–3452.
[12] Y. Marcus and G. Hefter, *Chem. Rev.* **106** (2006), 4585–4621.
[13] D. F. Evans and B. W. Ninham, *J. Phys. Chem.* **87** (1983), 5025–5032.
[14] M. Böstrom and B. W. Ninham, *J. Phys. Chem. B* **108** (2004), 12593–12595.
[15] M. Böstrom and B. W. Ninham, *Biophys. Chem.* **114** (2005), 95–101.
[16] B. A. Pailthorpe, D. J. Mitchell and B. W. Ninham, *J. Chem. Soc. Faraday II* **80** (1984), 115–139.
[17] W. W. Lucasse, *J. Am. Chem. Soc.* **51** (1929), 2597–2604.
[18] J. N. Butler and D. R. Cogley, *Ionic Equlibrium: Solubility and pH Calculations*. New York: John Wiley (1998).
[19] A. Voinescu, P. Bauduin, C. Pinna, D. Touraud, W. Kunz and B. W. Ninham, *J. Phys. Chem. B* **110** (2006), 8870–8876.
[20] H. D. B. Jenkin and Y. Marcus, *Chem. Rev.* **95** (1995), 2695–2724.
[21] K. D. Collins, *Methods* **34** (2004), 300–311.
[22] K. D. Collins, G. W. Neilson and J. E. Enderby, *Biophys. Chem.* **128** (2007), 95–104.
[23] J. Jiang and S. I. Sandler, *Ind. Eng. Chem. Res.* **42** (2003), 6267–6272.
[24] J. Setschenow, *Z. phys. Chem.* **4** (1889), 117–125.
[25] M. M. Ries-Kautt and A. F. Ducruix, *J. Biol. Chem.* **264** (1989), 745–748.
[26] N. Ni and S. H. Yalkowsky, *Int. J. Pharm.* **254** (2003), 167–172.
[27] F. A. Long and W. F. McDevit, *Chem. Rev.* **51** (1952), 119–169.
[28] P. H. von Hippel and T. Schleich, in *Structure and Stability of Biological Macromolecules*, ed. S. N. Timasheff and G. D. Fashman. New York: Marcel Dekker (1969).
[29] M. Lagi, P. Lo Nostro, E. Fratini, B. W. Ninham and P. Baglioni, *J. Phys. Chem. B* **111** (2007), 589–597.
[30] P. Lo Nostro, B. W. Ninham, S. Milani, L. Fratoni and P. Baglioni, *Biopolymers* **81** (2006), 136–148.
[31] S. Rossi, P. Lo Nostro, M. Lagi, B. W. Ninham and P. Baglioni, *J. Phys. Chem. B* **111** (2007), 10510–10519.
[32] P. Lo Nostro, B. W. Ninham, A. Lo Nostro, G. Pesavento, L. Fratoni and P. Baglioni, *Phys. Biol.* **2** (2005), 1–7.
[33] W. Kunz, J. Henle and B. W. Ninham, *Curr. Op. Coll. Interface Sci.* **9** (2004), 19–37.
[34] K. H. Gustavson, Specific ion effects in the behaviour of tanning agents toward collagen treated with neutral salts. In *Colloid Symposium Monograph*, ed. H. Boyer Weiser. New York: The Chemical Catalog Company (1926), and references therein.
[35] J. Loeb, *Science* **52** (1920), 449–456.
[36] S. Finet, F. Skouri-Panet, M. Casselyn, F. Bonnete and A. Tardieu, *Curr. Op. Coll. Interface Sci.* **9** (2004), 112–116 and references therein.
[37] M. Boström, F. W. Tavares, S. Finet, F. Skouri-Panet, A. Tardieu and B. W. Ninham, *Biophys. Chem.* **117** (2005), 115–122.

[38] A. W. Omta, M. F. Kropman, S. Woutersen and H. J. Bakker, *Science* **301** (2003), 347–349.
[39] L.-Å. Naslund, D. C. Edwards, P. Wernet, U. Bergmann, H. Ogasawara, L. G. M. Pettersson, S. Myneni and A. Nilsson, *J. Phys. Chem. A* **109** (2005), 5995–6002.
[40] D. Rosenfeld, *Science* **287** (2000), 1793–1796.
[41] B. W. Ninham and V. Yaminsky, *Langmuir* **13** (1997), 2097–2108.
[42] C. L. Henry, C. N. Dalton, L. Scruton and V. S. J. Craig, *J. Phys. Chem. C.* **111** (2007), 1015–1023.
[43] L. Onsager and N. N. T. Samaris, *J. Chem. Phys.* **2** (1934), 528–536.
[44] R. A. Stairs, *Can. J. Phys.* **73** (1995) 781–787.
[45] M. Boström, D. R. M. Williams and B. W. Ninham, *Langmuir* **17** (2001), 4475–4478.
[46] P. K. Weissenborn and R. J. Pugh, *J. Coll. Interface Sci.* **184** (1996), 550–563.
[47] A. A. Abramzon and R. D. Gaukhberg, *Russ. J. Appl. Chem.* **66** (1993), 1139–1146; 1315–1320; 1473–1480.
[48] R. Aveyard and S. M. Saleem, *J. Chem. Soc. Faraday Trans. 1* **72** (1976), 1609–1617.
[49] W. Kunz, P. Lo Nostro and B. W. Ninham, *Curr. Op. Coll. Interface Sci.* **9** (2004), 1–18.
[50] B. W. Ninham and V. A. Parsegian, *J. Theor. Biol.* **31** (1971), 405–428.
[51] D. J. Mitchell and B. W. Ninham, *Phys. Rev.* **174** (1968), 280–289.
[52] D. J. Mitchell and B. W. Ninham, *Chem. Phys. Lett.* **53** (1978), 397–399.
[53] M. A. Knackstedt and B. W. Ninham, *J. Phys. Chem.* **100** (1996), 1330–1335.
[54] P. Kékicheff and B. W. Ninham, *Europhys. Lett.* **12** (1990), 471–477.
[55] T. Nylander, P. Kékicheff and B. W. Ninham, *J. Coll. Interface Sci.* **164** (1994), 136–150.
[56] B. Beresford–Smith, D. Y. C. Chan and D. J. J. Mitchell, *J. Coll. Interface Sci.* **105** (1985), 218–234.
[57] R. M. Pashley and B. W. Ninham, *J. Phys. Chem.* **91** (1987), 2902–2904.
[58] J. N. Connor and R. G. Horn, *Langmuir* **17** (2001), 7194–7197.
[59] R. Manica, J. N. Connor, R. R. Dagastine, S. L. Carnie, R. G. Horn and D. Y. C. Chan, *Physics Fluids* **20** (2008), 032101 (1–12).
[60] R. G. Horn, M. Asadullah and J. N. Connor, *Langmuir* **22** (2006), 2610–2619.
[61] J. N. Connor and R. G. Horn, *Faraday Discuss.* **123** (2003), 193–206.
[62] R. Manica, J. N. Connor, S. L. Carnie, R. G. Horn and D. Y. C. Chan, *Langmuir* **23** (2007), 626–637.

4

The balance of forces

4.1 Forces in the DLVO theory of colloidal stability

For over 50 years the theory of the stability of colloid suspensions due independently to Deryaguin and Landau [1] and Verwey and Overbeek [2] held centre stage in the field. Following the derivation from quantum mechanics of the attractive force between two molecules, it was a straightforward matter to add these forces up to arrive at a force between two model colloidal particles. This potential of interaction per unit area behaves, for two particles modelled as planar interfaces a distance l apart as, $V(l) \approx -A/\left(12\pi l^2\right)$ where A is the Hamaker constant (see Chapter 2).

The double-layer force of repulsion between the (charged) surfaces behaves at large distances of separation as $V(l) \approx f \cdot \exp(-\kappa_{\mathrm{D}} l)$. Here κ_{D} is the Debye length, which depends on salt concentration, and the prefactor is a complicated function of the surface charge or potential (see Chapter 3). Then a combination of these forces gives rise to a predicted net force of attraction like that shown schematically in Fig. 4.1.

These forces depend on geometry assumed for the particles (see Chapter 2). The insertion of this potential into a theory of coagulation of particles predicted whether the particles would stay in suspension or flocculate. The particles could flocculate into a 'secondary minimum', with a barrier usually at $l \sim 1/\kappa_{\mathrm{D}}$.

Or in the absence of such a barrier they could (theoretically) adhere into a deep 'primary' minimum at a distance of a few molecular diameters. The rate of flocculation could be measured and the electrostatic potential correlated with measurements of single particle mobilities from electrophoresis or zeta potential measurements, standard tools for characterizing colloidal particles [4–6].

The interpretation of a surface potential measurement is not an independent one. It relies on the theory of the double layer itself. If the theory is deficient the interpretation is flawed, and this injects a consistent inconsistency into the entire field.

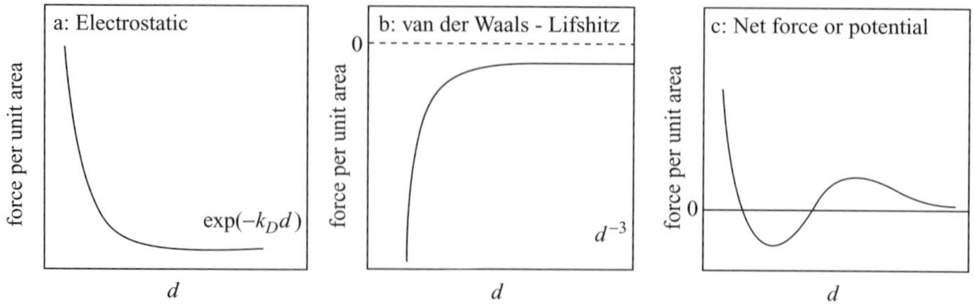

Fig. 4.1. The balance of van der Waals and double-layer forces. Schematic representation of van der Waals and double-layer forces per unit area between two planar media [3]. Adapted from Ref. 3, with permission. Copyright 2006 Springer-Verlag.

4.1.1 Direct measurements, assumptions and inadequacies

The assumptions used to estimate the attractive forces were that:

1 Van der Waals, dispersion forces, high-frequency induced dipole-induced dipole correlations in the visible to ultraviolet region dominate. A consequence of this assumption is that pairwise additivity of interatomic forces was reasonable.
2 An intervening liquid behaved as if it had bulk liquid properties up to a molecular distance of the surfaces. The dubiousness of such an assumption was emphasized explicitly by Hamaker and De Boer in their original papers. (They also first pointed out that van der Waals forces could be repulsive; see Chapter 1 in Ref. 7.) Decorations of the double-layer theory of electrostatic forces (Stern layers) to accommodate such problems were introduced by adding more unquantified parameters. These, for example, tried to take account of finite ion size and hydration via additional outer and inner planes near the surfaces that demarked regions of different dielectric constants. (Already this injects at least five extra assumptions!)
3 The fundamental ansatz was that electrostatic and quantum mechanical forces could be treated separately.
4 'Hydrophobic' colloids were explicitly excluded from consideration.
5 The problem of repeptization, how a colloid such as clay particles supposedly flocculated into a deep primary minimum could become resuspended on dilution of the background salt solution, remained unanswered. (Sometimes the flocculation was reversible, sometimes not.)
6 Surfaces were supposed to be molecularly smooth.

The theory of Deryaguin and Landau was published in Russian in 1941. The original paper can be found in the collected works of Landau or of Deryaguin. It is remarkable for the invective and scorn with which Landau dismisses an earlier attempt by S. Levine. Overbeek developed it independently during the war, while ostensibly manpowered for industry by the German occupation, and protected by his supervisor Verwey, and working all night secretly for the Dutch resistance. Overbeek's classic work on the *Theory of the Stability of Lyophobic Colloids* appeared in 1948 [2]. He had already been aware of Casimir's work on 'retardation', a weakening of van der Waals forces beyond about 4 nm which would affect matters. Both Deryaguin and Overbeek claimed validity for their theory for salt concentrations only up to at most $5{\cdot}10^{-2}$ M.

Deryaguin had from the beginning postulated and demonstrated the existence of an extra exponential repulsive force that he termed 'structural component of disjoining pressure'. We would now call this a hydration effect, or one due to surface-induced liquid structure.

Some progress was made in extending the theory to include surface charge regulation or ion binding. This resolved the unphysical issue of constant charge or constant potential boundary conditions (see Chapter 3).

Where experiments appeared to fit the theory it was found that constant potential boundary conditions worked best. This was not understood. Much later it makes sense as constant potential gives an upper bound to the repulsive double-layer forces [8]. A larger than expected 'effective double layer' force would be required if hydration forces are operating.

One apparently encouraging result of the theory was its prediction of the concentrations of salt required to precipitate 1:1, 2:2 and 3:3 salts. The Shultz Hardy rule predicted that the salt concentrations required were in the ratio $1{:}(1/2)^6{:}(1/3)^6$. But the experiments were incorrect. Further, linearization of the theory of the double-layer forces, necessary at the time to find analytic results, rendered the agreement even less credible.

4.1.2 Experiment and theory with the double layer

In innumerable studies in colloid science subsequently, the correctness of the theory was hardly questioned. The number of parameters required to fit data proliferated. Even so, with up to five free parameters, e.g., hydrated ion size, inner and outer variable planes in the 'Stern Layer' and genuinely molecularly smooth surfaces of mica, the theory could still not be made to work [9]. For example, in the careful work of Ref. 9, which measured forces between molecularly smooth mica surfaces in $LiNO_3$ solutions, a lithium counterion 'radius' of 0.5 nm was required to fit the data. This is absurd.

The work of Israelachvili and Pashley [10,11], who did the first direct measurements of double-layer forces via the pioneering Israelachvili Surface Force Apparatus, gave the correct long-range exponential decay. But the fitting of the data required the postulation of very large 'hydration' forces, with a range of 1, 2 and 3 nm for 1:1, 2:1 and 3:1 electrolytes, together with additional cation-specific binding constants. These 'secondary hydration forces' were extra-DLVO forces not explained by the theory. (The Russians under Deryaguin had long been comfortable with an additional 'structural component of disjoining pressure' [12].)

4.1.3 Ion specificity of double-layer forces

Equally careful measurements by Pashley et al. [13], and Parsegian, Evans and co-workers [14] on forces between insoluble cationic bilayers on mica surfaces, or osmotic forces in lamellar phases of these immersed in potassium bromide and potassium acetate electrolyte solutions, showed anion-dependent forces that differed, for the same salt concentration, by a factor of 50 in magnitude! (See Fig. 4.2.) The theory worked for acetate salts with no free parameters. These and other experiments were carried out with (1:1) salts at low concentrations with well-defined molecularly smooth surfaces where the theory ought to have worked. The long-range Debye–Hückel decay length was right, but nothing else made sense. To fit the theory to the experiments, one had to assume that 80% of the cationic surface groups 'bound' bromide ions, whereas no acetate ions were bound. The same was so for the standard characterization of charged colloidal particle surfaces by zeta potentials that depend on the Poisson–Boltzmann – Debye–Hückel equations. For a few simple alkali halide salts, matters seemed more or less reasonable, but not for others. The situation was much as for the problem of electrolyte activities and interfacial tensions. Differences in forces by a factor of 50 or more in magnitude for systems that ought to give the same result theoretically were attributed to unknown 'specific ion effects'.

The DLVO theory was and is still used routinely to describe interactions between biosurfaces and in applications to chemical engineering, soil and polymer science and nanotechnology. And it is used in high concentration ranges, in biology 0.15 M, where the assumptions of DLVO theory have broken down. But clearly something is missing if the key forces vary so much with specific anions. It is only in the last decade or so that Hofmeister's century-old experiments on specific ion effects have been taken seriously.

Apart from the anomalies of 'specific ion effects' which could hardly be ignored, but were, the theory is confronted by some other serious questions. Dissolved oxygen and nitrogen from the atmosphere were not considered. As we shall see in Chapter 8, removal of the gas gives us very different forces for the electrolyte system we were modelling. In such a case the forces sometimes change by another order of magnitude! The theory is apparently in tatters. The same remark applies to the standard characterization techniques of colloid science such as zeta potential measurements.

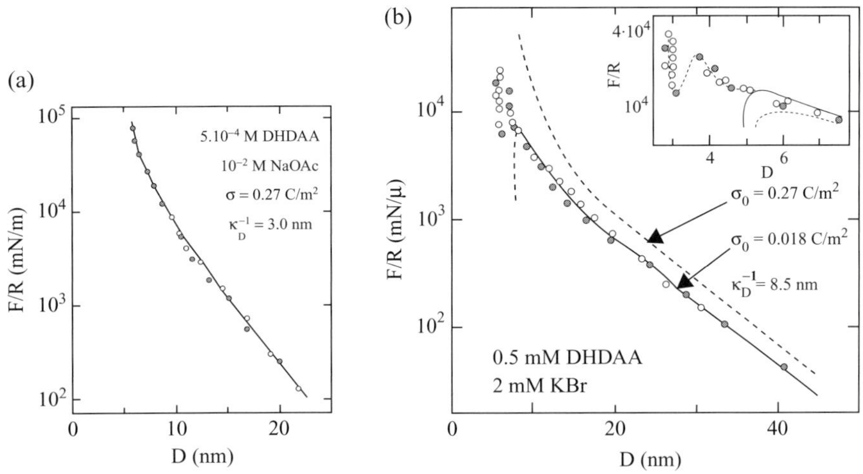

Fig. 4.2. Forces between cationic bilayers of dihexadecyl dimethyl ammonium acetate adsorbed onto molecularly smooth mica surfaces in an Israelachvili Surface Force Apparatus [11]. The bilayers interact across 10^{-2} M sodium acetate. At such low concentrations double-layer forces ought to dominate. The surface charge per unit area of the charged interacting surface is known independently and also measured. The mica surfaces have the geometry of crossed cylinders, equivalent to a sphere on a flat, which avoids alignment problems. The ordinate F/R is the measured force at distance apart *D*, divided by the radius of the cylinders, equivalent to the energy of interaction per unit area. The agreement with double layer based on the Poisson–Boltzmann is perfect, with no fitting parameters, down to the last 10 Å. At such close separations some liquid structural forces (not shown) decorate the force. The same results emerge from osmotic measurements on multilamellar phases of the surfactant bilayers of Evans and Parsegian [14], so that deformability of the surfaces near contact is not an issue. There is no ion binding. The corresponding results if the acetate ion is replaced by bromide fit to double-layer theory with the correct Debye length only if a 'bound' fraction of bromide of 80% of the surface charge is postulated. The forces for acetate are 50-times larger than those for bromide or chloride! There is no real, chemical binding. Adapted with permission from Ref. 58, Copyright 1987 American Chemical Society, and from Ref. 3, Copyright 2006 Springer-Verlag.

4.2 Forces of entropic origin

4.2.1 The ideas of Langmuir and Onsager

A different approach to colloid stability that predates the DLVO theory was taken separately by Langmuir and by Onsager. Langmuir was concerned with the stability of a suspension of Bentonite clays. He kept his samples for 30 years just to make sure of their stability! As did Faraday 100 years earlier in his observations on the stability of gold sols. Clays are aluminosilicate minerals made up of individual thin charged plates typically 200 nm × 200 nm × 2 nm. The surface charge depends on the very complex solid-state surface chemistry, which is still not understood.

Onsager was concerned with suspensions of tobacco mosaic virus particles. These are not planar, but are long and cylindrical, again charged.

The essential idea was that in any such system there will be a dilute supernatant that forms a separate phase. This phase contains a very few particles of high entropy. The free energy of the gas-like supernatant phase balances the free energy of the liquid-like condensed phase, particles of which are subject to steric hindrance as well as electrostatic forces of repulsion. The same ideas work for latex spheres in part.

References to Langmuir's and Onsager's work can be tracked through Refs. 15–17, or their collected works.

Extensive experimental and theoretical studies over many years using such colloid systems as models for molecular gases, liquids, liquid crystals, solids and their phase transitions were undertaken by Lekkerkerker, who succeeded Overbeek at Utrecht. There is in Utrecht a vast collection of such suspensions. Since many such systems have the dimensions or spacings of the wavelength of visible light they display extraordinarily beautiful opalescence.

The same ideas that focus on entropy vs. hard repulsive forces as the determinant of two- and three-phase behaviour are at the core of many theories of liquid crystals, liposomes and microemulsions. The many systems in Utrecht provide convincing model systems for the structure and theory of liquid-state physics [18].

The forces introduced by W. Helfrich, which enjoy much popularity in the area of liquid crystals, embrace essentially the same ideas, with the decoration of elasticity by the assignment of a 'bending modulus' to interacting planar bimolecular membranes or polymers.

4.2.2 Clays and agriculture

The swelling of clays depending on electrolyte and electrolyte concentration is of central concern in soil science. The plates, or dimers, trimers and other small aggregates of plates, can swell or coalesce rapidly under the influence of electrolyte to form an open floc-like stack of cards face to end; or else they can coalesce slowly into separated colloidal ordered aggregates. The last state makes the clay permeable. The first state of swelling clays makes the soil impermeable and is a huge problem for irrigation systems, or river systems with dams that interrupt natural flooding processes. After flooding, and then drainage of water containing salt into a separate disconnected water table, continuity is maintained and salt accretes in the surface soil layers with catastrophic agricultural consequences.

Flocculation and soil improvement is achieved by adding divalent salts, gypsum or limestone. The spacing of clay plates under different electrolyte conditions has been studied extensively over many years by X-ray analysis, and in direct force measurements on mica. In both there are at high electrolyte distinct strong potential wells that correspond to 1, 2 and even 3 hydrated ion radii. Attempts to model such systems with DLVO theory or decorations to include ion fluctuation forces (below) have simply not worked [19]. This is not surprising in view of what we now know of the limits of DLVO theory. (The concentrations of counterions between the flocculated plates is of the order of 6 M, so it all is really more like a hydrated salt.)

Perhaps related to the phenomena encompassed by Langmuir's ideas are the extensive experiments of N. Ise and co-workers [20] on very dilute suspensions of latex spheres. The spheres exist in extensive liquid or solid-like aggregates of colloidal dimensions. The separation of the spheres can be as high as 800 nm! The clusters of aggregates appear to be stable, unchanging in form, for months. Why this is so is not understood. The phenomenon has been invoked in support of the concept of very long-range water structure in biological systems; see the book by G. Pollack [21].

The systems are extensively dialysed, so that any excess salt due to the presence of surfactants used in the synthesis of the particles is removed. Then the range of electrostatic forces should be large, although the presence of atmospheric CO_2 and carbonic acid would lower the pH to around 5. The Debye length, if meaningful here, would then be at most 100 nm. Attractive ion fluctuation forces could be called in to explain matters. Another possible source of the Ise phenomenon that cannot be ruled out is polymer bridging between the particles due to unravelling of the polymeric latex particles; or else bridging due to polysilicic acid from the glass walls of the container.

Certainly bridging between silica particles (polywater) occurs over such large dimensions and seems to require no new ideas about long-range water for its explanation. But the system provides a challenge that remains unexplained [22,23].

4.2.3 Colloid science and genesis of ores and oil

A perennial idea that recurs has it that the origins of life-forming molecules were catalysed in and on the surfaces of clays (see Section 8.3 in Ref. 24). The advantage of such ideas is that they can be neither confirmed nor denied, so that the debate is without end.

More serious and suggestive is the work of McCollom [25] on the formation of meteorite hydrocarbons from thermal decomposition of siderite ($FeCO_3$). Thermal decomposition of siderite had been proposed as a source of magnetite in Martian meteorites. Laboratory experiments were conducted to evaluate the possibility that this process might also result in abiotic synthesis of organic compounds. Siderite decomposition in the presence of water vapour at 300 °C generated a variety of organic products dominated by alkylated and hydroxylated aromatic compounds. The results suggest that formation of magnetite by thermal decomposition of siderite on the precursor rock of the Martian meteorite ALH84001 would have been accompanied by formation of organic compounds and may represent a source of extraterrestrial organic matter in the meteorite and on Mars. The results also suggest that thermal decomposition of siderite during metamorphism could account for some of the reduced carbon observed in metasedimentary rocks from the early Earth [25].

The important, hardly noticed, point is that the addition of water to the inorganic iron carbonate rock produced a huge range of complex organic products that occur in oil reservoirs. It had been thought that such 'life' product molecules in oil reservoirs had to be the consequence of bacterial activity. It had been postulated by T. Gold in the 1990s, and confirmed by other model experiments done in 2004, that such processes can indeed be semi-infinite as yet untapped sources of natural gas and inorganic oil [26]. The matter is completely open and of extreme interest. It is also connected to the present interest in climate change.

The origin of rocks and mineral deposits has been meticulously researched and traced to physicochemical principles of colloid science over a 40-year period by John Elliston. The new geological theory is revolutionary and, if correct, has the same importance to that subject as Darwin's to biology. It gives, for example, a completely different explanation of the origin of granites involving colloidal interactions [27].

4.3 Effects of molecular size on forces in liquids

4.3.1 Oscillatory or depletion forces and hydration

Another kind of force that we wish to consider in further detail has been flagged in Section 2.2.

The molecular granularity of liquids gives rise to oscillatory forces that can stabilize dispersions. If the interacting surfaces are molecularly rough at a molecular scale the forces are smoothed and decay exponentially.

The idea of summing pairwise the attractive van der Waals forces between atoms of two colloidal particles interacting across an intervening liquid film to give a net force to oppose electrostatic double-layer forces led to the DLVO theory of colloid stability. In a seminal paper, Hamaker, or rather his thesis advisor De Boer, went to great pains to point out the approximations made. The key assumption was that of Laplace and Young (Section 2.2). That is, an intervening liquid such as water separating the interacting particles was treated as a medium of bulk liquid density up to a molecular distance between the 'interfaces'.

4.3.2 The notion of an interface

In any theory of interactions that involves surfaces the question arises: when does the bulk liquid approximation break down? That question presupposes that we understand and can define what we mean by the word surface.

For example, at an air–water interface, the system is clearly isotropic in the (horizontal) (x, y) plane. But in the z direction the water density changes from its bulk value, 55 mol/L, to a very low value (that of water vapour, and air, 1 atm or 1 M) as one crosses the boundary. Because of the granularity of matter, which is made up of molecules, this boundary cannot be infinitely sharp. There must be a change in density from the vapour to the bulk value inside the liquid. Likewise there must be a profile of density of air, at 1 atmosphere pressure, i.e. 1 mole per litre (and a profile of water vapour) as we move from the vapour into the liquid. The bulk solubility of oxygen in bulk water at 25 °C is about $5 \cdot 10^{-3}$ M. In oil it is 10-times greater.

At an oil–water interface, we will have two overlapping profiles of density, as water merges with oil, and vice versa (in fact three because the solubility of air in oil is 10-times greater than it is in water).

Generally there is also a profile of liquid density, orientation, hydrogen bonds, dipolar orientation and other characteristics of molecules as we approach such a surface. For water, hydroxide and hydronium ions are both supposed to be positively adsorbed. This adsorption excess is still in dispute, and gives rise to an electric potential at the interface. For a liquid in equilibrium with its vapour, the width of this interfacial region becomes infinite at the critical point! It drops to a few molecular layers as the liquid freezes, depending on the vapour pressure.

For globular proteins, made up of sequences of amino acids, or even for a phospholipid membrane or a surfactant–water interface, the abstraction of a surface that is smooth at a molecular level seems pretty absurd.

This problem of the definition of an interface seems arcane and too complicated to contemplate. But it cannot be dismissed and needs further comment and awareness.

The usual justification is to say that the thermodynamics of interfacial phenomena was explored by Gibbs in the second part of his much-quoted book *On The Equilibrium of Heterogeneous Substances*. All surface chemists pay obligatory obeisance to this. Gibbs's concept of dividing surfaces, with a clear geometrical image of the 'dividing surface' and the introduction of surface excess quantities, allowed a description of surface properties in the simplest way (see Fig. 4.3).

But it glossed over the problem of the structure and thickness of the surface layer. To do this requires statistical mechanics which was non-existent in Gibbs's time.

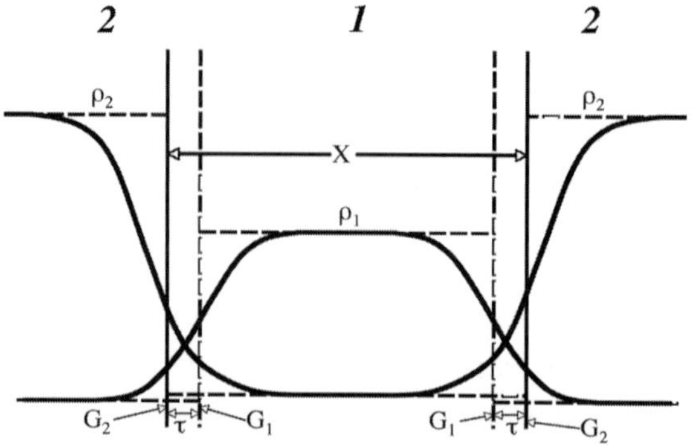

Fig. 4.3. Schematic model system comprising a thin film of liquid (1) separating an immiscible liquid (2). The Gibbs dividing surfaces for each liquid are separated by a 'vacuum' of width τ. Reprinted with permission from Ref. 34. Copyright 1980 American Chemical Society.

The problem is still unresolved. Gibbs's excess quantities (related to adsorption and other properties) depend on the dividing surface location. These are to be determined on the basis of maximum simplicity and convenience. For two adjoining liquids, he used two main positions of the dividing surface. One is the position for which the adsorption of one of the components is zero. This is called the equimolecular surface. The other is the position for which there is no explicit dependence of surface energy on curvature, called the surface of tension. Gibbs theory is mathematically simple but intuitively difficult. It is impossible to study Gibbs's work as a dilettante. The work is written concisely in very difficult language. It is so difficult in fact that Rusanov [28] reports that, according to the great Lord Rayleigh: 'this work is too condensed and too difficult, for most, I might say all readers'. He further reports that Guggenheim wrote also that 'it is less difficult to use Gibbs' formulae than to understand them', and further: 'I have found Gibbs' treatment difficult, and the more carefully I have studied it the more obscure it appears to me'. Rusanov and east European colleagues in the former East Germany and the Bulgarian School seem to be the only people to take such matters seriously. One can deduce that almost anyone who invokes the authority of Gibbs in support of his work has either not read the work or, if he has, has not understood it.

The matter seems incredibly arcane. The problem of, and confusion on, how to define a surface has bedevilled and continues to bedevil investigations in self assembly, membranes, electrochemistry, Langmuir Blodgett films and measurements of surface forces.

For a review of the connection between surface thermodynamics and forces see the review on colloid science from the Gibbs adsorption perspective [29].

4.3.3 Solid surfaces, surface-induced liquid structure

Returning to our problem, two things are a matter of assurance. Firstly, surfaces do exist, and can be defined in principle, even if we have difficulty in doing so. Secondly, the energy of interaction between surfaces is, from the point of view of thermodynamics, just the change in total interfacial energy due to their proximity. In so far as the interfacial tension is defined in terms of adsorption excesses, through the Gibbs adsorption equation, the interaction energy is rigorously a result of overlap of the adsorption profiles at each interface [29].[1]

Keeping these matters in mind, but not too much or we should make no progress at all, let us step back to an ideal molecularly smooth solid interface, like mica, or a crystal face of calcium fluoride, or any ideal colloidal particle immersed in a liquid. 'Molecularly smooth' means that the solid surface has the bumpiness of molecular spheres packed at the interface. If the molecules of an adjoining liquid had the same size and density (and therefore interactions) as in the solid there should be no profile of surface-induced order. But if the size and density of the liquid molecules are different from those of the solid, due to the different intermolecular forces that operate between the molecules of the solid and those of the liquid, there should be some profile of liquid order. The question is, how much of an effect is it?

This problem of surface-induced solvent structure and its influence on forces goes back to the dispute between Laplace and Poisson mentioned previously. Maxwell revisited the matter in a famous article on capillary forces [31]. He contrived by mean field theory, invoking renormalization and scaled particle theory etc., to study the forces due to surface-induced liquid structure. Although not correct, Maxwell got the range and size of the forces right. (Actually S. Marcelja published his paper on hydration forces in 1976 exactly 100 years later [32]. At the time this brought new concepts to colloid science, Marcelja and most colloid scientists at the time not knowing of Maxwell's work. His paper was exactly the same in approach and method to Maxwell's but less detailed.)

The range of these 'structural forces' and the profile of liquid order was exponential with a range of the size of the liquid molecule.

[1] A technical issue: even when an effective planar interface is defined thermodynamically, set by one or the other of the Gibbs dividing surfaces, this statement seems to conflict with theories of interactions between particles. Between surfaces separated by a liquid, or between two liquids, e.g. oil or membranes across water, treatment of the intervening liquid as a continuum as in the pairwise summation method or the later Lifshitz theory is a matter that is confusing. The matter is reconciled if we are aware that such theories of interactions are perturbation theories that give only the change of interfacial energies due to interactions. It turns out that to calculate free energies of interaction via thermodynamic perturbation theory [30], it is necessary to know the adsorption profiles correctly only to the zeroth order to get the interaction free energy correct to second-order approximation. The same is true for interactions between molecules in a solvent. This is why mean field theories based on continuum solvent approximation with Gibbs dividing surfaces work better than one might think. By contrast, in a first-principle statistical mechanical or simulation approach, it is necessary to calculate distribution functions to third order to get free energies correct to second order.

Forces due to liquid structure: oscillatory or exponential molecular profiles at interfaces?

For an extensive recent review of some of the matters below, in particular disjoining pressure in thin liquid foams and emulsions, see Ref. 33 and for earlier work with historical perspective see Refs. 34 and 35.

The first postulation and picture of oscillatory forces between molecules (or an oscillatory order profile at an interface) was due to Ruder Boscovic, a famous Croatian Jesuit natural philosopher working in Rome after whom the Ruder Boscovic Institute in Zagreb is named. There is a large book by Boscovic on his system of the world in Latin around 1600 which derives from this imagined molecular force. Boscovic's postulated force between 'atoms' oscillates and decays to a gravitational potential. While we can dismiss Boscovic's potential – it infuriated J. Clerk Maxwell – the question still remains: is any surface-induced solvent structure due to the molecular granularity of liquid matter a smooth function or not?

The electrical double layer is an example of an exponentially decaying order profile due to surface-induced liquid structure. To see this, consider a primitive model point ion electrolyte. It is a dilute 'liquid' adjoining a charged interface. (The water is here a background continuum.) Then we know that the electrolyte density profile is exponential, with a decay length given by the Debye length. The overlap of these density profiles gives rise to the double-layer force between two such surfaces. Again it is exponentially decaying. This is par excellence the prototype surface-induced liquid structural force. The charged surface induces the exponential density profile in the electrolyte. At much higher bulk electrolyte concentrations, where the hydrated ion size comes into play, what goes on?

The first statistical mechanical study of a simple model hard sphere, van der Waals liquid at and between smooth solid surfaces that predicted oscillatory forces was carried out 30 years ago [36–38].

The predicted forces and the order profiles were decaying, but oscillatory, with a period of the average molecular spacing. They merged into the usual asymptotic continuum theory (Lifshitz or Hamaker) result after about six oscillations. This was very surprising at first.

After these predictions which gave oscillations due to solvent structure and surface-induced solvent structure to surprisingly large distances, such forces were measured on a surface force apparatus (SFA) which uses molecularly smooth mica surfaces. The liquid chosen was octamethyl cyclotetrasiloxane, a molecule invented by Hildebrand early in the twentieth century for diffusion work. It is a van der Waals molecule, elliptical, 9 Å × 7 Å. Since the distance resolution of the SFA was 2 Å, it was ideal for the purpose (see Fig. 4.4). Theory and experiment [39] agreed! A review of subsequent experimental studies on continuum forces and these forces due to liquid structure is given in Ref. 40.

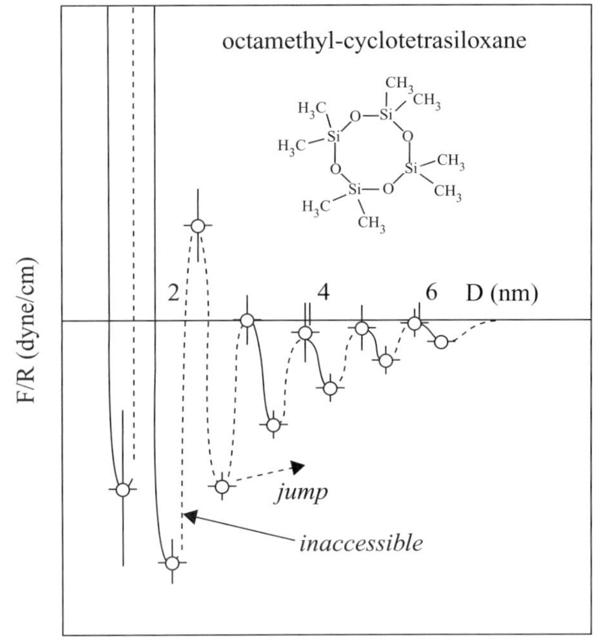

Fig. 4.4. Oscillatory forces between mica separated by a hard sphere liquid measured by Israelachvili Surface Force Apparatus (after Refs. 39 and 40). The abscissa is the energy of interaction per unit area between two planar media in 10^{-9} N/m. The distance of separation is in nanometres. The dashed line indicates inaccessible regions where the mica surfaces jump discontinuously due to the experimental set-up. Note that the oscillations have a periodicity about the size of the molecule illustrated (7 × 9 Å). Adapted with permission from Ref. 34. Copyright 1980 American Chemical Society.

This triggered a wave of computer simulations to arrive at the same result.

A lot of work followed on various systems and continues. Studies were made on the effects in propylene carbonate of dissolved salts, rather than in aqueous electrolytes [41]. The liquid molecule is a hard sphere with a permanent dipole moment and the liquid has a high dielectric constant of 20. The measurements showed the double-layer force on which were superposed three or four large oscillations at close separation. The idea was to gradually build liquids of increasing complexity, hard spheres, hard sphere dipoles, van der Waals forces, hydrogen bonds, to build an intuition about complex liquids such as water. Many other experiments can be found in Ref. 40. Another was on a molten salt, ethylammonium nitrate, to see how electrostatic interactions between liquid molecules might interfere [42].

Other work followed on differently structured liquid phases such as lamellar (one-dimensional) liquid crystals [43]. The oscillations are much larger than van der Waals forces and can be repulsive or attractive. One can understand the results intuitively as shown in Fig. 2.5.

Rather than use equations of liquid-state statistical mechanics there is an alternative approach that comes to the same prediction and derives them analytically from continuum theory.[2]

4.3.4 Oscillations due to liquid structure or exponentially decaying forces?

At this point, there remains a problem. On the one hand, it was verified that there were decaying oscillatory forces extending to quite large distances between surfaces. (Roughly six oscillations before the forces went over to predictions of continuum Lifshitz or Hamaker theory. This is reasonable. It corresponds to the overlap of two surface-induced profiles extending about three molecular layers from each surface. The forces will be repulsive on approach of the surfaces, riding over the maxima. They are attractive on pull off, passing through and over the minima.)

On the other hand mean field theories for 'hydration' forces due to surface-induced liquid structure gave a smooth exponential with a decay length of the order of a molecular diameter. The latter case includes the double layer for dilute electrolytes, where the range of the ionic interaction is the Debye length, which replaces molecular size. There is some liquid structure that shows up in forces at distances of a few hydrated electrolyte ion diameters. But direct measurements of hydration forces by Parsegian and Rand between phospholipid bilayers invariably gave exponentially decaying hydration forces with a range of the size of a water molecule.

So which was right?

4.3.5 Depletion forces stabilize colloidal suspensions and emulsions

The answer is both. The first kind of oscillatory force occurs only when the size of the liquid molecules is much greater than the surface bumpiness. In that case the liquid granularity and bulk liquid molecular packing at the surface dominates. That can be seen for measurements of forces between mica in surfactant solutions, above the critical micelle concentration (cmc). The 'liquid' across which the surfaces

[2] One can assign rigorously an anisotropic dielectric susceptibility to one, or two, or three planes of spheres or other objects and calculate the magnitude of the oscillatory forces precisely by an extension of Lifshitz theory of Chapter 5 [44,45]. The derivation of anisotropic dielectric susceptibilities which extends Lord Rayleigh's work on the Clausius Mossotti result can be found in Refs. 3 and 46. The analytic, asymptotic form of the forces, that is how they settle down to go over to the bulk continuum, isotropic liquid theory results after several oscillations, is reasonably straightforward to derive [34,36].

interact is now a system of micelles with their bound counterions (see Chapter 9). The radii of such 'pseudomolecules', comprising about 50 surfactant molecules, is about 1.5 nm. These are immersed in a continuum of water and free counterions. The force inside a distance less than a micellar size, about 30 Å, where micelles are excluded, is a double-layer force appropriate to an electrolyte composed of free counterions only. (The effective electrolyte density is much lower than would be expected from the bulk concentration of monomeric surfactants. The Debye length is longer and the repulsive force, which is exponentially screened, is much higher.) Beyond that distance, oscillations with periodicity equal to micellar size occur. Such forces are often called *depletion* or structural forces. Measurements of these forces between mica surfaces separated by a micellar solution are given in Refs. 47 and 48. They act to stabilize thin soap films, emulsions, microemulsions and colloidal suspensions [33].

The forces in the all-important small-distance regime are much more repulsive above the critical micelle concentration (cmc) of surfactants (cf. Chapter 9) than below the cmc due to this effect. This explains why it is that better stabilization of suspensions containing surfactants is often achieved with 'dirty' systems. The real-world 'dirty' systems contain hydrophobes, hydrophobic impurities that enhance micellization at concentrations lower than the cmc. Similarly the addition of salt to a suspension containing surfactant will lower the cmc and stabilizes the system in apparent defiance of thermodynamics.

The same occurs with suspensions of proteins and polymers such as RNA in cell biology. The depletion forces stabilize cell membrane interactions.

On the question of Debye length and range of coulombic interactions in dilute systems such as proteins and polymer salt mixtures, the theory has been verified by several direct force measurements [49–51].

4.3.6 Phospholipids and hydration

As opposed to this situation, when the surfaces are rough compared to the size of the molecules of the intervening liquid, or compared to the range of its molecular interactions (electrolytes) or surface fluctuations, as is the case with phospholipid membranes, the resulting surface inhomogeneities smooth out oscillations to give what can be fitted to exponentially decaying forces in agreement with mean field theories. Again surface inhomogeneities and macromolecular 'breathing' modes, oscillatory fluctuations in the infrared frequency regime that exist for proteins, lipid membranes, micelles and microemulsions, will also all smooth oscillations. Hydration forces between lipid membranes are usually exponentially decaying due to this and other effects, such as dipolar fluctuations of headgroups of the membrane lipid molecules.

Phospholipid membrane surfaces are not so rigid and smooth as that of mica. The hydrated headgroups of the constituent molecules bob and weave and rotate. And water and other hydrogen-bonded liquids are made up of not such rigid molecules or large micellar aggregates as those that display oscillatory forces. The surfaces of membranes are hugely 'bumpy' and dynamic compared with solid mica, and a water molecule is much smaller than the bumpiness. Here there is genuine hydration, surface-induced orientation of the vicinal water, more or less, 'hydrogen-bonded' to the headgroups at the membrane surface.

A very extensive literature on measurements of hydration forces using osmotic stress techniques with phospholipids exists. Much of this is associated with the work of Parsegian and colleagues.

There is no sign of oscillatory forces. The repulsive hydration forces – between like surfaces of membranes – are purely repulsive and apparently decay exponentially. The reason for this is clear. The phospholipid head groups are large compared to the size of a water molecule. The surface is rough and the water can no longer arrange in well-defined layers. The short-range hydration-desolvation force is not unique to water.

A major difference here and with any biomimetic surface is the fact that the zwitterionic headgroups have very different sizes, and different permanent dipole moments. They are more or less free to rotate but sterically hindered in the surface plane. Headgroup area itself depends on hydration. This changes with interactions between surfaces. The range of the exponentially decaying repulsive force is around 0.3 nm for the large headgroup (large dipole moment) common phosphatidylcholine, about the size of water molecule. This is what one would expect if liquid packing is the main determinant of the force. But for the smaller headgroup ethanolamine lipid it is 0.1 nm. The common plant lipids mono- and digalactodiglycerides also have much smaller hydration forces with a range from 0.6 to 1.2 nm. The large variation in range, different to that of the diameter of a water molecule, means that specific attractive and repulsive headgroup hydrations combine to give a formidable subtlety to 'hydration' forces. Attraction or repulsion depends on whether the hydrated water layers at the interacting surfaces are like or unlike.

In addition it can be shown that in-plane cooperative, sterically hindered long-range permanent dipole fluctuation forces are at least as large as other contributions like those from van der Waals or double-layer forces at small distances [52]. Their form is not exponential at all, but an inverse power law force. Their inclusion and different kinds of 'protrusion effects' may account for the differences in apparent decay lengths fitted to exponential forms. If only surface-induced liquid structure were involved one would expect a universal exponential decay with a range of 3 nm, the size of a water molecule.

4.3.7 The role of hydration forces in membrane interactions

The existence of strong short-range hydration forces between the lipids of biomembranes is reassuring. Otherwise cell membranes would all fuse and independent cells could not exist. In that connection the differences in the range of hydration forces must be an important source of communication and transport between cells. In any real membrane there are mixtures of lipids of different chain lengths and headgroups. With interactions between the surfaces of two membranes the in-plane distribution of lipids changes and phase separates in response to the hydration effects propagated from its approaching neighbour. Since the segregating headgroups have different sizes and intramolecular forces, the system will rearrange to change local curvature, so contributing to endocytosis and exocytosis. Such self-assembly processes, driven also by changes in physicochemical solution conditions, are augmented too by the capacity of lipids to flip from one side of a membrane to another to give local curvature changes.

Biophysical mechanisms that attend hydration of lipid membranes, and proteins and their capacity to respond to and communicate through hydration have been much explored by Parsegian and co-authors, to whose papers we recommend to the reader. A few that address some of the matters above are given in Refs. 53–57.

References

[1] B. V. Deryaguin and L. Landau, *Acta Phys. Chim. (U.S.S.R.)* **14** (1941), 633–642.
[2] E. J. W. Verwey and J. T. G. Overbeek, *Theory of the Stability of Lyophobic Colloids*. Amsterdam: Elsevier (1948).
[3] B. W. Ninham, *Progr. Colloid Polym. Sci.* **133** (2006), 65–73.
[4] R. J. Hunter, *Introduction to Modern Colloid Science*. Oxford: Oxford University Press (1993).
[5] D. F. Evans and H. Wennerström, *The Colloidal Domain – Where Physics, Chemistry, Biology, and Technology meet*. New York: VCH (1994).
[6] J. N. Israelachvili, *Intermolecular and Surface Forces*. 2nd edn. New York: Academic Press (1991).
[7] J. Mahanty and B. W. Ninham, *Dispersion Forces*. London: Academic Press (1976).
[8] B. Beresford–Smith, D. Y. C. Chan and D. J. J. Mitchell, *J. Coll. Interface Sci.* **105** (1985), 218–234.
[9] V. E. Shubin and P. Kekicheff, *J. Coll. Interface Sci.* **155** (1993), 108–123.
[10] J. N. Israelachvili and R. M. Pashley, *J. Coll. Interface Sci.* **98** (1984), 500–514.
[11] R. M. Pashley and J. N. Israelachvili, *J. Coll. Interface Sci.* **101** (1984), 511–523.
[12] B. W. Ninham ed. Deryaguin and his contributions. Three volumes In *Selected Works of B.V. Deryaguin, Progr. Surface Sci.* **40** (1992), 15–20.
[13] R. M. Pashley, P. M. McGuiggan, B. W. Ninham, J. Brady and D. E. Evans, *J. Phys. Chem.* **90** (1986), 1637–1642.
[14] Y. H. Tsao, D. F. Evans, R. P. Rand and V. A. Parsegian, *Langmuir* **9** (1993), 233–241.
[15] S. Marcelja, D. J. J. Mitchell and B. W. Ninham, *Chem. Phys. Lett.* **43** (1976), 353–357.

[16] P. A. Forsyth Jr., S. Marcelja, D. J. J. Mitchell and B. W. Ninham, *J. Chem. Soc. Faraday Trans. II* **73** (1977), 84–88.
[17] P. A. Forsyth Jr., S. Marcelja, D. J. J. Mitchell and B. W. Ninham, *Adv. Coll. Interface Sci.* **9** (1978), 37–60.
[18] V. J. Anderson and H. N. W. Lekkerkerker, *Nature* **416** (2002), 811–815.
[19] R. Kjellander, S. Marcelja, R. M. Pashley and J. P. Quirk, *J. Chem. Phys.* **92** (1990), 4399–4407.
[20] K. Ito, H. Yoshida and N. Ise, *Science* **263** (1994), 66–68.
[21] G. Pollack, *Cells, Gels and the Engines of Life: A new, unifying approach to cell function cells*. Seattle: Ebner and Sons (2001).
[22] V. V. Yaminski, B. W. Ninham, and R. M. Pashley, *Langmuir* **14** (1998), 3223–3235.
[23] D. T. Atkins and B.W. Ninham, *Coll. and Surf. A* **129** (1998), 23–30.
[24] S. T. Hyde, S. Andersson, K. Larson, Z. Blum, T Landh, S. Lidin and B. W. Ninham, *The Language of Shape. The role of curvature in condensed matter physics chemistry and biology*. Amsterdam: Elsevier (1997).
[25] T. M. McCollom, *Geochim. Cosmochim. Acta* **67** (2003), 311–317.
[26] F. J. Dyson, *Many Colored Glass: Reflections on the place of life in the universe* (Page Barbour Lectures). Charlottesville: University of Virginia Press (2007).
[27] J. Elliston, *The Origin of Rocks and Mineral Deposits: using current physical chemistry of small particle systems*. Castlecrag, Australia: Elliston Research Associates Pty Ltd. (2007).
[28] A. I. Rusanov, in *The Modern Theory of Capillarity: To the centennial of Gibbs' theory of capillarity*, ed. F. C. Goodrich and A. I. Rusanov. Berlin: Akademie-Verlag (1981).
[29] V. V. Yaminsky and B. W. Ninham, *Adv. Coll. Interface Sci.* **83** (1999), 227–311.
[30] L. D. Landau and E. M. Lifshitz, *Statistical Physics*. London: Pergamon Press (1958).
[31] J. C. Maxwell, *Capillary Action, Encyclopaedia Britannica*, 9th edn. (1876). See also his Collected Works.
[32] S. Marcelja and N. Radic, *Chem. Phys Lett.* **42**, (1976), 129–130.
[33] C. Stubenrauch and R. von Klitzing, *J. Phys.: Condens. Matter* **15** (2003), R1197–R1232.
[34] B. W. Ninham, *J. Phys. Chem.* **84** (1980), 1423–1430.
[35] B. W. Ninham, *Pure Appl. Chem.* **53** (1981), 2135–2147.
[36] D. J. J. Mitchell, B. W. Ninham and B. A. Pailthorpe, *J. Chem. Soc., Faraday Trans. 2* **74** (1978), 1098–1113.
[37] D. J. J. Mitchell, B. W. Ninham and B. A. Pailthorpe, *J. Chem. Soc., Faraday Trans. 2* **74** (1978), 1116–1125.
[38] D. Y. C. Chan, D. J. J. Mitchell, B. W. Ninham and B. A. Pailthorpe, *Chem. Phys. Lett.* **56** (1978), 533–536.
[39] R. G. Horn and J. N. Israelachvili, *J. Chem. Phys.* **75** (1981), 1400–1411.
[40] R. G. Horn and B. W. Ninham, Experimental Study of Hydration Forces. In *Micellar Solutions and Microemulsions*, ed. S. H. Chen and A. Rajagopalan New York: Springer–Verlag (1986), 81–99.
[41] H. K. Christenson and R. G. Horn, *Chem. Phys. Lett.* **98** (1983), 45–48.
[42] R. G. Horn, D. F. Evans and B. W. Ninham, *J. Phys. Chem.* **92** (1988), 3531–3537.
[43] P. Richetti, P. Kékicheff, J. L. Parker and B. W. Ninham, *Nature* **346** (1990), 252–254.
[44] V. A. Parsegian and G. H.Weiss, *J. Adhesion* **3** (1972), 259–267.
[45] V. A. Parsegian and G. H. Weiss, *J. Colloid Interface Sci.* **40** (1972), 35–41.

[46] B. W. Ninham and R. A. Sammut, *J. Theor. Biology* **56** (1976), 125–149.
[47] R. M. Pashley and B. W. Ninham, *J. Phys. Chem.* **91** (1987), 2902–2904.
[48] R. M. Pashley, P. M. McGuiggan, R. G. Horn and B. W. Ninham, *J. Coll. and Interface Sci.* **126** (1988), 569–578.
[49] P. Kékicheff and B. W. Ninham, *Europhys. Lett.* **12** (1990), 471–477.
[50] T. Nylander, P. Kékicheff and B. W. Ninham, *J. Coll. Interface Sci.* **164** (1994), 136–150.
[51] R. Waninge, M. Paulsson, T. Nylander, B. W. Ninham and P. Sellers, *Int. Dairy J.* **18** (1998), 141–148.
[52] P. Attard, D. J. J. Mitchell and B. W. Ninham, *Biophys. J.* **53** (1988), 457–460.
[53] R. P. Rand and V. A. Parsegian, *BBA Biomembrane Rev.* **988** (1989), 351–376.
[54] S. Leikin, V. A. Parsegian, D. C. Rau and R. P. Rand, *Annu. Rev. Phys. Chem.* **44** (1993), 369–395.
[55] V. A. Parsegian and R. P. Rand, *Langmuir* **7** (1991), 1299–1301.
[56] K. Gawrisch, D. Ruston, J. Zimmerberg, V. A. Parsegian, R. P. Rand and N. Fuller, *Biophys. J.* **61** (1992), 1213–1223.
[57] H. I. Petrache, S. Tristram-Nagle, K. Gawrisch, D. Harries, V. A. Parsegian and J. F. Nagle, *Biophys. J.* **86** (2004), 1574–1586.
[58] J. B. Evans and D. F. Evans, *J. Phys. Chem.* **91** (1987), 3828–3829.

5

Quantum mechanical forces in condensed media

5.1 Lifshitz theory and its extensions: an overview

5.1.1 Molecular recognition

The ideas behind the DLVO theory of long-range forces in colloidal particle interactions and of the electrical double layer that we have just outlined held centre stage in colloid science for at least 70 years. Quantitatively, as already remarked, agreement between experiment and theory was illusory, except at salt concentrations less than 10^{-2} M, or at most 10^{-1} M. It was illusory in the sense that while classical theory captured some essentials of the forces between ions and macromolecules, ion specificity was still missing. While the electrical double-layer forces decayed exponentially as predicted, the magnitude of the forces changed with a change in counterion or co-ion. Recall our typical examples. A change in counterion in a background electrolyte from Br^- to OAc^- could produce an increase in magnitude of the forces by a factor of 50 to 100! The same will be seen to occur with different surfaces with a change from Na^+ to Li^+ [1]. We will see much more of this specificity later. These differences depended on the nature of the charged interacting surfaces and, at the same electrolyte concentration, on the counterion. They occur even with a change in an apparently indifferent co-ion in an intervening electrolyte. The accommodation of such results could only be achieved by calling in more unquantified fitting parameters. This is unsatisfactory as predictability is lost. The attractive van der Waals forces were obscured by short-range effects due to solvent structure. The same problem with theory exists for pH measurements, buffers, electrochemistry, zeta and membrane potentials, electrolyte activities, interfacial tension of salt solutions, proton and ion pumps, enzymatic activities and many other phenomena that depend on so-called specific ion effects.

The entire intuition on forces in physical chemistry, electrochemistry and biochemistry all derives from the DLVO theory, double-layer, Debye–Hückel and Born electrostatic theories. The problem of the nature of specific short-range hydration around ions or zwitterions or at surfaces remained completely open. To make matters worse, additional much longer-range additional forces sometimes seemed necessary. Deryaguin called these 'structural component of disjoining pressure'. Others called them long-range 'hydrophobic forces'. Their existence remained an open question.

The validity of models based on the assumption that interacting surfaces were molecularly smooth and homogeneous was at best suspect. For complex macromolecules such as proteins and membranes that assumption seemed ludicrous.

All chemistry and all biology is specific. The specificity of association of molecules like enzymes and substrates due to their *geometry* (lock and key) seems to be easily comprehensible. But the question of the specificity of *long-range forces* that drive molecular recognition, and hence physical association of macromolecules, remained a mystery. Likewise the source of the exquisitely reproducible (chemical) energy that drives enzymatic catalysis was equally mysterious. That we should attempt to understand such forces was D'Arcy Thompson's plea, and that of the early founders of the cell theory of biology. It was the main aim of F. W. Ostwald, whose students included Arrhenius, van't Hoff, Nernst and W. Ostwald, the founder of modern colloid chemistry. Ostwald senior's vast interests and influence included electrochemistry and enzymes, which he recognized were catalysts. He did a vast amount too in electrochemistry.

The problem of how molecules 'feel their vibes', as it were, was in the air for practically 150 years. Such problems were made explicit in the work of Hofmeister in the 1870s, which is discussed below. It stands in the scheme of things as Mendel's did to genetics. Hofmeister's work was not ignored as was Mendel's work. We simply had no idea of how to fit it into the theoretical framework. Where problems occurred, and they did so increasingly, we have tended to argue them away. Specificity is captured in descriptive terms embodied in words such as kosmotropes, chaotropes, hydrophilicity, hydrophobicity, soft and hard ions, π–cation interactions, hydration and hydrophobic forces, ion binding and water structure.

The question remains: is it possible to quantify long-range specificity that guides short-range recognition? To tackle that question we put problems with the electrostatic forces to one side for the moment.

5.1.2 Lifshitz theory

There is an alternative approach to the problem of attractive forces operating between colloidal particles, to which we now turn.

Previously we had started with the van der Waals forces between individual constituent molecules and worked bottom-up. We could get an estimate of the attractive force between colloidal particles interacting in water through pairwise summation of these quantum mechanical intermolecular forces as in Chapter 2.

In a more sophisticated approach to condensed media interactions we can do better through a 'top-down' approach. The idea here is different. It begins with the observation that measured dielectric susceptibilities of a condensed medium as a function of electromagnetic frequency (static dielectric constant, infrared adsorption, visible refractive index, ultraviolet frequencies) contain, implicitly, all information on many-body molecular interactions. Much more is involved than the refractive indices and principal atomic adsorption frequencies that were used to estimate van der Waals force between atoms. The dielectric response of a condensed medium is specific to that medium. It is much too complicated to calculate or even to simulate on a computer. But we can measure it. We have such experimental information, at least in principle. Then our problem is how that information might be harnessed to arrive at a better idea of the specificity of forces between particles made up of different media.

The approach was articulated in 1894 when P. N. Lebedeff, a distinguished Russian biophysicist, wrote 'of special interest and difficulty is the process that takes place in a physical body when many molecules interact simultaneously, the oscillations of the latter being interdependent due to their proximity. If the solution of this problem ever becomes possible we shall be able to calculate the values of the intermolecular forces due to molecular inter-radiation, deduce the laws of their temperature dependence, solve the fundamental problem of molecular physics whether all the so-called "molecular forces" are confined to the already known mechanical action of light radiation, to electromagnetic forces, or whether some forces of hitherto unknown origin are involved . . .'. Lebedev was a friend of James Clerk Maxwell, whose great works had already appeared 20 years previously. One of these was on electromagnetism and the nature of light waves. The other was on capillary action and the molecular forces between thin films [2,3]. In 1900 Planck discovered and quantified the quantum nature of light. A few years later, in 1905, Einstein explained the photoelectric effect, both of which led on to quantum mechanics.

It was especially fitting that Lebedev's grand vision and speculations concerning the electromagnetic origin of molecular forces should have been confirmed by the Russians in the dramatic simultaneous advance, in theory by Lifshitz in 1955 [4,5], and in the first direct measurements of molecular forces between metal surfaces by Deryaguin and Abrikosova in 1956 [6]. It was especially so since the great Russian colloid chemist B. V. Deryaguin was Lebedev's stepson.

At that time, the 1950s, the molecular biology revolution was beginning. Solid-state physics had emerged in a new triumphant synthesis that led to the semiconductor and computer revolution. Chemists, following the work of Pauling and Coulson on the chemical bond, had adopted the new spectroscopies with a vengeance. Biochemistry took off.

The challenge of Lebedev that had come back into focus was to understand how it is that molecular forces of physical origin conspire, with the chemical bond and chemistry of solutions, and the size and shape of molecules, to play their role to biology.

5.1.3 Derivations of Lifshitz theory

In its original form, the theory that we are going to outline retains the assumptions of the original DLVO theory of colloid interactions. Recall that in first approximation, there are supposed to be two separate forces (see Fig. 4.1). These are the electrical double layer, with decorations to include surface dissociation equilibrium, and van der Waals forces. For both forces a continuum model is taken for an intervening solvent. No hydration, no surface-induced liquid structure is allowed for. The model colloidal particles are formed from dielectric media bounded by molecularly smooth surfaces separated by a bulk liquid.

Given those assumptions, the attractive van der Waals forces were then calculated from quantum field theory. This theory of Lifshitz includes all non-electrostatic forces. They are many-body dipole–dipole, dipole–induced dipole and induced dipole–induced dipole forces. They are temperature-dependent, and include effects due to the finite velocity of light, called retardation. These quantum mechanical forces are all accessible from measured bulk dielectric susceptibilities as a function of electromagnetic frequencies. We emphasize that the forces cannot be derived from summation of two-body molecular potentials.

The original Lifshitz theory dealt with interactions across a vacuum. Its extension to include interactions between the surfaces of colloidal particles separated by an intervening liquid medium by Dzyaloshinskii, Lifshitz and Pitaevskii [4] seemed to represent a triumph, and a tour de force of quantum statistical electrodynamics. The predicted large-distance behaviour of the forces between metals was confirmed by experiments of Abrikosova and Deryaguin [6,7].

The theory seemed to represent a complete solution of the interaction problem through its reliance on measured dielectric properties. Well and good. But the theory was extremely complicated, so complicated in fact that except for a few special limiting cases no one had the slightest idea of how to use it.

To illustrate that, the formal result for the free energy of interaction $F(l, T)$ per unit area between two dielectric media 1 and 3 separated by a medium 2 at a

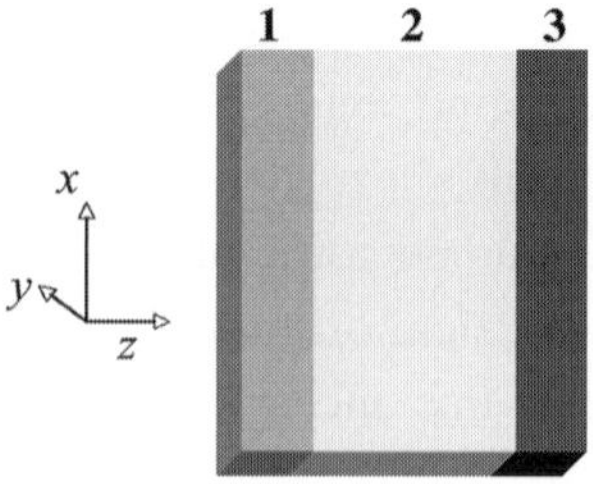

Fig. 5.1. A planar slab or material 2 separates two semi-infinite media of materials 1 and 3.

distance l is hardly transparent. It is:

$$F(l,T) = \frac{k_B T}{8\pi l^2} \sum_{n=0}^{\infty}{}' I(\xi_n, l) \tag{5.1}$$

$$I(\xi_n . l) = \left(\frac{2\xi_n l\sqrt{\varepsilon_2}}{c}\right)^2 \int_1^{\infty} pdp\left\{\ln\left[1 - \bar{\Delta}_{21}^R \bar{\Delta}_{23}^R \exp\left(-2p\xi_n l\sqrt{\varepsilon_2/c}\right)\right]\right.$$
$$\left. + \ln\left[1 - \Delta_{21}^R \Delta_{23}^R \left(\exp -2p\xi_n l\sqrt{\varepsilon_2/c}\right)\right]\right\} \tag{5.2}$$

$$\bar{\Delta}_{21}^R = \frac{s_1\varepsilon_2 - p\varepsilon_1}{s_1\varepsilon_2 + p\varepsilon_1}; \qquad \Delta_{21}^R = \frac{s_1 - p}{s_1 + p};$$
$$s_j = \sqrt{p^2 - 1 + \frac{\varepsilon_j}{\varepsilon_2}} \qquad \text{with } j = 1 \text{ or } 3; \quad \varepsilon = \varepsilon(i\xi_n) \tag{5.3}$$

In these expressions k_B is Boltzmann's constant and T the absolute temperature. The frequencies ξ are $2nk_BT/\hbar$ where $\hbar = h/(2\pi)$ and h is Planck's constant, c the velocity of light in vacuum and the term $n = 0$ in the sum is taken with a factor ½. The parameter p is defined as $p = c\rho_2\left(\xi_n\sqrt{\varepsilon_2}\right)^{-1}$, where ρ_2 is given by $\rho_2^2 = k^2 - \varepsilon_2\mu_2\xi_n^2\left(1 + \frac{i\sigma_2}{\varepsilon_2\xi_n}\right)$, μ is the magnetic permeability (1 for nonmagnetic media), σ is the electrical conductivity, and k is a vector perpendicular to the separating medium (z axis) as depicted in Fig. 5.1 [8].

With magnetic media the second factor is the same as the first but with the dielectric susceptibilities replaced by magnetic susceptibilities. The susceptibilities are to be evaluated on the *imaginary* axis in the complex frequency plane. This is a technical matter that we will not go into. The extension to include magnetic and conducting media gives a little more complex expressions [8].

The formula was derived from highly sophisticated quantum electrodynamics incomprehensible to non-experts, or indeed to experts. Apart from that, except for some limiting cases, from a practical viewpoint no one knew how to use dielectric data to extract any physical sense from such a complex expression.

The first problem was that the theory was incomprehensible! This was resolved later when it was recognized that the derivation that *appeared* to use the full formidable apparatus of quantum statistical electrodynamics was unnecessary. What had happened was this: technically, at a certain point in the development of the theory, a linear coupling constant integration in a so-called Dyson integral equation had been replaced by a linear coupling constant integration. (In less opaque language what this meant is that a complete theory still required a full statistical mechanical theory of the dielectric susceptibilities.) The result of this approximation was that the entire theory collapsed to a semi-classical theory! That is, it could be derived much, much more simply from the solution of Maxwell's equations for the fluctuating electromagnetic field with boundary conditions set by the dielectric properties of the different interacting media. This gave a set of independent allowed harmonic oscillator frequencies or modes. Assignment of a free energy to each mode with a weight given by Planck's hypothesis on the quantization of light and addition of these then gave the result [9,10].

So the claim that the Lifshitz theory and its extensions represented a complete solution to the many-body problem was too broad. Nonetheless it did represent the first attempt to implement, and captured the essence of, Lebedev's vision.

The second problem, how to use the theory, was resolved by a series of papers a decade later that dissected the formula and showed how to calculate forces from measured dielectric data [10–15].

5.1.4 Inferences from Lifshitz theory: molecular recognition

We repeat again that the approximations of DLVO theory remain. These are the assumption that surfaces of interacting particles are molecularly smooth, that an intervening liquid has bulk liquid properties up to the surfaces, and that electrical double-layer and quantum mechanical forces can be treated independently.

Nevertheless, analysis of the formulae enabled some of the subtleties of molecular forces in condensed media to be teased out.

The forces depended on the (in principle) measurable dielectric properties of the interacting materials, and those of an intervening medium. An extensive literature followed that explored how interactions depended on shape and size and isotropy of the interacting particles or molecules – spheres, cylinders, multiple layers. They depended on anisotropy of dielectric properties. Even more complicated formulae hold [10,11]. They depended on solutes and (later still, Chapters 6 and 7) on salt, temperature and the velocity of light through *retardation,* in calculable and very subtle ways [16].

All previous work emerged as special limiting cases.

5.1.5 Summary

In what follows we summarize some new concepts that emerged after Lifshitz theory began to be understood. For a physical, mathematical and quantitative account of these matters and how to calculate the forces explicitly we refer to a considerable literature [10–15]. These details are not so important to us as are the concepts. This is because, as we shall see, even this theory turns out to be flawed for real-world applications, even if the DLVO assumption of uniform media between two objects is accepted.

An analysis of the frequency dependence of the forces built into the dielectric susceptibilities of the interacting particles provides the first insights into a global picture of molecular recognition. To see this we continue to ignore the complication of electrolyte effects that we will come to later.

First consider only planar interactions between solid dielectric media like silica or mica. The simplest case is when the interactions occur across a vacuum. The dominant contributions come from frequencies in the ultraviolet and visible region. The result is then much as for pairwise summation of van der Waals forces, except that the inverse power law potential of interaction per unit area between two surfaces (Chapter 2) is now modulated by an exponential factor with a decay length of typically 20–40 Å. The forces across a vacuum are larger by an order of magnitude than when the interacting media are separated by a liquid.

The most interesting and more complicated case is when the intervening medium is water. For example, for water, its static dielectric constant ε_r is 78 at room temperature. This high value reflects the cooperative response of water molecule permanent dipole moments to an electric field. At infrared frequencies it relaxes to values $\varepsilon(\text{ir})$ around 5 with three or more relaxation bands. These reflect stretching, bending and torsional modes of the coupled dynamic water network. At visible frequencies the dielectric constant is $\varepsilon(\text{vis}) = n^2$ where n is the refractive index, and so on through the ultraviolet frequencies and beyond.

The complete formula for the interaction free energies (Equations (5.1–5.3)) is an infinite sum over different frequency contributions.

5.1.6 'Oil–water' systems: how molecules recognize each other

'Oil–water' systems are prototypes for those in biology. The temperature-dependent, zero-frequency, $n = 0$ term in the sum over frequencies is different, and due to classical correlations. It is especially important for oil–water systems, whose dielectric properties are similar to those of lipids and proteins. For these it provides 25–40% of the total interaction at small separations. Its contribution to the free energy of interaction falls off with an inverse square power law.

It dominates at large distances for all materials, where large will be defined below. This term is explicitly temperature-dependent. (It reflects cooperative permanent dipole interactions in water, the analogue of a sum of the Keesom interactions between two water molecules of Chapter 2, to which indeed it reduces in the low density limit. That sum of Keesom interactions would give a result *inversely* proportional to temperature. But in Lifshitz theory the zero-frequency contribution to the interaction in bulk water which includes all many-body dipole interactions is *proportional* to temperature! We make this remark to emphasize how qualitatively different many-molecular interactions are to expectation based on pairwise interactions [15].)

The remaining terms in the infinite sum for the free energy are loosely speaking due to quantum fluctuations alone. They have no explicit temperature dependence. All frequencies contribute, infrared, visible, and ultraviolet [14]. Each frequency provides a contribution. Each contribution has the form

$$F_n(l) \approx \frac{1}{l^2} \exp\left(-\frac{\xi_n l}{2c\sqrt{\varepsilon(i\xi_n)}}\right) \tag{5.4}$$

Here ξ_n is the nth frequency $2nk_BT/\hbar$ of the sum and ε the dieletric susceptibility of the intervening medium 2 at that frequency. That is, each term in the sum behaves in a way that mimics the pairwise summation result, but with *an effective cut-off distance* given by the exponential. For infrared frequencies the cut-off factor is at around several tens of nanometres. For visible frequencies it is typically 4 nm. For ultraviolet frequencies it is around 1 nm.

This provides some insight into how molecular recognition occurs: first molecules sense long-range microwave frequency interactions. At closer separations infrared frequency correlations kick in. At closer-still distances around 4 nm, visible frequencies take over. And in the last 1 nm ultraviolet high-frequency contributions, the domain of chemistry, take over.

For unlike particles or molecules some frequencies can provide positive contributions, others negative. So we have a picture where macromolecules successively sample their mutual electromagnetic vibrations, draw closer, and sample more frequencies, which can be repulsive, in which case they have no further use for each other. Or else they proceed to sample each other's properties more, and come closer still, until eventually coming into contact.

5.2 Measurements of forces

Direct measurement of forces between molecularly smooth mica across air with the surface forces apparatus began with Tabor and Israelachvili [10,17].

The first direct measurements were the work of Tabor and Winterton and of Israelachvili and Tabor on forces between molecularly smooth mica cylinders. They fitted their results to an 'effective' power law. The fit showed an apparent change from a potential of the Hamaker form $V(l) \approx Al^{-2}$ to a long-distance asymptotic 'retarded' form $V(l) \approx Bl^{-3}$. This latter form is incorrect. Theoretically it has to be as described above, i.e., a sum over all frequencies each term of which with an inverse power law form modulated by an exponential decay. When the full Lifshitz form was used later to compare theory with experiment no agreement could be reached. The matter was resolved by the realization that the measured radii of the interacting cylinders contained a systematic error of a factor of two! Theory and experiment were then consistent [18].

Extension to include forces across electrolytes and other liquid films followed. Other techniques such as atomic force microscopy (AFM) and systems have continued to expand an extensive force-measurement literature. A cursory perusal of this literature would leave the impression and confirm that the DLVO theory is correct. Certainly Lifshitz theory has been confirmed for forces between dielectrics across a vacuum. For forces across electrolytes the Debye length dependence of the electrostatic double-layer forces given by theory, even for multivalent electrolytes such as proteins, was confirmed (cf. Chapter 3).

But gradually flaws began to appear in the fabric that could no longer be denied. When force measurement experiments involving electrolytes as an intervening medium were compared with DLVO theory, the expected sum of double-layer plus van der Waals forces, large discrepancies often occurred. When theory could not fit experiment it became customary to invoke new 'extra-DLVO' forces. These began to proliferate without bound. The words hydration, specific ion effects, ion fluctuations, Helfrich, hydrophobic forces, and on to cation-π electron forces encompassed a veritable zoo of new, specific, unpredictable forces. The original simplicity of the DLVO theory and Lebedeff's vision was lost.

To anticipate: it turns out that classical theories, be it DLVO, or of electrolyte solutions or even of pH and buffers, were in serious trouble. This is so even with the continuum solvent approximation, and with the assumption of molecularly smooth surfaces. Happily, it turns out that when the faults are repaired, the revised theory does seem to accommodate the animals of the zoo. How it does so will be the subject of subsequent chapters.

5.3 Effects of unlike media, size, shape and anisotropy

Once it became clear that this complicated theory reduced to semi-classical physics, it was straightforward to explore a range of different situations besides planar objects. More subtleties emerged. And sometimes the theory worked very well.

Once again in considering these we are deferring the added complication of effects of electrolytes, our central interest.

5.3.1 Unlike media

If we have objects 1 and 3 of different dielectric properties interacting across a medium 2 some very subtle effects emerge.

An example is a medium '1', say air. This is separated by a medium '2', a thin film of oil of thickness l; medium '2' sits on a medium '3', water. We would like to know if the film spreads and wets or contracts to a drop, or exhibits more complex behaviour. (It depends on the alkane chain length. Pentane and heptane spread, octane does not. Decane and higher alkanes exhibit multifilm behaviour.) This was an old problem. It was Benjamin Franklin's famous problem, the spreading of a drop of oil on Clapham Pond, that enabled him to estimate the size of a molecule. It was also the subject of the first scientific paper, circa 1000 BC, in Assyrian cuneiform on clay tablets (cf. Chapter 13). Close to the saturated vapour pressure of the oil, this problem can be tackled, and answered by Lifshitz theory. There seemed to be no unknown liquid structure forces here.

Another might be the adhesion and spreading of a large oil drop or protein approaching a Teflon surface. Yet another is the adsorption of asphaltenes on rocks at the surface of oils in a reservoir. Another is the flooding of (microporous) oil rock reservoirs by saline solutions in oil recovery. This is a more complicated matter of some importance.

For the 'simplest' oil-spreading-on-water problem of the Assyrians and of Franklin, different frequency components to the thin film pressure as a function of thickness vary with alkane chain length refractive indices. They can give a net force across the film that either enhances or opposes the thin film thinning effect due to the vapour pressure. (Some frequency contributions can be repulsive, some attractive. Since each is exponentially damped at different distances, which contribution wins out depends on film thickness.) They can and do oppose each other in a very delicate manner that provides an exquisite test of the theory. The same complexity holds for the forces between macromolecules and their recognition. For a detailed exploration of the wetting problem, see Refs. 10 and 19.

The properties of thin films and adsorption and wetting is a vast subject in its own right. Different adsorption isotherms hold depending on vapour pressure, with the Lifshitz theory being useful here only very close to the saturated vapour pressure. A molecular theory that extends and includes Lifshitz theory gives out the Langmuir (monolayer regime) and BET (Brunauer, Emmett, Teller) isotherms used universally for surface area determination also [10,20–23].[1]

One of the more interesting confirmations of the theory is the work of Anderson and Sabisky on the thickness vs. film height of liquid helium on a molecularly smooth vertical barium fluoride substrate [10,24–26].

Another example that illustrates some of the subtlety of molecular forces in condensed media is the application of Lifshitz theory to the prediction of theta points of polymers in non-polar liquids [27].

So we can be fairly convinced that the Lifshitz theory does work – usually, and with some reservations with water at interfaces – under experimental conditions where it ought to.

5.3.2 Multilayered media

The theory can be extended to multilayered media [11,28,29]. One classical example is a soap film, a layer of water coated on each side by a thin molecular or multimolecular layer of surfactant (soap) molecules.

We can imagine a complex surface 1, say a glass substrate covered by a thin adsorbed protein layer 2 extruded by a bacterium that it generates to prepare a biocompatible substrate on which to adhere. The bacterial membrane 3 bounding its interior (4) interacts with the complex 1–2 substrate.

Then what comes out is that the bacterium 'sees' the substrate 1 at large distance. At closer distances it 'sees' the adsorbed protein. It is straightforward to write down corresponding formulae for free energies of interaction and formation of multiple-layered media such as liposomes, or multiple layers on substrates used for nanotechnological applications. Again, while such results have the same form as that which would be obtained by pairwise summation, when water is involved a much different interaction obtains for the (dominating) temperature-dependent terms [11].

5.3.3 Size, geometry and anisometry

For gravitational forces between two bodies long-range forces are proportional to the product of the masses, independently of their size and constitution. For molecular forces size and shape as well as material composition all count to give a rich diversity of behaviour. Thus the interaction potential between two point molecules is $V(r) \approx r^{-6}$. That between a molecule and a planar object is $V(z) \approx z^{-3}$. Between two planes it is $V(l) \approx l^{-2}$ per unit area to first approximation. The

[1] The matter is much more complicated and intriguing than these glib statements imply. See, for example, the discussion of thick films of water on quartz and polywater of Chapter 8 and Refs. 19 and 22.

different electromagnetic frequency responses of different media that are damped with different exponential decay add another overlay.

Between two parallel thin dielectric cylinders separated by a distance z much greater than their width, the free energy per unit length is $V(z) \approx z^{-4}$. If the cylinders are not parallel, there is an extremely strong torque that twists them into a parallel position as they come together.

To a rough approximation the results are a product of a geometric factor that would be obtained from pairwise addition of individual molecular forces multiplied by the Lifshitz (temperature-, material- and distance-dependent) factor for planes. The matter is not quite so simple. Thus for two spheres or a sphere and a plane, interacting across water with its peculiar dielectric properties at low frequencies, the zero-frequency temperature-dependent contribution is peculiarly long-range relative to the planar case. Likewise for thin cylinders – three-body forces are very strong and not additive.

For planar media with anisotropic dielectric properties, or separated by an isotropic medium, there are strong torques operating. The forces can go from attractive to repulsive depending on the alignment. A wide literature on such forces can be found in Refs. 10 and 11.

This is so also for molecules with anisotropic polarizabilities such as nitrates or hydroxide that associate with four water molecules in a planar arrangement. Anisotropy is an important factor in determining surface charge properties at air–water and other interfaces [10,11].

The theory was extended to different geometries, interacting spheres or cylinders, multilayered media such as membranes and materials with dielectric anisotropy and two- and three-body interactions. All had their own peculiarities; geometry and material properties reflected in dielectric data conspired to add new twists.

Even with spheres the relative importance to interactions of temperature-dependent, infrared-visible and UV frequency changes is different from that for planar particles. The temperature-dependent term is peculiarly long-range, hinting again at the connection between shape and recognition [30].

5.3.4 Conduction processes and peculiarities of cylindrical geometries

A further extension to include the effects of fluctuation forces due to salts and other sources of conducting processes adds another layer of subtlety (see Chapter 7 in Ref. 10). Such ion fluctuation or induction forces will be discussed further in Chapter 6. Especially and peculiarly is this so for cylindrical geometries, conducting polymers, polyelectrolytes or carbon fibres and model DNA molecules. The contributing fluctuation processes additional to those in dielectric media can be intrinsic (e.g. due to π electrons in conducting polymers). Fluctuation processes can occur in the sheath of adsorbed ions in the double layer surrounding charged cylindrical polyelectrolytes like DNA. Very long-range forces emerge here. The same results emerge from detailed model quantum mechanical calculations or from semi-classical theory. The existence of such forces was suspected by London as long ago as 1942, and suggested again by Froelich in the 1960s as a source of recognition for DNA molecules and viruses. They come about due to 'giant dipole' fluctuations which vary from system to system. For example the attractive potential of interaction for two parallel thin cylinders a distance R apart can be in different conditions as large as $V(R) \approx 1/R$, and $V(R) \approx \left[R^{3/2} \ln(R/a)\right]^{-1}$, where a is the cylinder radius. These forces are strictly non-additive [31,32].

They have both temperature- and salt-dependent contributions as well as those from infrared, visible and ultraviolet frequencies.

The role that such forces might play has hardly been recognized. But their existence is not in doubt. One might even speculate that one role for so-called indifferent sequences of identical nucleic acids in DNA, with specific frequency-dependent conducting properties not available to non-identical gene-determining sequences, might lie in such forces.

A summary of these forces is given in the Appendix to Chapter 6. They were derived by Parsegian and Ninham in work unpublished except in the book by Mahanty and Ninham [10].

To recapitulate, the theory of Lifshitz and its later extensions at first seemed impossibly abstruse. But it turned out to be in principle rather simple, at least conceptually. Results could be obtained by just solving Maxwell's equations with appropriate boundary conditions to determine allowed coupled modes. Then quantization of these modes with the Planck oscillator distribution gave the answer, and for different geometries. From such studies there emerged a picture of interactions and explicit confirmation of Lebedev's vision, even in the limiting case of the continuum solvent model. The picture shaped and confirmed the idea that structure and function, geometry and forces were intimately coupled.

The most subtle cases involve water, with its complicated frequency-dependent dielectric properties. These reflect many-body dynamic, dipolar and hydrogen-bonding fluctuations that are still difficult to capture by computer simulation. For example, the temperature-dependent contribution to long-range protein–protein interactions in water is proportional to temperature. A statistical mechanical angle averaged evaluation of this dipolar force would give a contribution proportional to the inverse of temperature (Keesom force).

Macromolecules, proteins and viruses approaching each other or surfaces do 'feel' and sample each other's electromagnetic ambience. The forces between them can be repulsive or attractive, change sign with distance, depend on material dielectric properties, geometry, size and shape, provide orienting forces that position them and allow them to come into contact in the correct lock and key arrangement. Our point is that a myriad complicated forces of different range and strength and sign are encompassed by the same formalism.

5.4 Dispersion self energy and atomic size

We now wish to remark on the concept of dispersion self energy of a molecule. For an ion this is the quantum analogue of the Born electrostatic free energy of Section 3.2. The contribution of quantum dispersion self energies to ionic free energies of transfer, passage of ions across membranes and other problems is ion-specific, depending as it does on ionic polarizabilty as function of frequency. It ranges from 25% to 50% of the Born energy or more.

The idea comes about as follows: theories of interactions discussed so far treat atoms or molecules as points. The interactions become infinite and diverge at zero separation. So workers in the liquid, solution and colloid sciences invoke hard cores, soft cores or some other artificial cut-offs to get around this. Unlike in solid-state physics where molecular spacings in a crystal are less ambiguous, these parameters – like 'hydration' – are fitting parameters that vary from situation to situation. 'Hydration numbers' deduced for a particular ion, say sodium, depend on and vary with the experimental conductivity, activity coefficients and so on.

To extend the Lifshitz theory of dispersion forces to include molecular size one needs first a proper definition of molecular size. This comes about if it is recognized that the polarization cloud of an atomic system is spread out over a region of the order of the 'size' of the atom or molecule. It can be defined formally in terms of the complete quantum mechanical spatial and frequency-dependent polarizability tensor of an atom (or an ion) in its nth state. It can then be shown that to a very good approximation, this can be approximated by, for example, a Gaussian atomic form factor in real space with a range of the order of the size of the atom, convoluted with the normal frequency-dependent response function of a point particle. The Gaussian form is not crucial and any peaked function consistent with the quantum state of the atom or molecule will do. The theory of dispersion self energy of an atomic system is developed in Refs. 10 and 21.

This is the precise analogue of the Born self energy of an ion, except that the size is defined in terms of the quantum state of the molecule. The Debye–Hückel correlation energy for an electrolyte is the change of the Born self energy of an ion due to its interactions with all of its neighbouring ions [10]. In the same way the dispersion interaction between two molecules is the change in dispersion self energy of the two isolated molecules which comes about due to the presence of surrounding molecules and the background equilibrium electromagnetic fields. In a vacuum the dispersion self energy is the change in the free energy of black-body radiation due to the presence of an atom. So, for a hydrogen atom, the dispersion self energy comes out to be the binding energy of the atom, but with the opposite sign, a nontrivial observation.

The Born and dispersion self energies in vacuum or solution are of the same order of magnitude. But the dispersion self energy has been completely ignored in electrolyte chemistry.

For a neutral atom the dispersion self energy is the chemist's standard chemical potential. It changes depending on the dielectric medium in which it is immersed. For non-polar molecules the dispersion self energy gives the free energy of transfer from gas to solution, or from one liquid to another; i.e., it can be used to make a respectable theory of solubility of gases in liquids.

We return to a more recent quantitative treatment of this problem in Chapter 7.

5.5 Connection to quantum field theory

The theory can be extended to include a background medium and temperature as for the original Lifshitz theory. This semi-classical theory can be used to derive a very respectable estimate of the Lamb shift [33]. The Lamb shift represents the ultimate triumph of quantum field theory. This and the further fact that the vacuum dispersion self energy is the same as the quantum mechanical binding energy for a hydrogen atom means that the analytical semi-classical theory of dispersion self energies we are discussing can be used quantitatively.

5.6 Interactions between molecules and hydration

In a condensed medium the dispersion self free energy of a molecule includes hydration but only in so far as the medium can be considered as a continuum. This can be seen as follows. For two point molecules the London or dispersion interaction energy becomes infinite on contact. For finite-sized molecules for which the polarization cloud is spread out the interaction energy is always finite. At zero separation the interaction energy, which is a measure of the change in self energy of the two isolated atoms, is quantitatively of the same order of magnitude as the binding energy of the molecule that they would form. Clearly this is indicative only, as the theory does not include real bound states. But when the molecules interact in a liquid medium, as they approach to atomic separations the intervening liquid is squeezed out. For isolated molecules or ions the self energy is modified by the dielectric properties of the medium in which they are immersed. The self energy takes some account then of 'hydration'. When they come together the hydration profiles around them overlap, and then disappear. The energies of interaction, the changes in self energy on contact, then give a first approximation to the changes in liquid structure or 'hydration' imposed by the individual ions on the liquid.

5.7 Bonds

But the difference now is that the London dispersion force between two molecules or atoms does not diverge on contact. Instead, when two atoms come together, the energy of interaction is the same as, or very close to, the covalent binding energy of a molecule that they form. As soon therefore as an atomic form factor is introduced into the formalism for energy that involves the polarizability tensor, here modelled as a Gaussian, the distinction between different kinds of bonds becomes blurred, as is appropriate in condensed media. The energies of interaction or bonding come out quantitatively about right.

The firm quantitative application of such theories to electrolyte activities and other phenomena such as changes in interfacial tensions at an air–water interface has been inhibited by the absence of quantitative frequency-dependent ionic polarizabilities. That problem has now been resolved in part by recent developments in quantum chemistry. (See e.g. Chapter 7, Ref. 34 and references therein. These papers go some way to defining ionic radii with electrostatic and dispersion radii consistently.)

5.8 Self energy changes in adsorption: remarks on formal theory

The same formalism that incorporates molecular size can also be used to write down the dispersion adsorption potential of an atom or an ion as a function of distance from a substrate across a medium. This leads to a reconciliation and bridge between the very different kinds of adsorption isotherms used in practice. These are: the Langmuir isotherm for monolayers, the Brunauer Emmett Teller (BET) isotherm, the most common isotherm widely used in practice for surface area determination, and an extension of the continuum theory of Lifshitz.

Lifshitz theory of thick film interactions and of wetting depends on both molecular size and on closeness of the vapour pressure of the adsorbate gas to the saturated vapour pressure. The energetics of adsorption – and consequent catalysis – depend very much on not just the polarizabilities of the adsorbate and substrate molecules but also on their relative sizes. A small adsorbate molecule can nestle into the spaces between substrate molecules that are much larger. It will experience a much large adsorption potential than an adsorbate molecule that is much larger than the substrate molecules and is so constrained to, as it were, flit over the substrate at a distance. (The matter is complicated and we refer the reader to Chapter 6 of Ref. 10 and Ref. 22.)

We will not here develop the formal theory that characterizes molecular size and the theory of dispersion self energy. The concepts are fairly straightforward as explained above. But the technicalities are mathematically extremely cumbersome. So we content ourselves here with providing the references above. The application of dispersion self energy to water of Andersson and Ninham accounts for its density maximum [35].

5.9 Dispersion and Born free energies

For electrolytes the Born self energy and free energy of transfer from one medium to another is a widely used fundamental concept. It ignores dispersion interactions of the ion with the medium in which it is immersed. It is only recently that these have been taken into account. Some preliminary estimates show that the dispersion contributions are ion-specific and can be of the same order of magnitude as Born energy itself [36–38].

The theory has been refined to take into account ion-induced solvent structure. More recently still, correct *ab initio* quantum calculations have allowed more accurate estimates. The more refined calculations give even larger specific ion dispersion contributions [34].

5.10 Cooperative substrate effects with adsorption: catalysis

The interactions of molecules physisorbed on to surfaces or interacting in the confines of slits or channels are very different from those forces operating in a bulk medium or vacuum. They are now affected due to their coupling to the dielectric or metallic conduction properties of the adsorbate, and its geometry. These effects mediate the forces and have a major role in catalysis that is explicit. Such forces cannot be modelled in general by two-body simulations. For some explicit results and their implications see Chapter 2 in Ref. 10 and Ref. 39.

5.10.1 Remarks on catalysis

What emerges from such studies contains some new conceptual insights. The interaction between two molecules in confined geometries, or at surfaces, is strongly mediated by the coupling of the free space electromagnetic modes of the two molecules to those of the substrate or channel. The interactions can be weakened or strengthened markedly. This is especially so for metal substrates where real conduction processes in the substrate contribute. The barriers to molecular formation that exist in free space to two adsorbed molecules diffusing on the surface can then be lowered. Alternatively a covalent molecule can be broken down into its components by the substrate–molecule interaction.

The important point is that the process is again cooperative.

A quantitive example is provided by considering catalysis in zeolites and reverse catalysis as a possible mechanism of dioxin poisoning.

(1) Catalysis in zeolites

Zeolites are bicontinuous molecular frameworks of predominately silica. They are honeycomb surfaces of extremely high surface area with average curvature (the sum of principal curvatures) equal to zero. The Gaussian curvature (product of principal curvatures) varies continuously over the microporous honeycomb-like surface. The pore diameters are of the order of 0.5–0.75 nm for hydrophobic zeolites and up to 1.3 nm for hydrophilic zeolites. Their industrial use in adsorption and as catalysts for 'cracking' oils is vast. The long-chained oils adsorb into the pores of the hydrophobic zeolite medium. The question is why the adsorbed oils at high temperature break down into smaller fragments. The self energy concept finds useful application here. Here the coupling of the substrate to the normal modes of vibration of the adsorbed oil molecule plays a role. The curvature of the interface plays an additional major role. Essentially in the environment of the zeolite interior, the longer-chained oil molecules shake themselves to bits – it is more favourable energetically say for hexadecane to break up into smaller hexane and decane fragments than to exist in its original form. For detailed quantitative calculations and demonstration see Refs. 40 and 41.

(2) Reverse catalysis: possible mechanism of dioxin poisoning

That the self energy concept and the coupling of modes of an extended molecule to those of a substrate and the substrate geometry apply to systems like zeolites seems to be correct quantitatively, and rendered more plausible by the fact that no other account of their catalytic action exists.

A reverse kind of coupling seems to occur with poisons like dioxin, potent in extremely small quantities. A convincing argument for the action of this highly toxic substance lies in the observation that it is inert to any possible biogenous manipulative action. At the same time it is loaded with mobile electrons and so constitutes an ideal target for barrel and twisted proteins. It is a false substrate that combines high chemical stability with high electron polarizability. On interception it can be expected to be immobilized instead of just passing through. This is much the same as for the strong specific ion binding in crown ethers, which depends not just on 'perfect fit' geometry but on the dispersion self energy of adsorption of the ion into the cavity, or for the formation of polypseudorotaxanes by and threading of polymers into the hydrophobic cavities of cyclodextrins (see Chapter 13).

With dioxin a relatively large amount of interactive energy is released by the protein–dioxin complex.

So in a kind of 'reverse enzymatic process', instead of the 'normal' situation where a protein successively catalyses say the assembly of a linear RNA molecule, here the high-frequency modes of the adsorbed dioxin molecule successively shake its targeted proteins to bits.

This observation is due to Zoltan Blum, whose exquisitely elegant Chapter 6 in Ref. 41 on the role of curvature in protein structure and function in membrane proteins, enzymatic action, geometry in hormone–receptor interactions, self–non-self recognition, DNA folding and self assembly and crystallization of proteins is extraordinarily insightful.

5.10.2 Other possible sources of specificity: defect substitution in drugs and DNA

The examples above illustrate an interesting point. What determines self and interaction energies is not only individual polarizabilities of the constituent atoms or molecular groups, or the geometry of enzymes and proteins that guide adsorption and then interactions and catalysis. The cooperative many-body vibrational modes of the interacting entities seem to provide an additional major source of specificity to forces operating. Some further speculations might be allowed. To illustrate, consider defect substitution in drugs and DNA and why enzymes are so big.

(1) Defect substitution in drugs and DNA

Small changes in large drug molecules, say replacement of an amine group (NH_2) by a quaternary ammonium $(CH_3)_4N^+$ group, can sometimes have large effects on the efficacy of the drug. Whatever the mechanism of action, it must involve interactions with a substrate of some kind. A part of this can be traced to compatibility of hydration of individual groups. Why such apparently minor substitutions can give rise to large effects is quite unclear. Similarly a defect in a large DNA molecule can result in a big change in expression.

One contribution to such effects can be seen from an illustrative model system. If we consider the collective many-body (vibrational) states of a long linear chain, e.g. CH_2 groups, the molecule will exhibit a continuum of collective infrared modes. But if we substitute a defect, depending on the defect and its place in the chain, this cooperative dielectric response can change. There can be specific long-lived metastable specific vibrational states that emerge from the band of infrared modes. This has been demonstrated explicitly in Ref. 42.

The interactions of two (macro)molecules that differ apparently only slightly from each other can then be very different.

(2) Why should enzymes be so big?

We might be allowed to speculate further that the same kind of many-body coupling may have a role to play in driving other processes such as some enzymatic processes. These are both highly specific and seem invariably to involve large macromolecular assemblies of particular, specific shapes and structure.

A case in point is horseradish peroxidase, probably the most studied enzyme of all. Yet the active site for the reactions has not been identified. The reactions take place on the enzymatic surface. This involves adsorption of reactants and reduction of the energy barrier confronting the reaction in bulk. Presumably this occurs, in fact physically must occur, due to coupling of the self energies of the reactants and their interactions due to the normal modes – if we like – the dielectric properties of the macromolecule. Sufficient flexibility in tuning the many-body cooperative modes of such macromolecular assemblies is only accessible if the assemblies have not just the required shape, but are sufficiently big. Apart from its constitution and surface area, which determine allowed modes, and hydration, which dictates specific adsorption of substrates, the enzyme must be big enough to provide an adsorption potential strong enough to overcome Brownian motion of the small substrate molecules.

5.11 Casimir–Polder and excited-state–ground-state interactions

5.11.1 Effects of temperature and the finite velocity of light on atomic interactions

We now revisit the atomic dispersion (London) forces between two atoms. The matters addressed are technical, but sufficiently important to deserve note. The zero-temperature form was derived from the Schrödinger equation. This approximate formulation of quantum phenomena assumes that the velocity of light c is infinite and takes temperature to be zero. We have seen that the interaction potential so derived is valid at distances l small enough that $2l \ll c\xi$, where ξ is the characteristic frequency of the atoms.

The effects of temperature can be taken into account explicitly via semi-classical theory. The full formula can be extracted from Lifshitz theory by a limiting process and is rather complicated [43]. The result is of interest in that it shows how temperature and distance are intimately coupled and shows how the correspondence principle comes about explicitly.

Of more interest is what happens when the finite velocity of light is taken into account. The potential calculated by Casimir and Polder first, and now a text book result, shows at large distance a weakening of the van der Waals potential from the London $V(R) \approx -R^{-6}$, to $V(R) \approx -R^{-7}$.

Explicitly this is the so-called retarded potential:

$$F(l,T) = -\rho^2\alpha^2(0)\left[\frac{23\hbar c}{120 l^3} + \zeta(3)\frac{\hbar c\pi}{4 l^3}\left(\frac{2k_B T l}{\hbar c}\right)^3 - \frac{7\hbar c\pi^4}{360 l^3}\left(\frac{2k_B T l}{\hbar c^4}\right) + \cdots\right] \tag{5.5}$$

The classical interpretation is that this ‘retarded’ form, a weakening of the interaction, is due to a reduction in correlations due to the finite velocity of light. This is in fact incorrect. The Casimir–Polder result can be shown to be correct *only* at zero temperature [43]. The weakening of the potential lies rather in the quantum nature of light. This can be reconfirmed by an explicit quantum statistical mechanical calculation that includes temperature [44].

Both the London ($c = \infty$), small distance, and Casimir Polder, large distance, forms revert to the explicitly temperature-dependent interaction given by Lifshitz theory at high temperature.

The proper result is that of Lifshitz theory, i.e. a sum of a contributions from different frequencies of the form

$$V(R) \approx k_B T R^{-6} \cdot \exp(-2R\xi_n n/c) \tag{5.6}$$

At large distance and ‘high’ temperature only the temperature-dependent, zero-frequency term remains.

5.11.2 Resonance or ground-state–excited-state interaction

That this classical accepted quantum mechanical result of more than 50 years standing and its interpretation is incorrect at any finite temperature is curious [45–47]. More curious still is the corresponding quantum mechanical result for excited-state–ground-state interactions, which turns out to be wrong even at zero temperature. The matter is not just academic but has an impact not just on physical but also on a number of biophysical problems, such as photon transfer in photosynthesis.

If one atom is in its ground state and another in an excited state, there is an interaction between them that involves transfer of a photon. If two different molecules are involved the process is called a Forster interaction. The interaction is at the core of many problems in biology. It is involved in photosynthesis, light harvesting and fluorescent light-emitting devices. Forster energy transfer was first discovered by Carlo and Franck in 1923. They exposed a mixture of mercury and thallium to a frequency that could only be emitted by the thallium, so demonstrating a transfer of energy between mercury and thallium atoms.

The mechanism behind the phenomenon has been proposed as a way to create entangled states for quantum logic using both molecules and quantum. This interaction furthermore underlies the interpretation of many problems in circular and optical dichroism and related techniques standard in biophysical chemistry.

A detailed knowledge of the nature of interaction energies and transition rates for diatomic systems in excited configurations is necessary to understand cold collision processes, laser cooling of atomic samples and coherent control of cold molecule formation. It is also relevant to possible future developments like the mooted molecular quantum computation. Efficient formation of long-lived cold molecules via photoassociation has been achieved experimentally. (There are remarkable developments in our understanding of the interaction of light with matter that involve laser techniques of trapping light for periods now as long as a second. Matters in this area are proceeding so fast that we refer the reader to the literature.)

The theoretical explanation involves the resonance interaction between excited and ground-state atoms. Atomic radiative lifetimes are sensitive tests of atomic-structure calculations, and experiments have been performed on molecules created from two atoms, one in its ground state and one in an excited electronic state.

Binding energies and lifetimes were determined using photoassociation spectroscopy of laser-cooled atoms and the results compared to the standard theoretical results.

The long-range, so-called retarded form of potentials of interaction and energy transfer derived from quantum field theory have been used to study exciton transfer in molecular crystals. This leads to a shape dependence of the physical properties of these crystals in so far as conditionally convergent series arise from use of the long-range potential. The retarded resonance interaction problem then has presented something of a puzzle of long standing, particularly since the basic theory from which it is derived appeared to be correct.

The classical quantum mechanical perturbation treatment of this problem gives an energy transfer interaction $\approx R^{-3}$.

At 'large' distances, again called the retarded case as for the Casimir–Polder interaction between atoms in their ground state, the classical result of 50 years standing is supposed to be oscillatory, of the form $\approx R^{-3}\cos(\omega R/c)$, where ω is the frequency of the photon transfer. It is therefore surprising to note that this result, like the Casimir–Polder result, is incorrect. The Casimir–Polder result for long-distance, retarded interactions is correct at zero temperature. But with the resonance problem even the classical zero-temperature result is quite wrong. The correct result can be derived easily by the same semi-classical techniques as for the Lifshitz theory [43,44]. (The incorrect result also emerges as due to too drastic an approximation.) The correct result is again a sum over frequencies exponentially damped with a prefactor $\approx R^{-4}$.

The excited-state–ground-state interaction involving photon transfer has applications to electron transport in biophysical applications. In the theory used to interpret such experiments the photon transfer part handled by the Forster expression and the consequent electron transfer are treated as independent events. In fact, as for the DLVO theory that relies on an ansatz that treats electrostatics and dispersion forces separately, the two processes are coupled and cannot be separated. This will be discussed later.

5.11.3 Speculations on insect pheromones and photon transfer

Interactions in molecular-sized cavities and at surfaces are of relevance to catalysis, and catalytic reactions. The resonance interaction can take on longer-range forms in both hydrophobic and metallic cavities, and give rise to bound states as for interactions at surfaces; e.g. in a hydrocarbon channel of dimensions 2 nm, for molecules with excited states in the infrared $\sim 10^{-14}$ rad/s, the interaction and formation of bound states is much enhanced [46]. It may be worth remarking and is suggestive that typical insect pheromones (for tiger moths, 2-methyl-heptadecane) have an excited state in this frequency range, which is also probably characteristic of vibration frequencies of pheromone-sensitive proteins. Likewise mosquitoes detect prey by sensing CO_2. In such a connection the enhancement of the interactions may or may not be coincidental.

A long-standing problem in biology is that of how pheromones work. Female insects emit extravagantly small numbers of species-specific molecules that attract males. Males seem to be able to pick up and track down females from extraordinary distances. Nearly all pheromones are very simple, uncharged hydrophobic molecules, typically say a linear (C_{12}) hydrocarbon chain that might have a double bond somewhere along the chain, and a terminal simple group such as an acetate. Present ideas have it that the pheromone molecules arrive and physisorb by van der Waals forces at the antennae of the male. These hard, hydrophobic antennae have along their surface stomata that provide access to proteins involved in recognition. The pheromones adsorbed on the hydrophobic antennae are supposed to diffuse along the surface of the antenna until they find and bind to such a protein. This causes a conformation change. The protein unbinds, traverses an aqueous lumen and binds to another protein at the surface of the interior nerve that runs down the length of the antenna. That in turn triggers a sequence of biochemical signals that are well understood.

There are several problems with this theory. There is no reason to believe that there is anything species-specific in the physisorption process. Indeed as far as van der Waals forces are concerned, there can be essentially no difference between all pheromones. Nor is any chemical interaction involved.

This is apart from the problem of too few pheromone molecules being available, by factors of at least millions; or why one might suppose that such molecules would actually physisorb and remain at the antenna surface. In fact if the story were as above, every other hydrophobic molecule in the atmosphere would also adsorb on the antenna. The unfortunate males would then be unable to fly, with their antennae weighed down like an iced-up Boeing aircraft. The biochemistry consequent on energy transfer to the initiating surface protein is not in doubt. But the process by which the information contained in the signalling pheromone molecule is transferred is not understood.

If we look more closely, we see that the only physical differences between different pheromone molecules must lie in the infrared frequency region. (All are essentially identical in the visible and ultraviolet regions that give rise to van der Waals interactions. There are usually no permanent dipoles, so no microwave or electrostatic frequency contributions are involved either.)

In the infrared, however, there are differences. The infrared (coupled mode vibrational) frequency spectra of such molecules are all different. They depend on the position of double bonds, bond strengths along the chain and the end group [42].

In fact they can have long-lived metastable excited modes, and could be emitted by the female in such a state. Quantum calculations of such modes confirm this.

It is of interest too that insects do not usually mate in wet conditions – water is a strong infrared adsorber. Similarly, depending on conditions, insects fly headlong to destruction into bush fires with intense IR radiation.

Given that situation it is then possible to speculate on this scenario: the pheromone molecule is emitted in an excited metastable state by the female. It has a specific frequency that corresponds to the IR vibration frequencies of the receptor protein that resides in the stomata of the male antenna. Transfer of a real photon via the long-range directional resonance interaction just discussed can then activate the recognition and trigger the necessary conformation change.

Although speculative, since such a physical recognition process exists it would be surprising if nature did not exploit it somehow. It seems significant that the distinguishing physical characteristics of pheromone and insect-repellent molecules all lie in the infrared. Unspecified electromagnetic origins for pheromone action have been advanced every decade or so and dismissed as anathema, or lunacy. In that connection, Erasmus Darwin established the Lunar Society, which met every month and discussed the theory of evolution, equally anathema.

The possibility of such a mechanism would be easy to test in principle by placing males inside a box with an IR window and using a tunable laser [48–51].

5.11.4 Casimir, mesons and nuclear interactions

The theory of Casimir (1948) for the force between two perfectly reflecting metal plates predates Lifshitz theory and is a special case of it, as is the Casimir–Polder force between two atoms [43,52–56].

The two ideal metal plate problem can be derived in many ways and the Casimir interaction or variants and adaptations thereof has been the subject of literally thousands of papers. (A historical aside: that the sum of London forces between two colloid particles decayed more rapidly than expected was first inferred by Overbeek, who gave a first estimate of retardation in his thesis [57]. He appears to have asked his friend Casimir to undertake the job. The Dutch School attempted to measure forces between glass surfaces in air but failed due to too large asperities and electrostatic charge effects. Deryaguin, the leader of the competing Russian School, independently asked Landau to develop the same theory of stability of colloids in 1941, and later still asked Lifshitz to develop his theory of interactions, which was confirmed by the experiments of Abrikosova and Deryaguin.)

We wish here to remark that extensions of the Casimir theory give rise to a theory of mesons and nuclear interactions that follow from electromagnetic theory, and consequences of this.

The low-temperature result for the potential of interaction per unit area between two perfectly conducting plates is [43]:

$$\begin{aligned} F(l,T) = &-\frac{\pi^2\hbar c}{720 l^3} - \frac{\zeta(3)(k_B T)^3}{2\pi(\hbar c)^2} + \frac{\pi^2 l (k_B T)^4}{45(\hbar c)^3} \\ &- \frac{(k_B T)^2}{\hbar c l}\left(1 + \frac{k_B T l}{\pi\hbar c}\right) e^{-\pi\hbar c/k_B T l} + O\left(e^{-2\pi\hbar c/k_B T l}\right) \end{aligned} \tag{5.7}$$

The leading term is the Casimir zero temperature result. The temperature correction terms form an asymptotic expansion. At very large distances (of the order of a micron) the formula goes back to a slowly decaying form.

$$F(l,T) = -\frac{k_B T}{8\pi l^2}\left[\zeta(3) + \left(\frac{8\pi k_B T l}{\hbar c} + 2\right) e^{-4\pi k_B T l/\hbar c} + O\left(e^{-8\pi k_B T l/\hbar c}\right)\right] \tag{5.8}$$

The leading term is due to current–current fluctuations in the metal plates (Nyquist noise). The correction terms are a sum of the same exponentially decaying terms we are now familiar with in interactions involving dielectrics.

It was this form that provided the first experimental verification of theory [6,7]. The first equation is of much interest. The first term is that of Casimir. The second is the chemical potential term, precisely that of an electron-pair plasma in the gap between the plates (such a plasma would be formed by the reaction $2\gamma \rightarrow e^- + e^+$ where γ is a photon) [58]. The third term is the free energy per unit volume in the gap of equilibrium black-body radiation.

It can be shown that this mathematical expansion is strictly valid only at absolute zero of temperature where the temperature corrections go to zero. At any other temperature the exponential form (Equation (5.6)) obtains. This has caused and continues to cause much confusion in the literature.

But there is another possibility that leads us into a connection between particle physics and electromagnetic fluctuation forces. If we assume we have available to us the Casimir energy and estimate the energy of interaction of two nucleons modelled as perfectly reflecting spheres of distance and radius one Fermi (10^{-13} cm) we find an energy of interaction of about 8 MeV. This is the binding energy of a nucleon in a nucleus which in the theory of weak interactions is mediated by mesons, not by electromagnetic radiation. How could mesons come out of such electromagnetic theory?

If we do equate the second and third terms of the expansion at distance l of one Fermi, we can find an equivalent back-body temperature. We can then exploit the photon radiation electron-pair equilibrium to calculate the density of such virtual electron pairs, which now form a plasma in the gap.

We can now reformulate the problem to consider the corresponding generalized Casimir interaction in the presence of a plasma whose dielectric response is known. The result gives an interaction that can be interpreted as a Klein–Gordon equation for π_0 mesons where the 'meson' is a plasmon, a collective mode in the electron–positron pair sea. It has the right mass and lifetime.

Once that connection is made, the existence of charged π^- and π^+ mesons is due to bound states of electrons or positrons to the plasmon. Similarly K mesons would emerge as double plasmon excitations. Whatever the status of such a theory, we would still have to account for the existence of such large energy available in the electromagnetic field. In any event the form of the Casimir interaction with macroscopic metal plates is analytically of a completely different form to the standard form, even for an infinitesimal plasma density in the gap. In any real system skin effects provide a source of such a homogeneous plasma in the gap. Continuing debates about the Casimir effect ignore this reality.

The connection between electromagnetic fluctuation forces and nucleon interactions and mesons, if correct, shows that meson theory is exactly equivalent to Onsager–Samaris and Lifshitz, i.e. has the same status and is a special case of that linear response theory. Then it suffers the same defects. All of nuclear physics would have the same problem as well. That is the electrostatics and mesons (here shown to come from quantum dispersion interactions) are separated and treated invalidly just like and exactly equivalent to the older DLVO theory of colloid science.

References

[1] V. E. Shubin and P. Kekicheff, *J. Coll. Interface Sci.* **155** (1993), 108–123.
[2] W. D. Niven (ed.), *The Scientific Papers of James Clerk Maxwell*, **2** vol. (1890, reissued in 1965).
[3] J. C. Maxwell, Capillary action. In *Encyclopaedia Britannica*, 9th edn (1876) updated by Lord Rayleigh in the 11th edn, building on ideas of van der Waals and Poisson.
[4] I. E. Dzyaloshinskii, E. M. Lifshitz and L. P. Pitaevskii, *Adv. Phys.* **10** (1961), 165–209.
[5] A. A. Abrikosov, L. P. Gor'kov and I. E. Dzyaloshinskii, *Quantum Field Theoretical Methods in Statistical Physics*. 2nd edn. London: Pergamon Press (1961).
[6] B. V. Deryaguin and I. I. Abrikosova, *J. Exp. Theor. Phys.* **30** (1956), 993–1006.
[7] B. V. Deryaguin, I. I. Abrikosova and E. M. Lifshitz, *Q. Rev. Chem. Soc. (London)* **10** (1956), 295–329.
[8] P. Richmond and B. W. Ninham, *J. Phys. C Solid State Phys.* **4** (1971), 1988–1993.
[9] B. W. Ninham, V. A. Parsegian and G. H. Weiss, *J. Stat. Phys.* **2** (1970), 323–328.
[10] J. Mahanty and B. W. Ninham, *Dispersion Forces*. London: Academic Press (1976).
[11] V. A. Parsegian, *Van der Waals Forces: a handbook for biologists, chemists, engineers, and physicists*. Cambridge: Cambridge University Press (2006).
[12] V. A. Parsegian and B. W. Ninham, *Nature* **224** (1969), 1197–1198.
[13] B. W. Ninham and V. A. Parsegian, *J. Chem. Phys.* **52** (1970), 4578–4587.
[14] B. W. Ninham and V. A. Parsegian, *Biophys. J.* **10** (1970), 646–663.
[15] B. W. Ninham and V. A. Parsegian, *Biophys. J.* **10** (1970), 664–674.
[16] B. W. Ninham and V. V. Yaminsky, *Langmuir* **13** (1997), 2097–2108.
[17] J. N. Israelachvili, *Intermolecular and Surface Forces: with applications to colloidal and biological systems*. Londaon: Academic Press (1985–2004).
[18] L. R. White, J. N. Israelachvili and B. W. Ninham, *J. Chem. Soc. Faraday Trans. I* **72** (1976), 2526–2536).
[19] P. Richmond, B. W. Ninham and R. H. Ottewill, *J. Coll. Interface Sci.* **45** (1973), 69–80.
[20] D. Beaglehole, E. Z. Radlinska, H. K. Christenson and B. W. Ninham, *Phys. Rev. Lett.* **66** (1991), 2084–2087.
[21] J. Mahanty and B. W. Ninham, *J. Chem. Soc. Faraday Trans. II* **70** (1974), 637–650.
[22] J. Mahanty and B. W. Ninham, *Disc. Faraday Soc.* **59** (1975), 13–21.
[23] D. Beaglehole, E. Z. Radlinska, H. K. Christenson, B. W. Ninham, *Langmuir* **7** (1991), 1843–1845.
[24] C. H. Anderson and E. S. Sabisky, *Phys. Rev. Lett.* **24** (1970), 1049–1052.
[25] E. S. Sabisky and C. H. Anderson, *Phys. Rev. A* **7** (1973), 790–806.
[26] P. Richmond and B. W. Ninham, *J. Low Temp. Phys.* **5** (1971), 177–189.
[27] D. Chan and B. W. Ninham, *J. Chem. Soc. Faraday Trans. II* **70** (1974), 586–596.
[28] B. W. Ninham and V. A. Parsegian, *J. Chem. Phys.* **52** (1970), 4578–4587.
[29] J. Lyklema and K. J. Mysels, *J. Am. Chem. Soc.* **87** (1965), 2539–2546.
[30] D. J. Mitchell and B. W. Ninham, *J. Chem. Phys.* **56** (1972), 1117–1126.
[31] B. Davies, B. W. Ninham and P. Richmond, *J. Chem. Phys.* **58** (1973), 744–750.

[32] R. F. Rajter, R. Podgornik, V. A. Parsegian, R. H. French and W. Y. Ching, *Phys. Rev. B* **76** (2007), 045417.
[33] J. Mahanty, *Il Nuovo Cimento B* **22** (1974), 110–120.
[34] D. F. Parsons and B. W. Ninham, *Langmuir* (2009) in press.
[35] S. Andersson and B. W. Ninham, *Solid State Sci.* **5** (2003), 683–693.
[36] M. Boström and B. W. Ninham, *Biophys. Chem.* **114** (2005), 95–101.
[37] M. Boström and B. W Ninham, *J. Phys. Chem. B* **108**, (2004), 12593–12595.
[38] M. Boström and B. W. Ninham, *Langmuir* **20** (2004), 7569–7574.
[39] M. Boström, J. J. Longdell and B. W. Ninham, *Phys. Rev. A* **64** (2001), 062702.
[40] Z. Blum, S. T. Hyde and B. W. Ninham, *J. Phys. Chem.* **97** (1993), 661–665.
[41] S. Hyde, S. Andersson, K. Larsson, Z. Blum, T. Landh, S. Lidin and B. W. Ninham, *The Language of Shape. The role of curvature in condensed matter physics, chemistry and biology*. Amsterdam: Elsevier (1997).
[42] B. W. Ninham, R. Nossal and R. Zwanzig, *J. Chem. Phys.* **51** (1969), 5028–5033.
[43] B. W. Ninham and J. Daicic, *Phys. Rev. A* **57** (1998), 1870–1880.
[44] H. Wennerström, J. Daicic and B. W. Ninham, *Phys. Rev. A* **60** (1999), 2581–2584.
[45] M. Boström, J. J. Longdell, D. J. J. Mitchell and B. W. Ninham, *Eur. Phys. J. D* **22** (2003), 47–52.
[46] M. Boström, J. J. Longdell and B. W. Ninham, *Europhys. Lett.* **59** (2002), 21–27.
[47] M. Boström and B. W. Ninham, *Phys. Rev. A* **69** (2004), 1–2.
[48] R. H. Wright, *Nature* **178** (1956), 638.
[49] R. T. Cardé and A. K. Minks, *Insect Pheromone Research: new directions*. New York: Chapman & Hall (1997), 111–163.
[50] S. Kanaujia and K. E. Kaissling, *J. Insect Physiol.* **31** (1985), 71–81.
[51] G. Chen, J. Wu, Q. Lu, H. R. Gutierrez, Q. Xiong, M. E. Pellen, J. S. Petko, D. H. Werner and P. C. Eklund, *Nano Lett.* **8** (2008), 1341–1346.
[52] B. W. Ninham and M. Boström, *Phys. Rev. A* **67** (2003), 030701.
[53] H. B. G. Casimir, *Proc. K. Ned. Akad. Wet.* **51** (1948), 793–796.
[54] H. B. G. Casimir, *J. Chim. Phys.* **46** (1949), 407–409.
[55] H. B. G. Casimir and D. Polder, *Nature* **158** (1946), 787–788.
[56] H. B. G. Casimir and D. Polder, *Phys. Rev.* **73** (1948), 360–372.
[57] E. J. W. Verwey and J. T. G. Overbeeck, *Theory of the Stability of Lyophobic Colloids*. Amsterdam: Elsevier (1948).
[58] L. D. Landau and E. M. Lifshitz, *Statistical Physics*. Oxford: Butterwoth-Heinemann (1951).

6

The extension of the Lifshitz theory to include electrolytes and Hofmeister effects

6.1 Inclusion of electrolytes and Hofmeister effects in the theory

We have seen that the classical theories of electrolytes are limiting laws that are strictly valid, if at all, at very low concentrations. That is so for the Born self energy, the correlation free energy of Debye–Hückel theory, the interfacial tension of electrolytes represented by Onsager's theory, and the double-layer theory of interactions between charged colloidal particles. These theories ignore water structure, hydration, induced by ions. They also ignore dispersion interactions between ions and between ions and surfaces. In reality, with moderate salt concentrations of interest in biology (> 0.1 M) strong specific ion effects emerge. These effects can be taken into account by introducing more (unquantified) parameters that attempt to build in hard core interactions, local 'water structure' and overlap of hydration shells. But even with such extensions, the parameters vary from one situation to another. For example, the double-layer forces between the cationic bilayers of Chapter 4 differ by more than an order of magnitude with sodium bromide or sodium acetate at concentrations as low as 10^{-3} M! This is entirely unsatisfactory. A theory of electrolytes that deserves the name ought to work predictively for all electrolytes.

As a first step to throwing some light on the problem, we here address how to build a theory that includes the missing dispersion forces together with electrostatics.

6.1.1 The effects of electrolytes and conduction processes

The theory of electrodynamic fluctuation forces between bodies that we have described so far has been restricted to spatially uniform dielectric media. The idea behind it was that instead of taking a bottom-up, molecular, approach via statistical mechanics or simulation, we recognize that all many-body forces are implicit in the measured dielectric susceptibilities as a function of frequency of the interacting media.

The totality of electromagnetic fluctuations between the two bodies can be Fourier analysed into a set of allowed frequencies or modes. The modes are determined by the dielectric boundary conditions at the surfaces. They can be treated as independent virtual harmonic oscillators. Then the sum of the free energies of the allowed harmonic oscillator frequencies leads to the total free energy of interaction. With conducting media, correlations between real current fluctuations add another level of complication. In the original Lifshitz theory [1,2] the forces between perfect low-temperature conducting media were handled by a limiting prescription. That gave out the earlier Casimir force (see Chapter 5), which had been derived differently. The Casimir ideal metal plate interaction-in-a-vacuum problem can also be dealt with in terms of real surface current–surface current fluctuation correlations [3]. With the Casimir problem we have seen in Section 5.11 that an intervening charged plasma between the metal plates, however dilute, changes the nature of interactions qualitatively.

This is so too for colloid particles interacting across electrolyte solutions, or with bodies that can themselves conduct electric currents. We turn now to this added complication.

6.1.2 The effects of electrolytes on dispersion interactions

First, recall that for a wide range of situations that concern us in colloid science and biology (oil, proteins, lipids and other macromolecules in water with or without electrolytes), roughly four different frequency regimes combine to give specificity in the dispersion interactions. The relative contribution of each allowed modal frequency depends on the characteristic properties of the interacting media reflected in their measurable dielectric susceptibilities. The theory tells us that the contribution of each frequency is modulated by an exponential cut-off factor. Its magnitude depends on the distance apart, and on the geometry of the particles. Interacting macromolecules sample these different frequency contributions and rearrange their orientation as they recognize each other and respond to them.

Temperature-dependent (zero-frequency), infrared, visible and ultraviolet frequency contributions, specific to each medium, are all sensed. On approach different frequencies lock in successively until the molecules eventually dock into place like two imagined space vehicles. In this picture the long-range zero-frequency, temperature-dependent contribution due to cooperative permanent dipoles is dominant at large distances. It seems rather boring.

With electrolytes, things become more interesting. There are several ways of tackling the problem of including conducting media. The formalism required to extend Lifshitz theory of interactions (and to include extensions of Born energies, Debye–Hückel and Onsager–Samaris interfacial tensions) to this complex case and applications thereof can be found in Refs. 4–7.

6.1.3 Extensions of Lifshitz theory and the Onsager limiting law

The difference to the pure dielectric case for interactions involving electrolytes is qualitative. The ions respond directly to the fluctuating fields and give rise to real current correlations that contribute to the forces.

The differences can be seen by comparing the forces with and without electrolyte. Without electrolyte, in the sum over frequencies that give the whole free energy of interaction in Lifshitz theory, the first contributing frequency is at around 10^{14} rad/s.

If we ignore the complication of retardation (take the velocity of light to be infinite) the net free energy per unit area between two dielectric slabs 1 across a dielectric medium 2 is:

$$F(l) = \frac{k_B T}{2\pi} \sum_{n=0}^{\infty}{}' \int_0^{\infty} \ln\left[1 - \Delta^2 \exp(-2kl)\right] \cdot k dk \tag{6.1}$$

The prime indicates that the term in $n = 0$ is to be taken with a factor ½, and the parameter Δ is given by: $\Delta(\xi_n) = [\varepsilon_2(i\xi_n) - \varepsilon_1(i\xi_n)] / [\varepsilon_2(i\xi_n) + \varepsilon_1(i\xi_n)]$.

The dielectric susceptibilities ε of the interacting media are evaluated at imaginary frequencies $i\xi_n$, where $\xi_n = 2\pi n k_B T/\hbar$.

For the usual concentrations n of electrolyte of interest, say 0.1 M, the ion plasma frequency $\omega_p = \sqrt{4\pi n e^2/m}$,where m is the mass of an ion and n the number density, is of the order of 10^{12} rad/s. At all frequencies higher than this, infrared and above, the ions cannot follow the fluctuating fields. To a first approximation then, we can ignore the effects of ionic motion on the dielectric responses of the media from all except the zero-frequency term. (A full treatment of the problem and to include hydronium ion responses has not been attempted. The hydrodynamic behaviour of the electrolyte can be modelled along the lines outlined in Ref. 5.)

So the high-frequency contributions are substantially unaltered. But for the first, $n = 0$ zero-frequency term of the sum, the dielectric responses are substantially modified by the electrolyte. This contribution to the interaction can be viewed as due to a correlation of electrostatic image interactions of the ions. The modified contribution, about half the full free energy of interaction for oil–water systems, now has the form

$$F_{n=0}(l) \equiv F_0(l) = \frac{k_B T}{4\pi} \int_0^\infty \ln\left[1 - \Delta_{12}^2(s)\exp(-2sl)\right] \cdot k dk \tag{6.2}$$

where $s = \sqrt{k^2 + \kappa_D^2}$, and $\Delta_{12}(s) = \dfrac{\varepsilon_2(s) - \varepsilon_1 k}{\varepsilon_2(s) + \varepsilon_1 k}$.

6.1.4 Where and how the theory breaks down

The corresponding contribution in the absence of electrolyte, which comes from a purely quantum treatment, cancels out identically.

As a technical aside, a careful analysis at small distances, $kl \ll 1$, shows that the interaction free energy from the $n = 0$ term modified by the presence of an electrolyte can be rewritten as an asymptotic expansion in concentration. The first approximation is the same original Lifshitz form for dielectric media in the absence of salt. The next term is a contribution which is twice the change in (electrostatic) interfacial free energies of the interacting particles due to dissolved salt in the linearized approximation of Onsager's treatment. The interaction can also be considered to be due to an osmotic effect. It can be considered to come about from an electrostatic double-layer interaction due to the overlap of ionic profiles set up at the interfaces. These inhomogeneous profiles are due to electrostatic image interactions of the ions with the two surfaces [6].

These observations demonstrate the explicit operation of the correspondence principle of quantum physics: at large distances or high temperature, classical physics takes over. In the opposite case, small distance or low temperature, quantum mechanics dominates. The point is that it is impossible in general to separate and treat independently double-layer electrostatic and quantum interactions.

At large distances when linearization of the ionic spatial distribution functions is allowed, some very interesting new forces emerge. For planar objects the predicted interaction energies are proportional to temperature and decay as $V(l) \approx -[k_B T \exp(-2\kappa_D l)]/l$, where κ_D is the Debye length of the salt solution.

6.1.5 Explorers in difficulty: what went wrong?

At this point in our development of theory, we are explorers in difficulty. We have put on our physicist's hat to study the dispersion forces. We have invoked the full apparatus of sophisticated quantum statistical physics, and simplified it to a semi-classical theory. We have used a description of matter via its dielectric susceptibility as a function of frequency that ought to include all cooperative many-body interactions. The alternative bottom-up descriptions of interactions between particles, starting from individual pairwise constituent molecular interactions and their simulations, are included as special cases.

But now, when we extend the theory to try to deal with an intervening electrolyte, it seems we are in trouble. This is because at small distances the Onsager–Samaris result reappears. That is, at small distances the zero-frequency contribution to the sum of frequencies of Lifshitz theory extended to include electrolytes is exactly equivalent to the linearized version of Onsager–Samaris theory for the change in interfacial tension with dissolved salt [6]. This we know is a limiting law valid, if ever, at impossibly low concentrations of salt. Even if we used the complete non-linear form of this electrostatic theory and included an artificial 'ion size', such a theory still does not work to describe interfacial tensions.

The predictions are not ion-specific. The experimental results are.

Worse still is this: it can be proved that this extension of the DLVO theory is seriously inconsistent. We began with an ansatz that separates interactions into attractive dispersion forces and electrostatic double-layer forces. This is inadmissible. The separation into two kinds of forces can be shown to violate both the Gibbs adsorption equation (which defines an interface) and the gauge condition on the electromagnetic field (conservation of current) [6].

So our basic ansatz of colloid science, the separation of forces, violates both thermodynamics and electromagnetic theory! Quantum fluctuation forces and double-layer forces are inextricably entangled and have to be treated at the same level. We cannot treat one, the quantum dispersion forces, via a linear theory, and the other, double-layer forces, by a non-linear theory as we do, for example, using a Poisson–Boltzmann distribution; cf. Chapter 3.

The same conclusions hold not just for colloidal particle interactions. The same is true for the Born and Debye–Hückel theories, when we attempt to combine long-range electrostatic forces with short-range (dispersion) interactions between ions, or with ions and the solvent. Fitting parameters to include short-range quantum effects with an effective 'ion size' can be used to try to paper over this difficulty. But these fitting parameters vary for the same ion from one situation to another, with temperature and nature of the interface and the mixture. Predictability is lost.

This is more than just a matter of fundamental esoteric interest. The entire theoretical framework and intuition we have on molecular forces and their impact on thermodynamic quantities in standard physical chemistry rests upon:

- the self energy of ions and other molecules in solution, the Born energy;
- correlation energies of electrolytes, osmotic pressures, Debye–Hückel theory and its extensions;
- adsorption phenomena of ions, ion binding, proteins, interfacial tensions of electrolytes and electrochemistry, pH, buffers, zeta potentials and membrane potentials in physiology;
- interactions between colloidal particles and proteins (DLVO theories).

All these basic concepts depend on classical electrostatic theories. These ignore dispersion forces entirely; or else they maintain the separation of forces ansatz as for the DLVO theory.

The matter is further compounded and confounded if we consider that the interpretation of measurements and the values assigned to measured quantities like pH depend solely on electrostatic-only theories that are flawed by the omission.

We will return subsequently to this problem of what went wrong and with one way to come to grips with it. For the moment, we pause to do a reality check and discuss the experimental challenges that a theory of interactions has to explain.

6.2 Hofmeister effects and their universality

By now we have run into specific ion effects often. The mere use of the words 'specific ion effects' means that we have no satisfactory theory.

Over a century ago Hofmeister, experimenting with precipitation, salting out of egg white proteins, showed that salts with the same cation but different anion precipitated the protein at different concentrations. The same occurred with some other colloidal suspensions such as Fe_2O_3. In terms of effectiveness in protein stabilization, the anionic sequence was [8]:

$$H_2PO_4^- > SO_4^{2-} > F^- > Cl^- > Br^- > NO_3^- > I^- > ClO_4^- > SCN^-$$

With fixed anion but varying cation the same phenomenon emerged. The usual sequence for the stabilization of proteins is [8]:

$$Me_4N^+ > NH_4^+ > Cs^+ > Rb^+ > K^+ > Na^+ > Li^+ > Ca^{2+} > Mg^{2+}$$

The variations for cations are not so qualitatively large as the variations of anions with fixed cation.

A translation of some of Hofmeister's original work can be found in Ref. 9. There are very many reviews on the Hofmeister effects; we refer the reader to Refs. 8, 10–14 and references therein.

In Chapter 3 we have run into the same Hofmeister phenomena with activity coefficients of electrolytes, with the interfacial tension of water induced by electrolytes and with direct force measurements between lipid bilayers. The first specific ion effect, with activity coefficients, is due to both interionic interactions and to the specific water structure induced by the ions of the electrolyte. (The nature of that water structure is itself a consequence of ion–water interactions.) The activity is a bulk effect, but it reflects local water structure around the ions.

The surface tension specific ion effect [15,16] is clearly a surface interaction phenomenon. Changes in interfacial tension by definition reflect changes in adsorption of ions at the interface through the Gibbs adsorption equation.

But again, the specific adsorption of ions reflects interactions of the ions with corresponding molecular moieties on the surface, via the local water structure around both.

6.2.1 Are surface or bulk effects responsible for Hofmeister phenomena?

Direct force measurements in salt solutions with different counterions apparently fit very well to predictions of electrostatic double-layer theory. They decay exponentially, with the theoretical Debye length. But the magnitudes of the long-range double-layer forces for Br^- vs. Cl^- or NO_3^- and CH_3COO^- differ by factors of 50 or more [17,18]! The required surface charge required to fit the double-layer theory prediction reflects different specific adsorption of the (hydrated) counterions (and co-ions) that are not accounted for by electrostatic forces. It is a surface effect. At least at low concentrations of electrolyte (e.g. 1–10 mM) that reflects what in conventional language is ascribed to 'ion binding'. (The effects here cannot be accommodated by postulating surface charge regulation–dissociation–association equilibria [19]. Charge regulation does occur, for example with amphoteric surface groups such as carboxylates or amines characteristic of proteins.)

In an elegant experiment on lamellar phases of double-chained cationic surfactant, it is shown that Hofmeister effects cannot be accommodated by postulating extra specific 'hydration' forces [20].

Short-range interactions of ions with neighbouring water molecules give rise to ion-specific hydration. Interactions between hydrated ions via all of electrostatic, hydration and ionic dispersion forces will give rise to specificity in bulk properties. Ion-specific interactions of ions with an interface are also clearly short-range.

Hofmeister himself, a pharmacologist, was bemused about the source of his phenomena, and speculated on whether they were due more to surface or to bulk water structure effects. On balance he seems to have favoured attribution more to a bulk effect, which he called the 'water withdrawing power of salts'. This is a view to which anyone given a dose of epsom salts, magnesium sulfate, would also subscribe. But to oppose that, the undoubted successes of the use of lithium acetate or chloride to control manic depression at a level of only a gram per day suggests rather that this particular effect is due to specific competitive ion adsorption of Li^+ with Na^+ at the surface of brain lipid membranes (the typical 5 litres of human blood contains around 100 grams of NaCl).

Sometimes, with differing colloidal or self-assembled particles, the order of the Hofmeister sequence becomes partially scrambled. (The surfaces with which the ions interact are different, and this has to implicate ion–surface molecular moiety interactions.) Sometimes the series is completely reversed. This depends on the colloidal particle, and its point of zero charge [21]. This observation again implicates a specific particle–ion surface interaction, rather than a bulk water effect. Hofmeister effects can reverse with buffer at the same pH.

The effects usually occur at 'high' concentrations of electrolyte, typically around and above about 0.15 M, when electrostatic forces begin to be strongly screened. This is precisely the biological milieu (the Debye length is around 0.8 nm at 0.15 M salt concentration, the characteristic physiological concentration in blood). But, depending on the surfaces of particles, they can also occur at low concentrations. The phenomena are ubiquitous and occur par excellence at a practical level in a myriad of completely different phenomena.

6.2.2 More examples of Hofmeister effects

They occur with self assembly of surfactants, micelles, microemulsions and emulsions and with polymers. This will be explored in later chapters.

They occur with optical activity of chiral solutes [22,23], with viscosity of aqueous electrolyte solutions [24], with cloud points of non-ionic or zwitterionic surfactants [25] and with formation of host–guest supramolecular complexes [26]. They occur with specific ion complexation of calixarenes at the air–water interface [27] and transitions involving surfactant lamellar phases [28].

With surfactants and their self assembly into micelles, lamellae, vesicles and other microstructures phases, with emulsions and microemulsions, and polymer solutions discussed in detail in later chapters of this book, specific ion effects are dramatic and legion.

Sometimes, and completely counterintuitively, Hofmeister effects show specific adsorption of cations at cationic surfaces [29]. They can be used to great effect to change self-assembled microstructures, as we shall see.

An example of considerable unexploited practical application are specific ion effects due to adsorption of salts in oil reservoir flooding for secondary oil recovery. This has huge economic implications for oil extraction via flooding with brines. It has only very recently been recognized that change of salt type, concentration and the nature of the reservoir rock surface changes interfacial tensions and wettability, and hence recovery, dramatically.

Hofmeister effects occur obviously in electrochemistry. A range of situations in which Hofmeister effects are discussed recently can be seen in a series of papers [11].

Importantly, the phenomena occur even with pH measurements. The Hofmeister series that rates the effectiveness of different ions in changing (apparent) pH can reverse with change of buffers [30]. This has an enormous impact especially on biological systems, where the use of different kinds of buffers and co-solutes is common.

6.2.3 Biochemical and biological examples of Hofmeister effects

With proteins (which are typically amphoteric with positive and negative dissociable surface groups due to carboxylates and amines) the series reverses above and below the isoelectric point of zero charge, which is assumed to be determined by pH. Ion–surface interactions missing from standard electrostatic theories must be operating here. The Hofmeister effects with bulk activity coefficients are not expected to undergo a reversal with a small change of pH. However, they do change. The reversal of the Hofmeister series above and below the point of zero charge of particles presented a continuing puzzle and the major challenge to theories from the beginning. Such an effect with proteins and its explanation are given in Ref. 21, and that this is a long-standing puzzle can be seen in Refs. 31 and 32.

Hofmeister effects with many naturally occurring surfaces are universal, but are rarely quantified. A nice illustrative example is the variation of uptake of water by natural fibres depending on the Hofmeister series [33].

Particularly in biology and biochemistry – based on systems that contain water and salts – the Hofmeister phenomena are ubiquitous. One such elegant example is the work of Vogel on conformational changes in rhodopsin induced by changes in anion [34,35].

Several examples are discussed explicitly further below. One in Section 6.4 concerns the specific ion dependence of restriction enzymes [36] and other enzyme activities [37–40].

Another of much interest is the Hofmeister specific ion effect in superactivity of horseradish peroxidase due to choline–sulfate specificity [41].

The literature on Hofmeister effects in biochemistry is vast. It is not even limited to just macromolecules such as enzymes or bio- and other polymers. It is at least comprehensible that with changes of type of background electrolyte, competitive ion adsorption for biosurfaces might affect conformation, and therefore recognition. It is not easy to see how changes in enzyme activities occur with changes in buffer.

At a higher level of organization a particularly striking example of Hofmeister effects occurs with whole cell growth. The growth of bacteria depends strongly on Hofmeister effects [42].

Another example in biology where specific ion adsorption effects have to be involved at some as yet unexplored level is in the central problem of 'ion pumps'. The maintenance of a large ionic differential in concentration of potassium and sodium to give a potential across a cell membrane is assigned to an active biochemical pump. (Sodium is 4-times higher outside the cell than inside, and vice versa for potassium.) Be that as it may, in a red cell, for example, the interior comprises densely packed haemoglobin molecules, proteins of high surface area. Preferential specific ion adsorption of cations undoubtedly exists, and must give rise to some contribution to the 'pump'.

Another whole animal effect, that of lithium acetate in controlling manic depression, we have already noted.

Specific ion binding, of potassium, even plays an integral role in dispersal of spores of particular fungal phyla [43], and in rhyzobial communication in algae with fungal mycelia and lichens.

The classical theories of forces in physical and colloid chemistry have had little to say on these phenomena, which were all well known long before better ideas on forces and bonds in physical chemistry developed after quantum mechanics began to impact.

It seems clear from this long litany that there must be some missing competitive adsorption forces acting on ions. This changes surface charge and the forces leading to coagulation. Theoretically, Onsager pointed out that with interfacial tensions of electrolytes there had to be some missing surface forces acting on ions to account for the facts as long ago as 1930.

6.3 Hofmeister effects with pH and buffers and implications

A problem of quite fundamental importance is that of specific ion effects on pH and buffers [44].

Measurements of pH are almost always made with standard glass electrodes. Figure 6.1 shows measurements of pH for two salt solutions with the pH fixed initially at 7.5 with two different buffers, phosphate and cacodylate.

The buffer concentrations are 5 mM. When salt of a strong electrolyte is added at fixed cation, here Na^+, the pH decreases, as much as a whole pH unit. This is a highly significant change in biological applications concerned with proteins. The Hofmeister series emerges for pH. The standard sequence is $H_2PO_4^- > SO_4^{2-} > F^- > Cl^- > Br^- > NO_3^- > I^- > ClO_4^- > SCN^-$. The left-side ions are considered strongly 'structure-making' (kosmotropic); merging into 'structure-breaking' or chaotropic ions. However, the standard ordering of effectiveness depends on substrate [45]. The sequence is reversed if phosphate buffer is replaced by cacodylate. The dashed curve shows pH calculated from the standard theory of electrolytes (extended Debye–Hückel theory). The trends are roughly correct. But to fit a particular electrolyte one would have to choose fitting parameters for buffer anion radii that vary from salt to salt and are absurdly large.

At these buffer concentrations carbon dioxide effects are suppressed – with no buffer the pH increases with added electrolyte. Similar effects occur for variation in cation.

Hofmeister effects on pH are seen with other buffers such as citrate and triethanolamine which buffer at pH 4.5 without salt. The first gives a decrease of apparent pH with added salts, the second an increase [45].

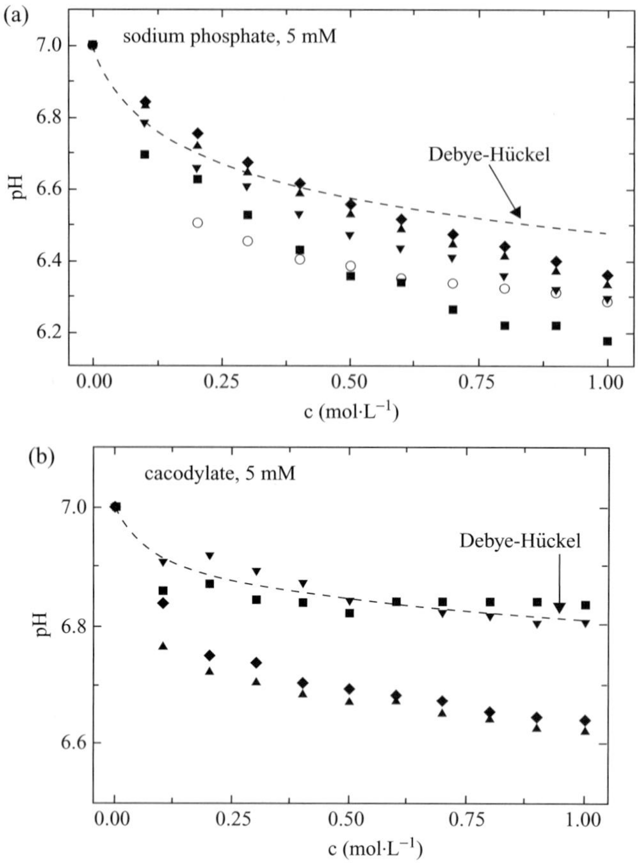

Fig. 6.1. Hofmeister series for pH measurements with glass electrode. Curves in (a) are for phosphate buffer at 5 mM, those in (b) for cacodylate. pH is fixed initially at 7.0 and salt added. The apparent pH sequence is reversed for the two buffers. With cacodylate and K^+ instead of Na^+, the sequence reverses again. NaCl (♦), NaBr (▲), NaI (○), $NaNO_3$ (▼) and $NaClO_4$ (■). Adapted with permission from Ref. 30. Copyright 2006 American Chemical Society.

Again one can attempt to fit results by the standard electrostatic theories of activities. The fits require again absurd parameters for buffer ion radii, and fail beyond 0.1 M.

6.3.1 Foundations that underlie pH measurement

A measurement of pH ultimately is a measurement of surface potential. The standard theory of interpretation of pH agreed by IUPAC convention is explained in Ref. 46. The interpretation of the surface potential is the question at issue. It rests intimately on two pillars. One is the extended Debye–Hückel theory (see Chapter 3) The other requires the theory of the double layer involving the Poisson–Boltmann distribution (see Chapter 3). Both are based on electrostatic forces alone.

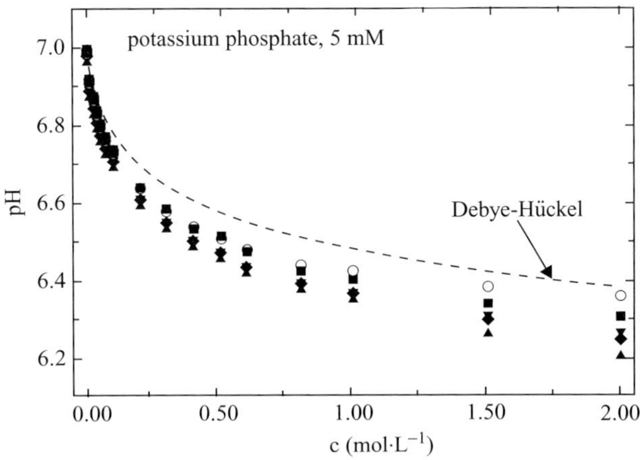

Fig. 6.2. Hofmeister series for pH measurements with a glass electrode and phosphate buffer at 5 mM. pH is fixed initially at 7.0 and salt added. The apparent pH sequence is reversed for the two buffers. KCl (◆), KBr (▲), KI (○), KNO_3 (▼) and KSCN (■). Adapted with permission from Ref. 30. Copyright 2006 American Chemical Society.

The classical theories ignore non-electrostatic ion–ion and ion–solvent interactions due to dispersion forces. We know that ignoring these quantum mechanical forces leads to non-trivial inconsistencies. These forces are crucial for a proper treatment of Born self energies, of activities, of interfacial phenomena and of interactions. The conventional electrostatic primitive model theories attempt to account for the Hofmeister pH effects with added salts in terms of calculated bulk activities. But since these theories are incorrect for bulk electrolytes, they are at best deficient. Buffer anion adsorption driven by dispersion forces competes with the counterions and co-ions for the electrode surface. If we make such an estimate it gives results of the right order of magnitude to do the job [47].

But the matter is still open. Its resolution depends on better dynamic polarizabilities of ions now accessible from ab initio quantum chemistry (Ref. 48 and D. F. Parsons and B. W. Ninham, *Langmuir* (2009), in press).

Concerning the possible adsorption effects of buffer anions it may be worth noting that electrically neutral zwitterionic vesicles formed from lipids will not respond to an applied electric field. On addition of phosphate buffer, however, they immediately charge to give an apparent zeta potential of 140 mV, indicating strong surface adsorption.

The glass electrode is a very complicated and not well understood beast. The best advice from IUPAC, the International Union of Pure and Applied Chemistry, is to avoid pH above 0.1 M added salt.

Figures 6.1 (top) and 6.2 compare pH measurements with change in cation from sodium to potassium in phosphate buffer. It is difficult to explain how the

Hofmeister pH series with anions reverses order when the sodium cation is replaced by potassium through via bulk activities alone.

6.3.2 Implications of the pH issue for zeta and membrane potentials

The uncertainties just discussed have some ramifications. The hydronium ion and sodium ion counterion concentrations at the glass electrode interface set the potential from which the pH is deduced via the classical theory of electrolytes. If the neglected ion-specific dispersion forces acting on ions are taken into account, there will be a further contribution to the surface potential component. This involves additional competitive adsorption of co-ions from the electrolyte and of buffer anions. Then the interpretation of the potential and the consequent pH deduced or assigned from a measurement is apparent, not real. This is a serious, unresolved issue. It impinges equally on measurements of surface (zeta) potentials in colloid science and on membrane potentials in physiology. These also have an inbuilt reliance on approximate theory that does not account for specific ion effects, either bulk activities or the nature of the substrates that affect competitive adsorption.

The effects just discussed open up the whole pH–Hofmeister question further. It may well be that in applications with multiple mixed electrolytes we have been working with a paradigm analogous to the Ptolemaic system of the planets in astronomy. The standard methods of physical chemistry to determine pH in buffers and salts are embodied in (for example) Ref. 49.

A difficulty is that with the complicated mixtures of real biological and industrial systems, this analogue to the Ptolemaic system breaks down. The whole meaning of pH that becomes a dependent variable becomes problematical. Thus Evens and Niedz [50] argue compellingly and show via a vast number of experiments that in real biological systems pH becomes a dependent variable resulting from a given ion formulation, not an independent variable to be controlled by buffering. Likewise, the ionic strength is taken as a dependent variable.

Another example of this compounding and confounding of pH effects occurs in sea water, where the solubility of carbon dioxide is a major issue in speculations on climate change [51].

The deficiencies of present theories mean that we can expect some interesting developments as our understanding of the meaning of pH and the interplay between pH and solutes and surfaces becomes better quantified.

6.4 Hofmeister effects with restriction enzymes and speculations on mechanisms

A remarkable example of Hofmeister effects in real biology is that of the efficiency of restriction enzymes as a function of salt concentration and type. Restriction enzymes are an important tool in molecular biology. They are catalysts that

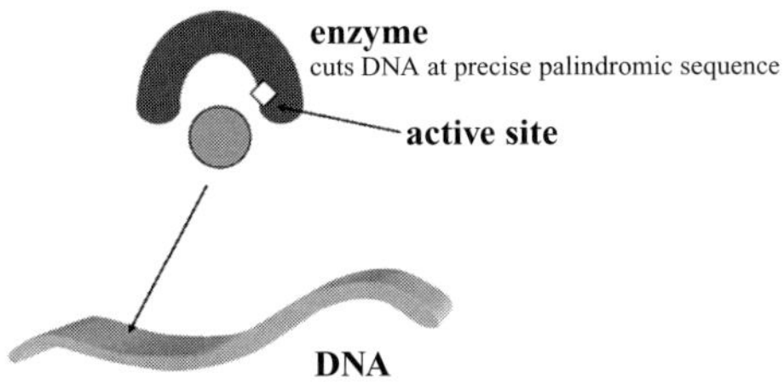

Fig. 6.3. Schematic picture of a restriction enzyme, cutting a DNA strand at a specific sequence of nucleotides. Adapted from Ref. 7, with permission. Copyright 2006 Springer-Verlag.

cut linear DNA at a particular palindromic small sequence of nucleotides (see Fig. 6.3).

The enzyme has a hydrophobic pocket, an 'active site ' formed by a sequence of nucleotides. The dimensions are about 10–20 Å. To fold up in the right conformation one or two members of the sequence need to bind specifically a Mg^{2+} ion. Other divalent ions such as Mn^{2+} can sometimes substitute.

The enzyme attaches to DNA, the association driven by molecular forces that we are familiar with. It then dimerises and diffuses along the (linear) DNA until the active site of the enzyme finds the required stereochemical (lock and key) sequence that it has to cut. Then it cuts the DNA at a phosphate bond. It does so with exquisite, repeated regularity.

The mechanism by which this occurs is attributed to an oxygen free radical. But a question remains, as for all catalysis and surface chemistry: what produces this putative free radical? What is the source of the energy that drives the process? The weak physical molecular forces of association are somehow harnessed, presumably cooperatively, to produce the high, chemical, energy acquired to produce the free radical. A buffer is usually used, and is assumed to be necessary.

When the cutting efficiency experiment [36] is performed in the presence of different salts, we obtain a sequence for anions (with fixed cation) and one for cations (with the same anion). And when two different buffers (sodium phosphate or sodium cacodylate) are used, the series that reflect the efficiency are inverted:

phosphate buffer:

$$Br^- < I^- < NO_3^- < SO_4^{2-} < ClO_4^- < Cl^-, OAc^- < H_2PO_4^-$$

cacodylate buffer:

$$ClO_4^- < SO_4^{2-} < I^- < NO_3^- < Cl^- < OAc^- < (CH_3)_2AsO_4^- < Br^-$$

This ought to be a problem of colloid science of polymer (DNA)–protein (enzyme) interactions. Nothing in the classical theories of physical chemistry has anything to say on the matter.

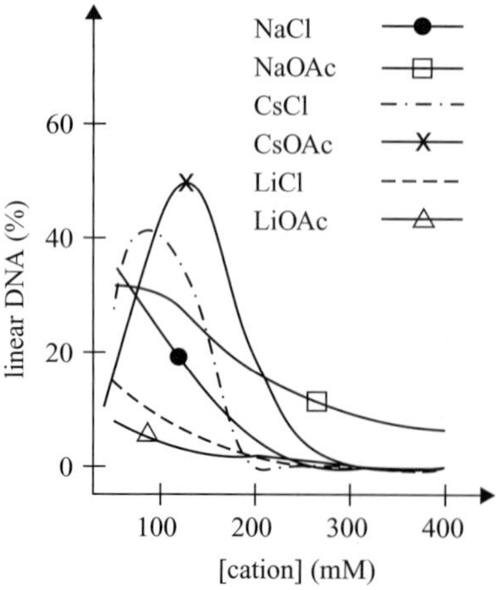

Fig. 6.4. Specific cation effect on the restriction enzyme *Eco*RI and supercoiled plasmid pBR322 DNA. They were diluted to 10^{-4} in a 5 mM sodium phosphate buffer (left) or in a 5 mM cacodylate buffer (right) at pH 7.5 with different concentrations and types of cations, in the presence of either 4 mM magnesium chloride or magnesium acetate, depending on the co-ion used. Cation concentration was varied between 50 mM, 80 mM, 125 mM, 175 mM, 200 mM, 300 mM and 400 mM. Details in Ref. 36. Adapted from Ref. 36, with permission. Copyright 2001 Springer-Verlag.

But there are strong hints that the mechanism calls into play all of the specific ion effects, hydrophobic interactions and cavitation, as well as dissolved gas, which we will subsequently discuss in Chapter 8. To see this consider Figs. 6.4 and 6.5.

Figures 6.4 and 6.5 show the cutting efficiency of a particular restriction enzyme (called Hindi 2) as a function of salt concentration. Figure 6.4 is for fixed anion (chloride or acetate) in two different buffers at pH 7.5. Figure 6.5 illustrates the trend with fixed cation (Na^+) and varying anion, with the same buffers and at the same pH.

For each ion pair the efficiency of cutting increases, more or less, to a maximum at the magic 0.15 M concentration, at which the bubble–bubble coalescence phenomenon discussed in Chapter 8 occurs. The results make some sense in terms of a reduction of classical electrostatic interactions between enzyme and the DNA polyelectrolyte with increasing salt. Reduction of electrostatic repulsive forces ought to enhance association and therefore cutting efficiency. Beyond 0.15 M, the efficiency decreases, with negligible efficacy by about 0.5–1 M.

At any fixed concentration, the efficiency of cutting follows a familiar Hofmeister sequence. But the sequence is reversed on change of buffer, as has been seen for pH measurements with glass electrodes. For the enzyme problem this reversal can only be due to competitive adsorption of buffer anion with salt anion in the

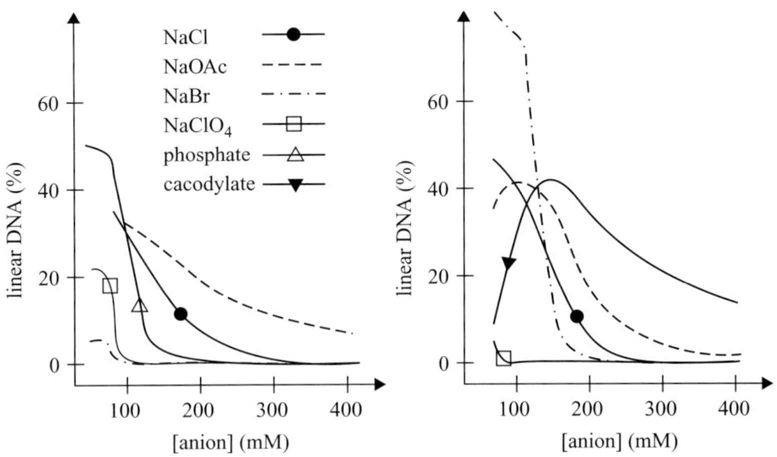

Fig. 6.5. Specific anion effect on the restriction enzyme *Eco*RI and supercoiled plasmid pBR322 DNA. The electrolytes were diluted from 1 M stock solutions into 5 mM sodium phosphate (left) or cacodylate (right) at pH 7.5 in the presence of the corresponding magnesium salt (4 mM). Sample preparation and reaction procedures are the same as for Fig. 6.4. Details in Ref. 36. Adapted from Ref. 36, with permission. Copyright 2001 Springer-Verlag.

hydrophobic active-site region. The effects cannot be due to pK_a changes, as there are no nucleotide or DNA sites affected by pH. The DNA does not change in conformation over the pH range 5–8 [52].

The same phenomena are repeated if we fix the anion and allow the cation to vary.

One tenuous clue to what might be going on is that we know that hydrophobic interactions between solid surfaces are reduced by addition of salt.

There is another clue. Addition of vitamin C, ascorbic acid, to the system stops the enzyme in its tracks. Vitamin C is a well-known free-radical scavenger. So free radicals must be involved. Where does this leave us in understanding the mechanism? A reasonable hypothesis might be as follows:

1 The enzyme attaches to linear DNA, dimerizes and diffuses along it until it finds the stereochemically favoured site.
2 In the active site, spontaneous hydrophobic cavitation then occurs (Section 8.10), creating an instantaneous microbubble joining active site to DNA. The microbubble probably contains dissolved atmospheric oxygen.
3 It is known from sonochemistry that within such small cavities the instantaneous pressure and temperature are such that free radicals form.
4 The free radical created then cuts the phosphate bond.

In such a scenario it is not known whether dissolved gas is necessary or not. An experiment to test this is difficult for prosaic reasons only.

In any event we see here a mechanism by which nature might harness all the weak physical forces acting in water cooperatively, via a micro or pseudo phase transition to do the job. Other enzymes probably have different energy sources, as discussed in Chapter 5. Whatever the status of such theorizing there is no question that Hofmeister effects loom large.

6.5 Clues to the Hofmeister problem

6.5.1 Resolution via bootstrapping: partial insights into the phenomena

In Hofmeister's original work on the relative efficacy of salts in precipitation of proteins there was no recognition of pH or of its salt dependence, which can affect surface charge. All systematic experiments in physical chemistry dealing with specific ion effects (practically all experiments involve the effects implicitly) 'fix' pH with a buffer. Studies as a function of electrolyte concentration use fixed cation and vary the anion, or vice versa. In very few cases has the specific ion dependence of pH itself or of buffers with salt concentration been taken into account. 'pH' itself becomes a dependent variable, measurements of which show a Hofmeister series, as we have seen. The anion series can reverse in order with change of cation. The series can reverse in order above and below the 'point of zero charge' of a protein or colloidal article. This is mysterious enough. Enzymatic activity depends strongly on Hofmeister effects. The activity depends on pH, which is assumed to be the determining variable. The experiments of Evens and Niedz [50] show that pH in a real situation is a dependent variable. Those on restriction enzyme activity do not depend on pH. And apparent pH itself as conventionally determined depends on background electrolyte. We have discussed whole-cell Hofmeister effects in bacteria that affect growth rates, perhaps due to their influence on a rate-determining enzyme.

So we have in all something of a muddle. About the only firm clue we can cling to comes from the observation that the effects are stronger with anions than for cations. Anions have a much broader spread in electron number and therefore polarizability than do the cations. Whatever the situation, there is a universal correlation between the polarizability of ions (and therefore in dispersion forces) and the Hofmeister effects. And since we know the conventional theories do not include dispersion forces acting on or due to ions, this has to be remedied as a first step.

We remind the reader again that we use the term 'dispersion forces' as a mnemonic to describe all the non-electrostatic forces.

These can be direct forces: between ions in solution, or between ions and an interface; or they can be indirect, specific ion–solvent interactions that affect hydration around ions or at interfaces.

6.5.2 Dispersion forces involving ions in the continuum solvent approximation: origins of ion specificity

We can pursue this clue further [6]. To recapitulate: the classical theory that underlies physical chemistry and colloid science has it that forces between particles or molecules are of two kinds, electrostatic and quantum mechanical dispersion forces. The basic assumption is that they can be treated independently.

The first are treated in a non-linear theory, the Poisson–Boltzmann distribution or its extensions in statistical mechanics. The second are treated via the linear Lifshitz theory and its extensions. The sum of these gives the net force between two particles or surfaces. The same ansatz underlies the entire field from pH to surface and solution chemistry generally, where the dispersion forces due to ions are simply ignored. The separation of forces assumption violates two basic laws, the Gibbs adsorption equation and the continuity of electromagnetic current.

Lifshitz theory of the attractive forces includes some specific ion effects in the dependence of refractive indices on salt and salt type. But this dependence is extremely weak.

As a first step towards remedying the inconsistencies, we have to treat both electrostatic and dispersion forces on the same equal *(non-linear)* footing. When that is done many of the Hofmeister effects begin to fall into place.

We illustrate this schematically for a single ideal interface. Explicitly consider the limiting law of Onsager–Samaris for the change in interfacial tension due to a dissolved electrolyte at concentration $c = \sum_i c_i$.

The Gibbs adsorption equation gives

$$d\gamma = -\sum_i \Gamma_i d\mu_i \tag{6.3}$$

where the chemical potential $\mu_i = \mu_i^0 + RT \ln c_i$, and Γ_i is the adsorption excess of species 'i'. Therefore

$$d\gamma = -k_B T \sum_i \Gamma_i \frac{dc_i}{c_i} \tag{6.4}$$

and

$$\Gamma_i = c_i \int_0^\infty \left[e^{\frac{-w_i(x)+z_i e\phi}{k_B T}} - 1 \right] dx \tag{6.5}$$

x is the distance of the centre of an ion from the interface, and the image potential $w_i(x)$ is given by:

$$w_i(x) = \frac{z_i^2 e^2}{4x\varepsilon} \Delta e^{-2\kappa_D x} \tag{6.6}$$

where κ_D is the inverse Debye length, and $\Delta = (\varepsilon - 1)/(\varepsilon + 1)$. A self-consistent potential $\phi(x)$ is set up near the interface because the size of ions is different and therefore so is their distance of closest approach.

The expression for the electrostatic image potential becomes infinite at the interface for point ions, but the adsorption excess does not. No cut-off in ion size is necessary. The only length scale is the Bjerrum length, which is independent of ion size (and electrolyte). The expression forms the basis of Onsager's treatment [4,6].

However, the ions also experience a dispersion potential corresponding to the electrostatic image term.

This has the form

$$w_i^d(x) = \frac{\hbar}{4\pi x^3} \int_0^\infty \alpha_i^*(i\xi) \frac{\Delta_{21}}{\varepsilon_2(i\xi)} d\xi \tag{6.7}$$

where $\Delta_{21} = (\varepsilon_2 - \varepsilon_1)/(\varepsilon_2 + \varepsilon_1)$, $\varepsilon_2 = \varepsilon_2(\mathrm{i}\xi)$, and $\varepsilon_1 = 1$. $\alpha_i^*(i\xi)$ is the excess polarizability of the ion, medium 2 is water and medium 1 vacuum. This continuum theory expression holds to a very good approximation for $x \geq a_\mathrm{i}$ (the ion radius). At smaller distances, or at the interface, the more complete expression is finite and does not diverge. It can be extended in principle to include induced permanent dipole interactions. Essentially the potential equation is the change in dispersion self-energy of the ion in medium 2 due to its interaction with the interface. The difference in dispersion self-energy of the ion in medium 1 or 2 is a measure of solubility differences in the media. To obtain some estimate of how large this potential might be, take:

$$\alpha_i^*(i\xi) = \frac{\alpha(0)}{1 + \left(\xi^2/\omega_1^2\right)} \tag{6.8}$$

$$\varepsilon_w(i\xi) = 1 + \frac{n^2 - 1}{1 + \left(\xi^2/\omega_2^2\right)} \tag{6.9}$$

where $\alpha(0)$ is the static effective polarizability of the ion, $\hbar\,\omega_1$ its unknown electron affinity, n the refractive index of water and ω_2 a typical ultraviolet relaxation potential [6].

Very detailed explanations of how to calculate forces using dielectric data and for different geometries have been given [4,19,53].

If we compare the dispersion potential with the image potential it can be shown that the dispersion potential is always larger than the electrostatic term for $\kappa_D^{-1} <$ 1 nm, i.e. salt concentrations >0.1 M. Even at quite low concentrations the dispersion potential can dominate in determining adsorption excesses. Thus at

$\kappa_D^{-1} \sim 30$ Å corresponding to 10^{-2} M, the dispersion potential can dominate at very small distances (∼5 Å) or very large distances (∼100 Å).

Although only qualitative, the arguments suggest that the dispersion potential can play a strong role in setting ion specificity.

Specificity should be accessible from bulk solution properties such as partial molal volumes and refractive indices. The excess polarizabilities can be much greater than the value 1 $Å^3$ chosen for the illustrative estimates above. Thus Br^-, with 80 electrons, can be expected to have a much higher polarizability, whereas the OAc^- ion with roughly the same electron density as water can be expected to have essentially zero excess polarizability. In general, we can expect excess polarizabilities and electron affinities (in solution) to change through the Hofmeister series of alkali metal halides and halide cations. The way that they change should be experimentally accessible, but in fine detail is not so obvious as the differences between Br and carboxylates. The inference is that current theories of interfacial tensions for which adsorption excesses are at the air–water interface and which are based on electrostatic potentials alone must be inadequate. We note two further matters:

1 Provided the excess polarizability is positive, the sign of the dispersion potential is the same as that of the electrostatic potential at the air–water interface. Both give rise to negative adsorption. We can expect much enhanced negative adsorption, larger $\Delta\gamma(c)$, for NaBr than for NaOAc. Awareness of this effect may explain the ion pair correlation effects and its additivity for surface activity as evidenced by, e.g. bubble–bubble interactions [54]. The qualitative differences for different ion pairs occur at concentrations ≥ 0.1 M, where dispersion effects dominate.
2 Positive adsorption, i.e. reduction in γ with added salt, can occur if the excess polarizability is negative. This can occur and depends on partial molal volumes, which reflect water structure around the ions. The notion of hydrophobic hydration can be rationalized on these lines.

6.5.3 Oil–water interfaces

We use the term 'oil' loosely to mean any surfaces – phospholipids, Teflon, silica, etc. – different from air. The integral which occurs in Equation (6.7) for the dispersion potential can now change sign. It depends on the dielectric properties of the two media, and their adsorption frequencies in the UV:

$$I = \int_0^\infty \frac{\alpha^*(i\xi)}{\varepsilon_w} \left(\frac{\varepsilon_w - \varepsilon_0}{\varepsilon_w + \varepsilon_0} \right) d\xi \qquad (6.10)$$

It can be estimated by taking in the denominator:

$$\varepsilon_w + \varepsilon_0 = 2 + \frac{n_w^2 - 1}{\left(1 + \xi^2/\omega_w^2\right)} + \frac{n_0^2 - 1}{\left(1 + \xi^2/\omega_0^2\right)} \approx 2 \tag{6.11}$$

then we have:

$$I \approx \frac{\pi}{4}\alpha(0)\, w_1 \left[\frac{\left(n_w^2 - 1\right)\omega_w}{\left(\omega_1 + \omega_w\right)} - \frac{\left(n_0^2 - 1\right)\omega_0}{\left(\omega_1 + \omega_0\right)}\right] \tag{6.12}$$

with limiting forms:

$$\begin{cases} I \approx \dfrac{\pi}{4}\alpha(0)\,\omega_1\left(n_w^2 - n_0^2\right), & \text{with } \omega_1 < \omega_w, \omega_0 \\ I \approx \dfrac{\pi}{4}\alpha(0)\,\omega_1\left[\left(n_w^2 - 1\right)\omega_w - \left(n_0^2 - 1\right)\omega_0\right], & \text{with } \omega_1 > \omega_w, \omega_0 \end{cases}$$

Clearly, and exactly as for the problem of spreading of alkanes on water [55], the sign of the dispersion potential is delicately poised. It depends on the interface. It can be repulsive, so enhancing electrostatic image effects. Or it can be attractive, leading to positive or reduced negative adsorption excesses. Thus for KI, $\partial\gamma/\partial c$ changes from positive ($\Gamma < 0$) at the water–air interface to negative ($\Gamma > 0$) at the water–octane interface [56]. The expression corresponding to Γ of Equation (6.5) is now schematically:

$$\Gamma_i = c_i \int_0^\infty \left\{\exp\left[-\left(\frac{A}{x}e^{-2\kappa_D x} \pm \frac{B}{x^3} + z_i e\varphi\right)/k_B T\right] - 1\right\} dx \tag{6.13}$$

1 There can now be a significant potential set up at the interface with, e.g., positive adsorption for the anion and negative for the cation because the magnitude of the dispersion potentials acting on cation and anion will be very different.

2 The potential Bx^{-3} can be expected to be much larger for halide ions than for, say, carboxylates or acetates with low electron density. Note that $|\Gamma_{KCl}| > |\Gamma_{KBr}| > |\Gamma_{KI}|$ are all negative at the air–water interfaces, and this apparently is a convenient way to systematize the Hofmeister series. The maximum effect at Na^+ in the cation series should be reflected in bulk data.

3 What may be more of significance is that it would appear that no inferences of ion binding to, say, proteins at the air–water interface can be carried over to binding at an oil– or phospholipid–water interface.

4 There is no reason to expect that the dispersion potential will not act at a charged interface and even dominate when local potentials felt by an ion are highly screened, and this would explain a number of puzzling effects, outlined in the introduction, to do with counterion dependence of double-layer forces. Ion-exchange effects occur at concentrations orders or magnitude lower than for interfaces of water with air, octane or

Teflon. Nonspecific electrostatic interactions here are modulated by dispersion effects which vary between ion pairs and interfaces.

5 The hydronium ion, with no dispersion electrons, is clearly very different. Pure paraffins and air bubbles charge in water. With an acid we may expect enhanced adsorption of a halide ion due to dispersion effects and no such effect on H^+ (a reverse situation occurs at the water–air interface). This sets up a potential which may bear on the matter of what we mean by pH measured by electrodes.

So the potential felt by an ion is both ion- and substrate-specific, depending on ionic polarizabilities and their frequency dependence and those of substrate and medium in which the ion is dissolved.

That is, if the theory is done correctly by putting dispersion and electrostatic forces on the same footing, even within the primitive model we have captured some essentials of ion specificity. Both ionic interactions in bulk solution and ion interactions with and between surfaces have to include dispersion forces at the same level as the double-layer theory. If the distribution function for ions is linearized with respect to the dispersion forces, one recovers the older DLVO theory that misses the Hofmeister effects.

This kind of approach to the inclusion of the missing direct dispersion forces has been applied to numerous examples, interfacial tensions, to estimate additional ion-specific contributions to Born energies, to ion specificity with electrolyte activities, in micellization, microemulsions, in polyelectrolytes, pH buffers, colloidal interactions and other systems discussed in later chapters of this book [15,17,18,28, 57–73].

These references that demonstrate Hofmeister series effects are proof-of-concept only. This is because the polarizability data for ions are subsumed into a (guessed) single frequency response function and the ionic radii are not properly handled. However, they do show that if dispersion forces are included consistently, the main features of Hofmeister effects are included qualitatively.

The essential conclusion is that with the proper inclusion of dispersion forces the Hofmeister effects do make sense. The intuition drawn from electrostatics alone can be quite misleading. Once ionic dispersion forces are taken into account, it becomes comprehensible, for example, that co-ions of the same charge sign as an interface with which they interact can be attracted to, not repelled from, an interface.

6.6 Indirect effects of ionic dispersion forces: ion–solvent interactions, chaotropic and kosmotropic ions

Hofmeister's bemusement as to whether his phenomena on the specificity of salting in or salting out of proteins were due to electrolyte-induced changes in bulk water or to surface-specific ion adsorption (so changing effective surface charge and electrostatic interactions between particles) is still not resolved by the considerations above. What can be deduced is that with inclusion of the missing ionic dispersion forces, surface adsorption alone can be sufficiently large to accommodate specificity. Which dominates, in a given situation, surface or bulk effects, depends on the system and the substrate. Most theoretical and simulation work that attempts to account for phenomena within the primitive model has been hampered by several factors. These are (see Chapter 7 for developments):

1 One is the absence of quantitative knowledge of ionic excess polarizabilities and their frequency dependencies.
2 Another is the problem of definition of a surface – at an air–water interface the gas concentration changes from 1 M at normal atmospheric pressure to about $5 \cdot 10^{-3}$ M across the relevant interface.
3 The meaning of ion size – it is inconsistent to use an effective radius of an ion deduced from fitting to Born free energies of hydration for electrostatics, and another for dispersion forces using an arbitrary Lennard-Jones potential or hard-core potential from crystallographic ion–ion distances.
4 In interfacial problems too, a full account for phenomena has to include changes of self energies on adsorption at an interface.
5 The modelling of a colloidal interface as a smooth plane or sphere or cylinder seems as absurd to the biologist dealing with a protein made up of a complex range of amino acid units as the proverbial 'consider a spherical horse' of the physicist.

Nonetheless such models that isolate one or another factor that influences events have been valuable in gaining insights into building a better theoretical intuition.

Most work done to elucidate ionic dispersion effects within the primitive model has also skirted around some of these questions by guessing the frequency dependence of polarizabilities and choosing an arbitrary ionic radius. Quantitative predictions are therefore indicative but unreliable. The manifestations of direct dispersion interactions with interfaces are probably more important than for ionic dispersion contributions in solution. This is because the first are longer-range ($\sim r^{-3}$) than the latter ($\sim r^{-6}$).

6.6.1 *Indirect dispersion forces: chaotropic vs. kosmotropic ions*

An approach to the problem of ionic specificity has been taken by Collins. This appears at first sight to be different to that we are following. Collins first developed a phenomenological characterization of ion specificity inferred from the standard heats of solution of electrolytes and the heats of hydration of its cation and anion. This produces a classification of ions into chaotropic (structure-breaking) and kosmotropic (structure-making) ions that span from one end of the Hofmeister sequence to the other [12,13].

This led further to the notions of compatibility of hydration shells (the old Gurney potential of classical electrolyte chemists) and a set of rules that covered a range of observations. For example, choline and sulfate ions were highly compatible and this leads to a better understanding of matters such as enzymatic superactivity that are well known in biochemistry [41].

In a later important development Collins, Neilson and Enderby via extensive neutron scattering work have shown that 'hydration' of ions is invariably limited to a single or at most two layers of water surrounding them [10]. This is not new. It is consistent with the extensive work of Parsegian and colleagues on hydration forces between lipid bilayers. These decay exponentially with a range of the order of the dimensions of a single water molecule. (The precise range depends on the dipole moment of the lipid bilayer headgroup and its flexibility.) What is new is that they conclude that

> The continuum electrostatics model of Debye and Hückel[1] and its successors utilize a macroscopic dielectric constant and assume that all interactions involving ions are strictly electrostatic, implying that simple ions in water generate electric fields strong enough to orient water dipoles over long distances. However, solution neutron and X-ray diffraction indicate that even di- and tri-valent ions do not significantly alter the density or orientation of water more than two water molecules (5 Å) away. Therefore the long range electric fields (generated by simple ions) which can be detected by various resonance techniques such as fluorescence resonance energy transfer over distances of 30 Å (about 11 water diameters) or more must be weak relative to the strength of water–water interactions. Two different techniques indicate that the interaction of water with anions is by an approximately linear hydrogen bond, suggesting that the dominant forces on ions in water are short range forces of a chemical nature.

The concept of kosmo- and chaotropicity has been developed by several other authors in the past, based on different parameters. Among these, for example, are the works of Samoilov and Krestov on the mobility of water around the ions (reflected in the change of entropy, ΔS_{II}) [75,76], and the work of Marcus on the viscosity of salt solutions (Jones–Dole B coefficient) [24].

The inferences are drawn from experiment and are therefore sound. But the intuition still relies on theory that depends on electrostatic forces alone. This we have shown is incorrect, and in the absence of dispersion forces it is natural that the effects should be assigned to short-range forces of a 'chemical' nature. We have seen in Chapter 4 that at solid interfaces forces due to surface-induced liquid structure can

[1] P. Debye and E. Hückel, *Phys. Z.* **24** (1923) 185–206.

decay and extend to span at most 6–12 oscillations. (The liquids are not hydrogen-bonded.) In this case, ions in water, the interaction between ions proceeds and is mediated via intramolecular cooperative water molecule interactions.

The first hydration layers around an ion will certainly be affected by geometry (size and shape and charge). But the additional effect due to dispersion forces acting between ions and the water can be expected to give rise to cooperative longer-range specificity due in hydrogen-bonded water orientations.

In more picturesque language, the ions might be imagined to be like defects of different size and strength within a dynamic connected zeolite or polymeric material. Between two such defects, fluctuating 'cracks' in the solid that connect them give rise to forces between them. For real solids the defects are responsible for the strength of the material as in the classical theory of solid mechanics of Griffiths. The same occurs in liquids – the strength of liquids containing dissolved atmospheric gas is a thousand times lower than that one would calculate without defects. So the analogy may not be too far-fetched. In a harmonic solid, defects either attract or repel each other depending on their coupling to the normal modes of oscillation of the solid. This coupling of collective modes leads to phase separation in solids, and presumably precipitation of salts from electrolyte solutions. This view comes back closer to the ideas of Bernal and of Langmuir on water, who viewed it more as a giant fluctuating molecule than a molecular liquid.

The weak 'cracks' in the glass, as it were, that stretch over distances of up to 30 Å and connect solute molecules in water involve ion–solvent interactions that depend on specific dispersion forces. In water containing dissolved gas from the atmosphere the same defects occur over the same range. The 'cracks' are propagated from and percolate from one solute molecule to another and, as we shall see in the Chapter 8, this mechanism seems to be responsible for long-range hydrophobic interactions.

The presence of dissolved atmospheric gas will be seen to have quite dramatic effects on colloidal interactions and this raises more complications, complexity and interest.

In any event there is no conflict between the chaotropic–kosmotropic classifications and the theory of forces that includes all electrodynamic fluctuation forces properly. We come to a quantitative treatment and inclusion of dispersion forces in Chapter 7.

6.6.2 Some consequences of anisotropy and anisometry

Effects of dielectric anisotropy in colloidal interactions have not been much explored. There are several situations that certainly are of much interest.

1 Anisotropy of dielectric properties of interacting particles: van der Waals interactions can become repulsive, not attractive! It depends on the relative orientation of the dielectric axes [4,20,53,77–83], and can lead to strong orienting torques. With membrane mimetic bilayers or other liquid crystals or biomembranes the necessary form birefringence of the oriented lipids of the membrane affects the magnitude of forces between them.
2 Anisotropy guides membrane protein recognition forces: protein interactions and binding – recognition – to proteins embedded in a (chain-oriented) lipid bilayer will experience a different interaction with the lipid background substrate from those with its target protein. The anisotropy of the lipids can effect a situation where the protein is repelled from the lipid region, by the anisotropy effect, and hydration, and guided to the target.
3 Anisotropic polarizabilities of ions such as OH^-, SCN^-, NO_3^- and $H_2PO_4^-$ clearly give a competitive edge to adsorption and specific binding of one kind of ion over another as for the air–water interface apart from specific forces due to differences in their polarizabilities, and due to the anisotropy of an interfacial layer.
4 Effects of geometrical anisotropy (cylinders) and non-additivity.

Anisometry of the interacting particles exemplified by the thin cylinder–cylinder interaction provides another source of orientation and different, sometimes very strong, kinds of interaction, as we have seen. That is a feature of all interactions involving particles of different length scales like polymers.

6.6.3 Anisotropy in the interfacial tension of water

One difficulty remaining in the way of more progress is that of the interfacial tension of water. When dispersion forces alone are operating as in the oil–water interface, i.e. dipolar forces and hydrogen bonding are absent, the interfacial tension can be calculated by using Lifshitz theory at a distance of the order of a molecular spacing. The free energy so calculated is twice the dispersion part of the free energy. Thus the free energy of two semi-infinite media containing water interacting across a vacuum is given by:

$$F(l) = -\frac{k_B T}{8\pi l^2} \sum_{n=0}^{\infty \prime} \left[\frac{\varepsilon_w(i\xi) - 1}{\varepsilon_w(i\xi) + 1} \right]^2 \tag{6.14}$$

the dispersion part is given by:

$$F_{n \neq 0} \simeq -\frac{\hbar}{16\pi^2 l^2} \int_0^\infty \left(\frac{\varepsilon_w - 1}{\varepsilon_w + 1} \right)^2 d\xi \tag{6.15}$$

Taking l to be the mean interaction spacing in water, we get a contribution to the surface energy of about 20 mN/m, which is much smaller that the experimental value of about 72.8 mN/m at 20 °C. That the dispersion part is given correctly can be inferred from a number of experiments. But, as expected, Lifshitz theory or additivity theories cannot be used to give the whole story down to intermolecular distances for highly polar liquids like water, where complications due to short-range forces (e.g. 'hydrogen bonds') are involved. At interatomic distances, the term in $n = 0$ in Equation (6.15) gives negligible contributions when compared with the dispersion part. On the other hand, if we consider two water molecules at a distance r apart, the whole interaction energy can be expressed as:

$$\begin{aligned} V(r) &= V_{\text{Keesom}} + V_{\text{Debye}} + V_{\text{London}} \\ &= -\left[\frac{\mu^4}{3k_B T} + 2\mu^2\alpha(0) + \frac{3}{4}\alpha^2(0)\hbar\omega_{uv}\right] r^{-6} \end{aligned} \tag{6.16}$$

Taking the values $\mu = 1.85 \cdot 10^{-18}$ esu, $\alpha(0) = 1.48 \cdot 10^{-24}$ cm^3, $\omega_{uv} = 1.9 \cdot 10^{16}$ rad/s, the relative ratio of the orientation energy, V_{Keesom}, to the dispersion energy is about 3 to 1. Hence we might expect the relative ratios of these contributions to surface free energy to remain roughly in the same proportion, and that the surface tension will decrease roughly as T^{-1} with increasing temperature. This is in fact so, but the above model is oversimplified. Unfortunately, a decomposition of the measured surface-free energy into enthalpic and entropic terms shows immediately that both the enthalpy and entropy increase with temperature.

The reason for this is probably that dipolar and hydrogen bonding conspire to give an anisotropic molecular dipolar ordering at the interface that stretches at the most about three molecular layers from the interface.

The dielectric properties of such an anisotropic layer can be estimated and given a simple analytic expression which is a generalization of Lord Rayleigh's extension of the Clausius Mossotti formula for isotropic media [84].

A better estimate of the dipolar contribution to the interfacial tension then requires the van der Waals interaction across a vacuum of a triple film: bulk water–anisotropic molecular film–vacuum–anisotropic molecular film–bulk water. This can be done analytically but has not been explored for the water interface [53,78–80].

The problem of how to define an interface here remains one of great difficulty. The matter is further complicated by the fact that there are no measurements of the interfacial tension of water, with or without electrolyte without atmospheric gas. The solubility of nitrogen in water at 1 atmosphere is around $5 \cdot 10^{-4}$ M; for O_2 it is $2.8 \cdot 10^{-4}$ M. It drops to zero by around 1 M salt. Carbon dioxide also dissolves in water to change pH to 5.6, which further complicates matters. CO_2 solubility in water at 25 °C is about 1.45 g/L (0.033 M). In any event, in the interfacial region, the amount of atmospheric gas drops from 1 M to about $7.8 \cdot 10^{-4}$ M over a distance of a few surface water layers. This affects ionic adsorption.

These difficulties also affect computer simulation of interfacial effects. For example, the best water molecular model gives an interfacial tension of about 1/3 the actual value. But the simulations do not include effects of dissolved gas.

6.6.4 Ion fluctuation or induction forces

Induction forces come about due to a coupling between kinetic fluctuations of ionic motion with an induced polarizability of neutral molecules or ions. Known results are at the same level as the linear Debye–Hückel or Born theories, but they do appear to capture some qualitatively new aspects of interactions that emerge from ion fluctuation correlations. The results below have not been previously published except in Chapters 6 and 7 of Ref. 4. Some further results on long-range fluctuation forces in different geometries of interest are given there. A net attractive force between rod-like molecules is reminiscent of the long-range force between conducting polymers peculiar to thin cylinder geometries of Chapter 5. The strength of this force depends on the presence of electrolytes, and on the charge located on the molecule (for example in polyelectrolytes). However, the main point is that theory predicts long-range attractive interactions that basically depend on the concentration of these dispersed objects in the surrounding medium, and on the different susceptibility of the solute and of the solvent.

A confirmation of such a model has been recently established in the formation and precipitation of pseudopolyrotaxanes, i.e. inclusion complexes produced by cyclodextrins and long PEG-based chains in water [85]. The attractive interaction predicted by the present models explains the behaviour of these supramolecular assemblies, and the correct temperature dependence of the process [86].

The attractive interaction peculiar to cylinders shows up in the interfacial tension of microemulsions discussed in Chapter 12. With polyelectrolytes where the interacting cylinders are charged the forces would have the same form and might well be a key factor in recognition with DNA. Even with uncharged cylindrical molecules the additional adsorption of ions due to Hofmeister effects gives essentially the same, but probably much larger, attractive forces. Such forces were first suggested by London as long ago as the 1940s.

We list here some surprising results that include effects due to ions on the temperature-dependent ion fluctuation contributions. These are given here because they have never been published except in Ref. 4.

1 There is a zero frequency dispersion contribution to the Born electrostatic self energy of an ion. In an electrolyte solution for a spherical ion this is

$$\Delta F_{\mathrm{SO}} \approx -\frac{2k_{\mathrm{B}}T}{\sqrt{\pi}a}\kappa_{\mathrm{D}}^{2}\alpha(0) = -\frac{8}{\sqrt{\pi}}\cdot\frac{\alpha(0)}{a\varepsilon(0)}\left(\sum_{\nu} n_{\nu}e_{\nu}^{2}\right) \tag{6.17}$$

where n_v is the mean density of ions of species n, $\varepsilon(0)$ is the static dielectric constant, a is the ionic radius, and $\alpha(0)$ the polarizability at zero frequency. The numerical factor is of not much significance, depending on the polarizability model. This gives a significant temperature-dependent contribution to the hydrophobic free energy of transfer of an ion or ionic headgroup in a surfactant that seems confirmed by experiment [87].

2 For two electrically neutral molecules in an electrolyte solution, the temperature-dependent contribution to dispersion interactions is

$$F_0(R) = -\frac{\alpha^2(0)\,k_B T}{2\varepsilon^2(0)\,R^6}\left[6 + 12\kappa_D R + 10\kappa_D^2 R^2 + 4\kappa_D^3 R^3 + \kappa_D^4 R^4\right]\cdot e^{-2\kappa_D R} \tag{6.18}$$

The interaction is weakened by the factor $\exp(-2\kappa_D R)$ but decays also as R^{-2} rather than R^{-6}.

3 For small spherical particles that can contain electrolyte such as polyelectrolytes or proteins, or salt porous colloidal particles at large distances, the results for these additional ion fluctuation forces are

$$f_0(r) = -\frac{k_B T a^6}{r^6}\exp(-2\kappa_m r)\cdot\Big[3E_{sm}^2\left(1 + 2\kappa_m r + 2\kappa_m^2 r^2 + \frac{4}{3}\kappa_m^3 r^3 + \frac{2}{3}\kappa_m^4 r^4\right) + E_{sm}(N_{sm} - E_{sm})\kappa_m^2 r^2\left(1 + 2\kappa_m r + 2\kappa_m^2 r^2\right) + \frac{1}{2}(N_{sm} - E_{sm})^2\kappa_m^4 r^4\Big] \tag{6.19}$$

where $N_{sm} = (n_s - n_m)/(3n_m)$ and $E_{sm} = (\varepsilon_s - \varepsilon_m)/(\varepsilon_s + 2\varepsilon_m)$.

The form for the interaction $r^{-2}\cdot\exp(-2\kappa_D r)$ was first predicted by Kirkwood in 1952.

In the absence of salt, n_s and $n_m \to 0$, $\kappa_m \to 0$, and $f_0(r)$ reduces to the usual result for two dielectric spheres:

$$f_0(r) \to -3k_B T\left(\frac{a}{r}\right)^6\left(\frac{\varepsilon_s - \varepsilon_m}{\varepsilon_s + 2\varepsilon_m}\right)^2 \tag{6.20}$$

for $\varepsilon_s = \varepsilon_m$ (the electrolyte-free case) this formula predicts no attraction between the spheres. Instead, for high salt concentrations, $\kappa_m r \to \infty$, and we obtain:

$$f_0(r) \to -k_B T\kappa_m^4\frac{a^6}{2r^2}(N_{sm} + E_{sm})^2\cdot\exp(-2\kappa_m r) \tag{6.21}$$

with an attractive energy given by

$$E_{attr} = -\frac{k_B T}{2}\left(\frac{a}{r}\right)^6 N_{sm}^2\kappa_m^4 r^4\cdot\exp(-2\kappa_m r) \tag{6.22}$$

When the mean salt concentration is uniform, $n_s = n_m$, but $\varepsilon_s \neq \varepsilon_m$, and the force is given by

$$f_0(r) = -\frac{3k_B T a^6}{r^6}\left(\frac{\varepsilon_s - \varepsilon_m}{\varepsilon_s + 2\varepsilon_m}\right)^2\left(1 + 2\kappa_m r + \frac{5}{3}\kappa_m^2 r^2 + \frac{2}{3}\kappa_m^3 r^3 + \frac{1}{6}\kappa_m^4 r^4\right)\cdot e^{-2\kappa_m r} \tag{6.23}$$

4 With cylindrical molecules there are some extremely interesting long-range effects.

Let us consider now a rod-like object with a radius a, and different susceptibility ($\varepsilon_{r\perp}$ and $\varepsilon_{r||}$) in the longitudinal and transverse direction, respectively. The dielectric constant of the embedding medium is ε_m. The concentration parameter for the rods is $n_r = \sum \nu^2 n_\nu(r)$, while the concentration parameter for the medium is n_m. For skewed rods with an inclination θ and at minimum separation distance R, the zero-frequency interaction free energy is given by:

$$\begin{aligned} F_0(R,\theta) = &-\frac{k_B T \pi^3 a^4}{4R^4 \sin\theta} \cdot \Bigg\{ \frac{3}{4}\Bigg[\Delta^\perp \left(\Delta^\perp + \Delta^{||}\right) \\ &+ \frac{\left(\Delta^\perp - \Delta^{||}\right)^2}{16} \cdot (1 + 2\cos^2\theta)\Bigg] \cdot \Bigg[1 + 2\kappa_m l + \frac{(2\kappa_m l)^2}{2!} \\ &+ \frac{(2\kappa_m l)^2}{3!} + \frac{(\kappa_m l)^2}{2}\Bigg] \cdot \Bigg[N\frac{\left(3\Delta^\perp + \Delta^{||}\right)}{2} - \frac{\Delta^\perp(\Delta^\perp + 3\Delta^{||})}{2} + \\ &- \frac{(\Delta^\perp - \Delta^{||})^2}{8} \cdot (1 + 2\cos^2\theta)\Bigg] \cdot (1 + 2\kappa_m l) \\ &+ (\kappa_m l)^4 \Bigg[N^2 - N(\Delta^\perp + \Delta^{||}) + \Delta^\perp \Delta^{||} \\ &+ \frac{(\Delta^\perp - \Delta^{||})^2}{8}\Bigg] \cdot (1 + 2\cos^2\theta) E_1(2\kappa_m l)\Bigg\} \end{aligned} \quad (6.24)$$

where $N = (n_r - n_m)/(2n_m)$, $\Delta^\perp = (\varepsilon_r^\perp - \varepsilon_m)/(\varepsilon_r^\perp + \varepsilon_m)$ and $\Delta^{||} = (\varepsilon_r^{||} - \varepsilon_m)/(2\varepsilon_m)$. For the salt-free system, N and κ_m are 0, and the previous equation reduces to the pure dielectric result:

$$F_0(R,\theta) = -\frac{3k_B T}{16R^4} \cdot \frac{\pi^3 a^4}{\sin\theta} \cdot \left[\Delta^\perp \left(\Delta^\perp + \Delta^{||}\right) + \frac{\left(\Delta^\perp - \Delta^{||}\right)^2}{16} \left(1 + 2\cos^2\theta\right)\right] \quad (6.25)$$

Just as for spheres, when $n_m \to 0$ and $\kappa_m \to 0$ (all mobile charges are confined inside the rods), the leading terms of $F_0(R, \theta)$ become the interaction energy for the ion-free plus the terms

$$-\frac{k_B T}{R^4} \cdot \frac{\pi^3 a^4}{\sin\theta} \cdot \left[\left(\frac{8\pi e^2 n_r}{\varepsilon_m k_B T}\right) \frac{n_r}{R^2} \left(\Delta^\perp + \Delta^{||}\right) + \left(\frac{8\pi e^2 n_r}{\varepsilon_m k_B T}\right)^2 \ln\left(2\kappa_m R\right)\right] \quad (6.26)$$

and for $\kappa_m \to 0$ this energy – due to correlation of ionic displacements along the rod – is dominated by the term $\ln(2\kappa_m R)$, and predicts an infinitely strong pairwise attraction at large distances of separation. On the other hand, for high salt concentrations in the suspending medium, the interaction goes as

$$F_0(R,\theta) \approx -\frac{\exp(-2\kappa_m R)}{R} \tag{6.27}$$

which is a quite different result from the R^{-4} dependent van der Waals free energy for the salt-free case.

Eliminating the angular dependence, i.e. for parallel rods, and at high ion concentration ($2\kappa_m R \to 1$) we have the asymptotic expansion:

$$F^{||}(R) = -\sum_{j=0}^{\infty} \frac{b_j \exp(-2\kappa_m R)}{(2\kappa_m R)^{\frac{3}{2}+j}} \cdot (\kappa_m a)^4 \sqrt{2\pi}\,(\kappa_m k_B T) \tag{6.28}$$

the calculation of the first coefficients provides:

$$b_0 = (A+B+C)/2, \qquad b_1 = (54A + 22B - 5C)/32,$$
$$b_2 = (1089A + 97B - 36C)/256,$$

where

$$A = \Delta^{\perp}\left(\Delta^{\perp} + \Delta^{||}\right) + \frac{3\left(\Delta^{\perp} - \Delta^{||}\right)^2}{16},$$
$$B = \frac{N\left(3\Delta^{\perp} + \Delta^{||}\right)}{2} - \frac{\Delta^{\perp}}{2}\left(\Delta^{\perp} + 3\Delta^{||}\right) - \frac{3\left(\Delta^{\perp} - \Delta^{||}\right)^2}{8},$$
$$C = \frac{1}{4}\left[N^2 - N\left(\Delta^{\perp} + \Delta^{||}\right) + \Delta^{\perp}\Delta^{||} + \frac{3\left(\Delta^{\perp} - \Delta^{||}\right)^2}{8}\right].$$

On the other hand, for $\kappa_m R \to 0$ with n_r finite we obtain

$$F^{||}(R) = -\frac{\pi k_B T}{32R}\left(\frac{\varepsilon_r}{\varepsilon_m}\right)(\kappa_m a)^4 \tag{6.29}$$

Although the model sounds artificial, the very long-range force predicted when $n_r \gg n_m$ suggests that in very long thin rod-like conducting molecules – such as highly charged polyelectrolytes – there may very well be forces operating which are of much longer range than ordinary van der Waals forces. More extensive results for anisotropic media are given in the book by Parsegian [88].

References

[1] I. E. Dzyaloshinskii, E. M. Lifshits and L. P. Pitaevskii, *Usp. Fizich. Nauk.* **73** (1961), 381–422.
[2] I. E. Dzyaloshinskii, E. M. Lifshitz and L. P. Pitaevskii, *Adv. Phys.* **10** (1961), 165–209.
[3] D. J. Mitchell, B. W. Ninham and P. Richmond, *Am. J. Phys.* **40** (1972), 674–678.

[4] J. Mahanty and B. W. Ninham, *Dispersion Forces*. London: Academic Press (1976).
[5] B. Davies and B. W. Ninham, *J. Chem. Phys.* **56** (1972), 5797–5801.
[6] B. W. Ninham and V. V. Yaminsky, *Langmuir* **13** (1997), 2097–2108.
[7] B. W. Ninham, *Progr. Coll. Polymer Sci.* **133** (2006), 65–73.
[8] P. M. Wiggins, *Physica A* **238** (1997), 113–128.
[9] W. Kunz, J. Henle and B. W. Ninham, *Curr. Op. Coll. Interface Sci.* **9** (2004), 19–37.
[10] K. D. Collins, G. W. Neilson and J. E. Enderby, *Biophys. Chem.* **128** (2007), 95–104.
[11] W. Kunz, P. Lo Nostro and B. W. Ninham, *Curr. Op. Coll. Interface Sci.* **9** (2004), 1–18.
[12] K. D. Collins and M. W. Washabaugh, *Q. Rev. Biophys.* **18** (1985), 323–422.
[13] K. D. Collins, *Methods* **34** (2004), 300–311.
[14] Y. Zhang, S. Furyk, D. E. Bergbreiter and P. S. Cremer, *J. Am. Chem. Soc.* **127** (2005), 14505–14510.
[15] M. Boström, D. R. M. Williams and B. W. Ninham, *Langmuir* **17** (2001), 4475–4478.
[16] C. L. Henry, C. N. Dalton, L. Scruton and V. S. J. Craig, *J. Phys. Chem. C* **111** (2007), 1015–1023.
[17] M. Böstrom and B. W. Ninham, *J. Phys. Chem. B* **108** (2004), 12593–12595.
[18] M. Böstrom and B. W. Ninham, *Biophys. Chem.* **114** (2005), 95–101.
[19] B.W. Ninham and V. A. Parsegian, *J. Theor. Biol.* **31** (1971), 405–428.
[20] H. I. Petrache, T. Zemb, L. Belloni and V. A. Parsegian, *Proc. Natl. Acad. Sci. USA* **103** (2006), 7982–7987.
[21] M. Boström, F. W. Tavares, S. Finet, F. Skouri-Panet, A. Tardieu and B. W. Ninham, *Biophys. Chem.* **117** (2005), 115–122.
[22] P. Lo Nostro, B. W. Ninham, S. Milani, L. Fratoni and P. Baglioni, *Biopolymers* **81** (2006), 136–148.
[23] S. Rossi, P. Lo Nostro, B. W. Ninham and P. Baglioni, *J. Phys. Chem. B* **111** (2007), 10510–10519.
[24] H. Donald, B. Jenkins and Y. Marcus, *Chem. Rev.* **95** (1995), 2695–2724.
[25] M. Lagi, P. Lo Nostro, E. Fratini, B. W. Ninham and P. Baglioni, *J. Phys. Chem. B* **111** (2007), 589–597.
[26] P. Lo Nostro, J. R. Lopes, B. W. Ninham and P. Baglioni, *J. Phys. Chem. B* **106** (2002), 2166–2174.
[27] B. Lonetti, P. Lo Nostro, B. W. Ninham and P. Baglioni, *Langmuir* **21** (2005), 2242–2249.
[28] P. Lo Nostro S, B. W. Ninham, M. Ambrosi, L. Fratoni, S. Palma, D. Allemandi and P. Baglioni, *Langmuir* **19** (2003), 9583–9591.
[29] S. Murgia, F. Portesani, B. W. Ninham and M. Monduzzi, *Chem. Eur. J.* **12** (2006), 7889–7898.
[30] A. Salis, M. C. Pinna, D. Bilaničová, M. Monduzzi, P. Lo Nostro and B. W. Ninham, *J. Phys. Chem. B* **110** (2006), 2949–2956.
[31] K. H. Gustavson, Specific ion effects in the behaviour of tanning agents toward collagen treated with neutral salts. In *Colloid Symposium Monograph*, ed. H. Boyer Weiser. New York: The Chemical Catalog Company (1926), 79–101.
[32] J. Loeb, *Science* **52** (1920), 449–456.
[33] P. Lo Nostro, L. Fratoni, B. W. Ninham and P. Baglioni, *Biomacromolecules* **3** (2002), 1217–1224.
[34] R.Vogel, *Curr. Op. Coll. Interface Sci.* **9** (2004), 133–138.
[35] M. Boström, D. R. M. Williams and B. W. Ninham, *Europhys. Lett.* **63** (2003), 610–615.

[36] H.-K. Kim, E. Tuite, B. Nordén and B. W. Ninham, *Eur. Phys. J. E* **4** (2001), 411–417.
[37] P. Bauduin, F. Nohmie, D. Touraud, R. Neueder, W. Kunz and B. W. Ninham, *J. Mol. Liq.* **123** (2006), 14–19.
[38] M. C. Pinna, A. Salis, M. Monduzzi and B. W. Ninham, *J. Phys. Chem. B* **109** (2005), 5406–5454.
[39] D. Bilaničová, A. Salis, B. W. Ninham and M. Monduzzi, *J. Phys. Chem. B* **112** (2008), 12066–12072.
[40] A. Salis, D. Bilaničová, B. W. Ninham and M. Monduzzi, *J. Phys. Chem. B* **111** (2007), 1149–1156.
[41] M. C. Pinna, P. Bauduin, D. Touraud, M. Monduzzi, B. W. Ninham and W. Kunz, *J. Phys. Chem. B* **109** (2005), 16511–16514.
[42] P. Lo Nostro, B. W. Ninham, A. Lo Nostro, G. Pesavento, L. Fratoni and P. Baglioni, *Phys. Biol.* **2** (2005), 1–7.
[43] F. Trail, I. Gaffoor and S. Vogel, *Fungal Gen. Biol.* **42** (2005), 528–533.
[44] A. Salis, M. C. Pinna, D. Bilaničová, M. Monduzzi, P. Lo Nostro and B. W. Ninham, *J. Phys. Chem. B* **110** (2006), 2949–2956.
[45] A. Voinescu, P. Bauduin, C. Pinna, D. Touraud, W. Kunz and B. W. Ninham, *J. Phys. Chem. B* **110** (2006), 8870–8876.
[46] R. P. Buck, S. Rondinini, A. K. Covington, F. G. K. Baucke, C. M. A. Brett, M. F. Camoes, M. J. T. Milton, T. Mussini, R. Naumann, K. W. Pratt, P. Spitzer and G. S. Wilson, *Pure Appl. Chem.* **74** (2002), 2169–2200.
[47] M. Boström, V. S. J. Craig, R. Albion, D. R. M. Williams and B. W. Ninham, *J. Phys. Chem. B* **107** (2003), 107, 2875–2878.
[48] D. F. Parsons and B. W. Ninham, *J. Phys. Chem. A* **113** (2009), 1141–1150.
[49] J. N. Butler and D. R. Cogley, *Ionic Equlibrium: solubility and pH calculations.* New York: John Wiley (1998).
[50] T. J. Evens and R. P. Niedz, *Scholarly Res. Exch.* (2008), 818461.
[51] P. Y. Tishchenko, R. V. Chichkin, E. M. Il'ina and C. S. Wong, *Oceanology* **42** (2002), 27–35.
[52] V. P. Antao, D. M. Gray and R. L. Ratliff, Robert, *Nucleic Acids Res.* **16** (1988), 719–738.
[53] V. A. Parsegian, *van der Waals Forces: a handbook for biologists, chemists, engineers, and physicists*. Cambridge: Cambridge University Press (2006).
[54] V. S. J. Craig and B. W. Ninham, *J. Phys. Chem.* **97** (1993), 10192–10197.
[55] P. Richmond, B. W. Ninham and R. H. Ottewill, *J. Coll. Interface Sci.* **45** (1973), 69–80.
[56] R. Aveyard and S. M. Saleem, *J. Chem. Soc. Faraday Trans. I* **72** (1976), 1609–1617.
[57] M. Boström, D. R. M. Williams and B. W. Ninham, *Langmuir* **18** (2002), 6010–6014.
[58] M. Boström, D. R. M. Williams and B. W. Ninham, *Langmuir* **18** (2002), 8609-8615.
[59] M. Boström, D. R. M. Williams and B. W. Ninham, *J. Phys. Chem. B* **106** (2002), 7908–7912.
[60] M. Boström, D. R. M. Williams and B. W. Ninham, *Biophys. J.* **85** (2003), 686–694.
[61] M. Boström, D. R. M. Williams and B. W. Ninham, *Europhys. Lett.* **63** (2003), 610–615.
[62] M. Boström, D. R. M. Williams and B.W. Ninham, *Eur. Phys. J. E* **13** (2004), 239–245.
[63] M. Boström, D. R. M. Williams, P. R. Stewart and B. W. Ninham, *Phys. Rev. E* **68** (2003), 041902.

[64] M. Boström, D. R. M. Williams and B. W. Ninham, *Curr. Op. Coll. Interface Sci.* **9** (2004), 48–52.
[65] M. Boström and B. W. Ninham, *Langmuir* **20** (2004), 7569–7574.
[66] M. Boström, W. Kunz and B. W. Ninham, *Langmuir* **21** (2005), 2619–2623.
[67] V. Deniz, M. Boström, G. Franks and B. W. Ninham, *Adv. Coll. Interface Sci.* 123–126 (2006), 5–15.
[68] M. Boström, V. Deniz and B. W. Ninham, *J. Phys. Chem. B* **110** (2006), 9645–9649.
[69] M. Boström, F. W. Tavares, B. W. Ninham and J. M. Prausnitz, *J. Phys. Chem. B* **110** (2006), 24757–24760.
[70] L. A. Moreira, M. Boström, B.W. Ninham, E. C. Biscaia and F. W. Tavares, *J. Braz. Chem. Soc.* **18** (2007), 221–230.
[71] M. Boström, F. W. Tavares, D. Bratko and B. W. Ninham, *Prog. Coll. Polymer Sci.* **133** (2006), 74–77.
[72] V. Deniz, M. Boström, D. Bratko, F. W. Tavares and B. W. Ninham, *Coll. Surfaces A* **319** (2008), 98–102.
[73] M. Boström, E. R. A. Lima, F. W. Tavares and B. W. Ninham, *J. Chem. Phys.* **128** (2008), 135104 1–4.
[74] S. Murgia, F. Portesani, B. W. Ninham and M. Monduzzi, *Chem. Eur. J.* **12** (2006), 7889–7898.
[75] O. Ya. Samoilov, in *Water and Aqueous Solutions; structure, thermodynamics, and transport processes*, ed. R. A. Horne. New York: Wiley Interscience (1972), 597–612.
[76] G. A. Krestov, *Thermodynamics of Solvation*. New York: Ellis Horwood (1991).
[77] E. R. Smith and B. W. Ninham, *Physica* **66** (1973), 111–130.
[78] V. A. Parsegian and G. H. Weiss, *J. Adhesion* **3** (1972), 259–267.
[79] V. A. Parsegian and G. H. Weiss, *J. Coll. Interface Sci.* **40** (1972), 35–41.
[80] V. A. Parsegian, *J. Chem. Phys.* **56** (1972), 4393–4398.
[81] R. Podgornik and V. A. Parsegian, *J. Chem. Phys.* **121** (2004), 4767–4773.
[82] R. Podgornik and V. A. Parsegian, *Phys. Rev. Lett.* **80** (1998), 1560–1563.
[83] R. F. Rajter, R. Podgornik, V. A. Parsegian, R. H. French and W. Y. Ching, *Phys. Rev. B* **76** (2007), 0454717.
[84] B. W. Ninham and R. A. Sammut, *J. Theor. Biol.* **56** (1976), 125–149.
[85] A. Becheri, P. Lo Nostro, B. W. Ninham and P. Baglioni, *J. Phys. Chem. B* **107** (2003), 3979–3987.
[86] P. Lo Nostro, L. Giustini, E. Fratini, B. W. Ninham, F. Ridi and P. Baglioni, *J. Phys. Chem. B* **112** (2008), 1071–1081.
[87] E. Z. Radlinska, B. W. Ninham, J. P. Dalbiez and T. N. Zemb, *Coll. Surfaces* **46** (1990), 213–230.
[88] V. A. Parsegian, *Van der Waals Forces: A Handbook for Biologists, Chemists, Engineers, and Physicists*. Cambridge: Cambridge University Press (2006).

7

Specific ion effects

In collaboration with Drew F. Parsons.

(This chapter is more or less self-contained. It can be read independently of earlier chapters.)

7.1 Hofmeister effects in physical chemistry

7.1.1 Introduction

It will be clear by now that the theoretical edifice that deals with molecular forces in physical chemistry, built painstakingly over the past 150 years, is flawed. That there are defects and limitations in any theory we can live with. Of course there are. But the emerging picture from the preceding chapters is more disturbing. It seems that our intuition has become so seriously flawed that what ought to be the enabling discipline of physical chemistry has become impotent in the face of challenges posed by the biological sciences. To readers familiar with and schooled in the classical textbook literature, this claim is not easy to accept. And the burden of proof of a challenge to an established discipline lies with the proponent. So, in this chapter we recapitulate the standard literature ideas, and outline where they went off course. Then we revisit and pull together a number of examples from biology, some already mentioned, where it is clear that the standard theories fail. Then we outline our views on how it is that a reconciliation is on the way to being effected. How to fit Hofmeister effects into the scheme of things is a first issue.

In the following chapter we shall have to confront even more dramatic assaults on classical ideas on molecular forces due to the neglect of the effects of dissolved gas. But first things first.

7.1.2 Specific ion effects: the classical picture

We begin by rehearsing textbook ideas on specific ion effects in electrolytes. In tracing out those ideas we flag where approximations are made, and remark on them (see the footnotes).

Any salt contains one or more cations (positive charges), and one or more anions (negative charges). The overall charge is zero. In equilibrium conditions, cations cannot be separated from anions, due to the electroneutrality condition, unless an external electric field is applied, as in electrophoresis. Electrostatic forces give rise to a solid which is a crystalline lattice.[1] In aqueous solutions of electrolytes the presence of the solvent lowers the electrostatic interactions because of the large dielectric constant of water. However, it is not simply a decrement in the attraction between particles of opposite charge. In fact water molecules 'hydrate' the ions, i.e. establish different kinds of interactions (hydrogen bonds, ion–dipole interactions, etc.) that become part of the 'hydrated ion', depending on its charge, size and nature.[2] This results in dissolution of the crystalline structure. Ions diffuse freely in the solution. The process of solubilization depends on a competition between the free energy of the crystal lattice, the Gibbs free energy of hydration of the ions,[2] interactions between ions in solution[3] and their entropy. The properties

[1] The electrostatic energy of formation of a simple salt crystal can be calculated by summing up the Coulomb interactions between the ions of the crystal (Ewald sums), with the interionic distances determined by X-ray crystallography. This can be used as a reference state for estimating the free energy of transfer of an ion into water (Born energy). There is already a problem here. The procedure ignores the dispersion energies of interactions of the ions, cf. Chapter 2; no set of electrostatic charges can be in equilibrium. Quantum (dispersion) interactions are necessary and missing. Their inclusion gives rise to different ionic radii and very different self free energies of transfer. Usually one thinks of a hard sphere radius for the ions, a sort of metallic sphere. But the electron clouds are smeared and the idea of an ionic radius becomes complicated. A consistent definition is developed below.

[2] The Born energy is the electrostatic self free energy of transfer from a standard state (which can be chosen, e.g., to be that in the crystal or in vacuum) to bulk water. It can be used to get some estimates of solubility of a solid or of partitioning and transport of ions into oils and across membranes. The Born energy involves the interaction of an ion with a continuum solvent characterized by a bulk static dielectric constant. It requires the idea of an ionic 'size' or radius. With a small radius the electrostatic field at the surface of the ion and its self energy is larger than for an ion of large radius. In a decoration of this idea, the interactions of the ion with water change water 'structure'. This is reflected in partial molal volumes and heats and enthalpies of solution. The now hydrated or 'dressed' ions possess radii different to ions in vacuum or in a crystal. Rarely do any two experiments yield the same hydrated ionic radii. Ionic size is somewhat arbitrary and its definition involves a confusing literature. In fact dispersion interactions of the ion with the solvent are specific and important even in the continuum solvent model. Preliminary estimates give (20–50%) corrections [1–5]. More refined work has defined ionic electrostatic and dispersion radii consistently. The latest *ab initio* data for ionic polarizabilities and their frequency dependence give even larger corrections to classical estimates [6]. So hydration depends on dispersion interactions as well as electrostatics. The two cannot be separated. The same is true for dipolar interactions and ion-induced dipole effects can play a role too (see Chapter 7 of Ref. 7) [7–9]. To say that this is a problem for classical physical chemistry is an understatement.

[3] The next stage in development of theory is the correlation energy of an electrolyte solution. This is the change in self energy due to the presence of other ions. This is a many-body interaction because of the long range of the Coulomb interaction. Debye–Hückel theory is a limiting law valid at best up to 10 mM. Ion size is unimportant at such low concentrations. (But dissolved atmospheric gas, which has big effects, is, and is always ignored;

of salt solutions, such as osmotic pressure, lowering of the vapour pressure of water, freezing point depression, boiling point elevation, electrical conductivity and viscosity, depend upon the number density of the cations and anions in the system.[4] Ion pairing, charge transfer and other specific interactions between the ions with the assistance of the solvent contribute to modify their physicochemical behaviour in solution. However, these properties depend not only on the concentration and on the valence of the ions, but also on their specific nature. That is to say: if we measure the viscosity of a 1 M aqueous solution of a 1:1 electrolyte (NaCl, KBr or $LiNO_3$), we will obtain different values depending on the particular salt used. The term 'specific ion effects' signifies the different effects produced by a particular pair or mixture of cations and anions at the same concentration. These effects are not encompassed by classical theory.

There is a plethora of phenomena in which specific ion effects occur. Among the most familiar, we recall: water activity [12], pH [13], buffers [14], weakly associated vs. strongly associated electrolytes, heats and entropies and partial molal volumes of solution [15–17], self-diffusion coefficients of ions in water, and water in electrolytes [18–20], viscosity of aqueous salt solutions [21], interfacial tensions [22,23], adsorption at interfaces [24], ion-exchange chromatography [13], electrophoresis [25], surfactant and polymer cloud points and critical micellar concentration [26–30], polymer swelling and gelation [31,32], protein solubility and denaturation temperatures [33,34], degree of protein aggregation, coacervate

cf. Chapter 8). At any concentration of interest to biology, greater than 0.1 M, the hydrated ionic sizes are required to characterize distances of closest approach of anions and cations that come into the theory. More ion-specific parameters are required for each ion pair to describe the hydrated radii and the interpenetration of shells to fit data such as activities. Electrostatics, which dominates our intuition, becomes irrelevant at high concentrations and at interfaces where ion concentrations become high. Linearization of the statistical mechanical distribution functions for ions fails for asymmetric electrolytes – the range of electrostatic force length is not the Debye length (cf. Chapter 3) and non-linear theory has to be used. The theory as described above is inconsistent. The dispersion and electrostatic forces taken together are necessary to describe hydration and the Hofmeister series in activities. Note, and this is a serious problem, that the interpretation of measurements with buffers, surface potentials and pH all depend on classical Debye–Hückel theory or its extensions in double-layer theory. Given the presence of many different multicharged protein ions in biology and a concentration above 0.1 M, it is hardly surprising that the classical intuition based on electrostatic notions alone can be misleading.

4 Weakly associated electrolytes such as acetic acid and buffers and pK_as for ionizable groups are characterized by specific association constants. The changes in pH or pK_as that occur in the presence of salts are all ion, electrolyte type, and mixture and concentration specific. Classical text book theory attempts to handle these effects by the same linear Debye–Hückel or double-layer theory (for surfaces) decorated to include hydrated ion size. Fitting parameters vary from situation to situation and become meaningless. The same is true for pH, which (cf. Evens and Niedz in Section 6.3 [10] and Section 7.1.9) becomes a *dependent* variable. The same is true for interpretation of ion binding measured by NMR. This leads to a proliferation of arbitrary parameters. It is understandable because of our lack of theory of water, or for that matter of any liquid. The problem is brought to order somewhat by the recognition of missing quantum mechanical dispersion forces just as the Newtonian system did for planetary motion. An ultimate absurdity is the proliferation of constants for 'predicting' activities of mixed electrolyte solutions by the Pitzer tables. The matter is of extreme practical interest in industry and in geochemistry, and even in astrophysics, where equations of state for dilute plasmas are relevant to the theory of evolution of stars [11].

behaviour [35],[5] microemulsions [36], enzyme activity [37–42], growth rates of bacteria [43], optical activity of chiral entities [44], water absorption in natural fibres [45], bubble–bubble coalescence [46], ion complexation of calix[n]arenes at the air–water interface [47], formation of host–guest complexes [48], transition in surfactant lamellar phases [49,50] and even more.

This brief list shows that practically all of solution behaviour is specific. Chemists are used to dealing with 'specificity'. The periodic table itself was drawn up by Mendeleev by recording different classes of 'specific' behaviour of the elements. Actually similarities in specific behaviour were considered by Mendeleev to be more important than the regular increment in atomic weight when ordering the then-known elements into groups (e.g. for Te/I and Co/Ni).

Each atom possesses a nucleus with protons (atomic number) and neutrons (mass number), surrounded by an electron cloud according to the particular electronic configuration. The atomic number and the electron configuration provide a sort of 'identity card' for each element. Atomic properties such as the ionization potential, the electron affinity, the electronegativity and the polarizability depend strictly on the position of an element in the periodic table. Their properties change progressively from left to right along the periods and from top to bottom within each group.

Atoms lose or acquire electrons to produce cations or anions. By reacting, they attain thereby a more stable electron configuration. This process usually affects only the external electron shell, while neutrons and protons of the nuclei remain unaltered. When fluorine or iodine acquire an electron and form F^- or I^-, respectively, the total number of electrons in the external shell is the same (isoelectronic ions). However, they differ in their atomic numbers and in the distribution of their electronic clouds. The same holds for isoelectronic cations such as Mg^{2+} or Ba^{2+}, and for polyatomic species such as ClO_4^- or SO_4^{2-}. All such specificity, valency and the formation of molecules are accounted for by quantum mechanics.

These intrinsic differences between isoelectronic species are exploited routinely. For example, the different solubility of silver halides is used in analytical chemistry. If a desiccating agent is needed, there is a vast selection of candidates, from Na_2SO_4 to $MgCO_3$, from KOH to P_4O_{10}, depending on the particular case. No one would ever use – for example – caesium iodide as desiccating agent. The same holds when one has to a prepare an environment with a desired relative humidity by using a saturated solution of a specific salt (e.g. $CaCl_2$, $Mg(NO_3)_2{\cdot}6H_2O$ or KNO_3). Again, the crucial tools of ion-exchange chromatography, electrophoresis and separation technologies involve specific 'hydration' effects of ions.

[5] A 'coacervate' is a small droplet of organic molecules (e.g. lipids) that are supposed to interact through 'hydrophobic forces' and freely diffuse in the surrounding aqueous medium.

Traditionally, the systematization of the chemistry of solutions containing salts has focussed on electrostatic forces and so far relatively ill-defined concepts of hydration. It has not involved quantum mechanics at the same level that we know is essential to explain the chemistry of gases and molecules, or of solids.

Where quantum mechanics appears explicitly, in van der Waals forces, hydrogen bonding, acceptor–donor coupling or π-ion interactions, the concepts come from calculations that involve a two-molecule interaction in vacuum. They are grafted, uncomfortably, into a background theory that ignores effects imposed by water, the molecular solvent medium. This situation guarantees that the reflection and specific expression of such concepts, while valid in vacuo, now become many-body, not two-body, interactions. A strong covalent bond remains a covalent bond in solution. But a weaker bond like a hydrogen bond is now shared indistinguishably over several water molecules.

7.1.3 Further exploration of classical ideas

We dissect the problem further. In a simple salt crystal, such as NaCl, cations and anions are assumed to interact mainly through electrostatic forces.[1] These forces depend on the 'size' and charge of the interacting species via Coulomb's law.[2] The fact that specific ion effects show up in solutions or in dispersed colloidal systems indicates that it is not just a matter of electrostatics.[2] Electrostatic theories of electrolytes activities, the Debye–Hückel model and its extensions[3] and the DLVO theory of colloidal particle interactions do not predict specific ion effects: in principle sodium chloride or lithium thiocyanate should work in the same way. They are both 1:1 electrolytes containing monovalent ions (apart from differences in hydrated ion size).[6]

[6] The Onsager electrostatic theory of interfacial tension change due to dissolved electrolyte is the analogue for surfaces of the Debye–Hückel limiting law for activities of electrolytes. Onsager himself said that some then-unknown forces were operating.

The electrostatic theory of the double layer for charged surfaces is at the heart of all colloid and surface electrochemistry. It ignores specific ion effects which are central to pH, pK_as, membrane and zeta potentials (or else accounts for them through arbitrary fitting parameters).

The DLVO theory of colloid stability set up to describe the forces in colloid science is actually the last stage in the development that can be schematically illustrated as:

solid → dissolution → solvent-ion interactions (self energies) → correlation energies → interfacial energies and adsorption → interaction energies between colloidal particles.

It is only at this last stage in the development of classical ideas that dispersion–van der Waals–Lifshitz forces appear. While this was the first step in the right direction, they are treated in a linear theory as opposed to the non-linear double-layer theory. They do not handle specific ion effects that can differ by an order of magnitude for the same electrolyte conditions but different electrolyte type.

The theory is fundamentally inconsistent and the forces cannot be separated [51]. The situation is then that we have built physical chemistry on what seemed reasonable assumptions. But the foundations were flawed, and the defects have propagated until the whole house of cards falls apart.

Instead:

1 The specific ion effects depend on the particular nature of the electrolyte (the cation and anion species), and on its interactions with the solvent molecules, and with any particular substrate (e.g. a polymer or a surfactant or buffer anion, or protein). Heats of hydration and solvation numbers are different for different ions and ion pairs [52]. Solvation numbers are in general a nebulous concept [53].
2 'Local' interactions between an ion and a solute (or a part of it as, for example, for a protein) or with an interface show that ions can be specifically adsorbed. This is not predicted by electrostatics. 'Hydration', interactions between hydration shells around ions or at surfaces, is involved. But so too are dispersion forces that depend on the frequency-dependent dielectric properties of the ion and of the substrate. The 'hydration' associated with an ion depends on both its electrostatic, dispersion interactions, and charge-induced dipole interactions with local water molecules. A consequence of this is that – especially with complicated systems (membranes, supramolecular structures, cells) – the overall effect is due to several different identified processes taking place simultaneously, so that the occurrence of anomalies such as displacements in the classical Hofmeister series (and even inversions) is often recorded. The combined effects are not always additive.

We are familiar with some characteristic molecular parameters. A permanent dipole moment in a molecule responds to a static external electromagnetic field, applied or due to neighbouring molecular dipoles and their orientations. Such effects are strictly non-additive. Again, the existence of molecular polarizability reflects the fact that with ions and neutral molecules the electron cloud can move and respond to an external oscillating electric field, a field that includes electromagnetic fluctuations due to its neighbours. In other words, the electron cloud is more or less soft, or deformable. Since the atomic number, the ion size and particularly the electronic configuration determine the softness of the electron cloud, these factors certainly play a role in specific ion effects. As a matter of fact fluoride ions – which possess relatively few internal electrons in fully occupied orbitals – are very hard, with the highest electronegativity and very low polarizability (the nucleus attracts the external electrons very strongly). On the other hand, another halide such as iodide, with the same external configuration but with several internal electrons that shield the nuclear attractive force from the outer electrons, is much softer, with relatively low electronegativity and higher polarizability. It is clear that in soft ions such as I^- or SCN^-, the high polarizability implies that these species are more sensitive than hard ions to external electric fields, and therefore to other ions and dipolar molecules. Depending on their composition and structure, ions, neutral molecules and macromolecules interact through dipolar interactions, hydrogen bonding, π-ion interactions, acceptor–donor coupling and van der Waals forces. Strictly speaking, while we use such words as mnemonics, in water such a dissection into separate forces is not generally valid. They are all coupled many-body effects. They can only be accessed from the in principle measurable complicated dielectric susceptibilities as a function of frequency of the media and moieties involved; or else through *ab initio* quantum mechanics. The forces depend mainly on the properties of ions and molecules such as charge, size, anisotropy, number density, ionization potentials and dynamic polarizabilities.

How to take all these measurable properties of ions together in a predictive framework is our task.

In passing, it is very interesting to recall that electrophoretic mobility does not require the existence of net charges on the surface of the migrating particles. Very recently it has been shown that anisotropic dipole orientations exist in neutral oil droplets, and that the interpretation of electrophoresis in terms of purely continuum effects must be reconsidered [54].

Shape is another key issue. Halides (F^-, Cl^-, Br^-, I^-), calcogenides (O^{2-}, S^{2-}, ...) or most of the cations (Rb^+, Ca^{2+}, Al^{3+}) are well approximated by spheres, meaning that their interactions are isotropic. Certainly this is not true for polyatomic species such as $NO_3{}^-$, SCN^-, $H_2PO_4{}^-$, OH^-, etc. that possess a significantly asymmetric structure. Due to the existence of symmetry axes, these ions will possess different values of polarizability depending on the direction in which it is measured. Moreover, the interaction between an asymmetric ion with another molecule (solvent, co-solute, substrate or interface) will depend on the direction of approach. This is well known in inorganic coordination chemistry, where the same donor (for example OCN^-) can produce different complexes depending on the donating atom (cyanates and isocyanates).

An extreme example is the anisotropy that exists in long conducting polymers or polyelectrolytes such as DNA. This gives rise to very long-range forces between them.

Similarly, cooperative giant dipole fluctuations in linear molecules with a permanent or induced dipole moment in water or electrolytes can give rise to large effects (cf. Chapter 5).

7.1.4 Specific ion effects in electrolyte solutions

In the case of simple aqueous solutions of salts, specific ion effects show up in very consistent and interesting ways. Although Hofmeister phenomena that deal with proteins, colloidal stability, microorganisms and enzyme activity are dramatic and have important practical ramifications, the specificity of electrolytes in changing some basic physicochemical quantities takes us straight to the core of chemistry proper.

7.1.5 Activity coefficients

We have discussed in Chapter 3 the activity coefficients ($\gamma_{\pm}$) of electrolytes in water solution. We consider the matter further. At fixed high concentration, say 0.5 M, when the range of electrostatic forces drops to around about 6 Å, these forces are screened. Their range is of the order of the distance of closest approach of an anion and cation. Other forces must come into play. The activities at the same concentration follow a Hofmeister series as illustrated in Fig. 7.1: for the cations (a, b, c, d), and for the anions (e, f) (see also Fig. 3.2 for lithium and sodium salts). See Equation (7.6).

The sequences for $\mathrm{Log}\gamma_{\pm}$ are:
for monovalent chlorides (Fig. 7.1a)

$$H^+ > Li^+ > Na^+ > K^+ \approx NH_4^+ \approx Rb^+ > Cs^+$$

for divalent chlorides (Fig. 7.1b):

$$Mg^{2+} > Ca^{2+} > UO_2^{2+} > Ni^{2+} > Co^{2+} > Sr^{2+} > Mn^{2+} > Ba^{2+} \approx Cu^{2+} > Zn^{2+} > Cd^{2+}$$

for trivalent chlorides (Fig. 7.1c):

$$Al^{3+} > Cr^{3+} > Sc^{3+} > Y^{3+} > La^{3+} \approx Ce^{3+}$$

for sulfates (Fig. 7.1d):

$$Cs^+ > Rb^+ > Na^+ \approx K^+ > NH_4^+ \text{ (monovalent cations)}$$
$$Be^{2+} > Mg^{2+} > UO_2^{2+} > Mn^{2+} > Zn^{2+} > Ni^{2+} > Cu^{2+} \text{ (divalent cations)}$$

for potassium salts (Fig. 7.1e):

$$OAc^- > OH^- > F^- > I^- > Br^- > Cl^- > SCN^- > ClO_3^- > BrO_3^- > NO_3^- > H_2PO_4^-$$

for caesium salts (Fig. 7.1f):

$$OAc^- > OH^- > Cl^- > Br^- > I^- > NO_3^-$$

for lithium salts (see Fig. 7.1g):

$$I^- > ClO_4^- > Br^- > Cl^- > NO_3^- > OAc^- > OH^- > SO_4^{2-}$$

and for sodium salts (Fig. 3.2):

$$OAc^- > I^- > SCN^- > Br^- > OH^- > Cl^- \approx HCOO^- > ClO_4^- > ClO_3^- > F^- > NO_3^- > BrO_3^- > H_2PO_4^-$$

The Debye–Hückel theory, extended to include fitted hydrated ion sizes, does account, roughly, for the observed trends, a decrease in activity at low concentration,

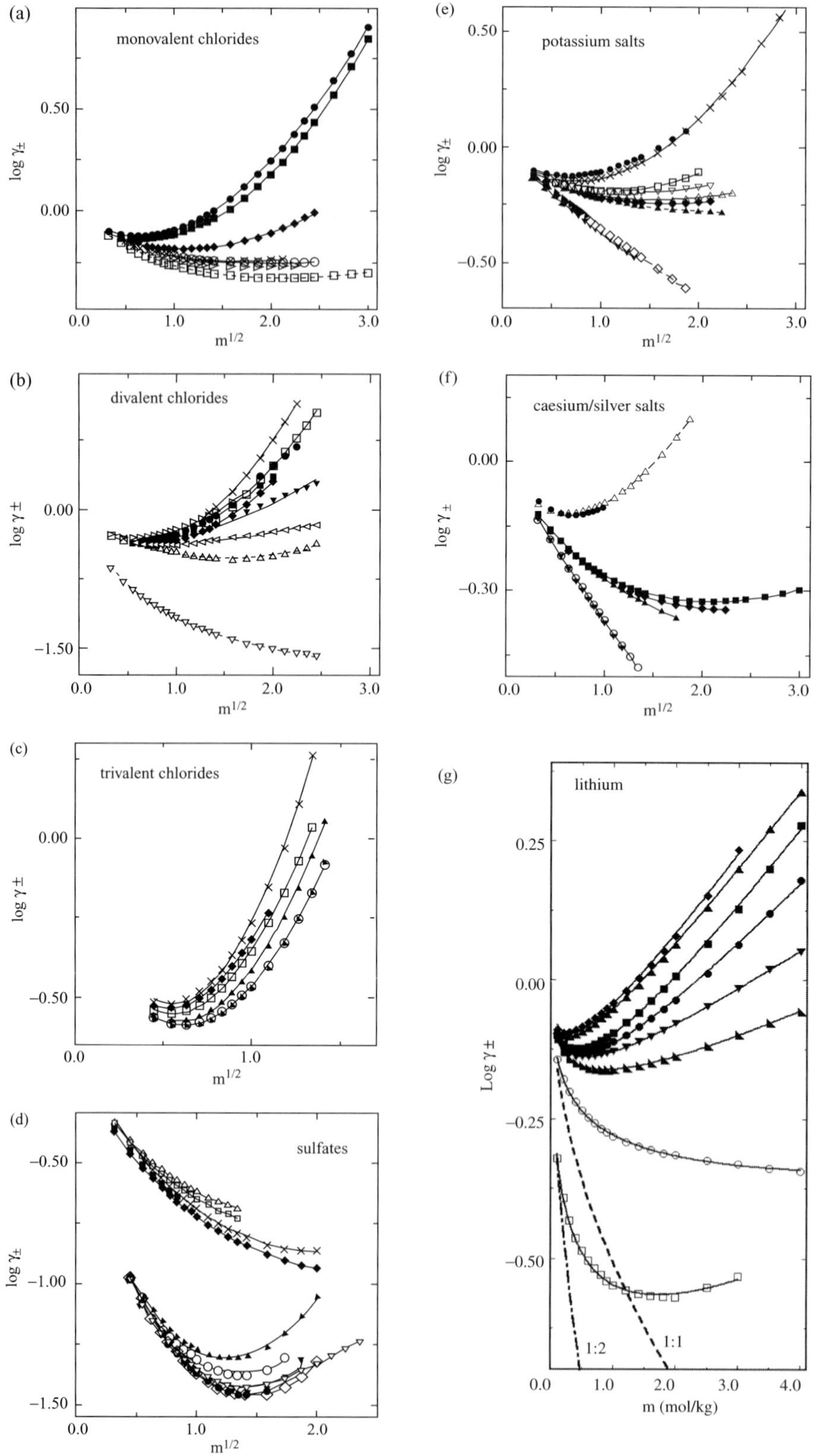

(a)
monovalent chlorides
log γ±
0.50
0.00
0.0
1.0
2.0
3.0
m1/2
(b)
divalent chlorides
log γ ±
0.00
−1.50
0.0
1.0
2.0
3.0
m1/2
(c)
trivalent chlorides
log γ ±
0.00
−0.50
0.0
1.0
m1/2
(d)
sulfates
log γ±
−0.50
−1.00
−1.50
0.0
1.0
2.0
m1/2
(e)
potassium salts
log γ±
0.50
0.00
−0.50
0.0
1.0
2.0
3.0
m1/2
(f)
caesium/silver salts
log γ ±
0.00
−0.30
0.0
1.0
2.0
3.0
m1/2
(g)
lithium
Log γ ±
0.25
0.00
−0.25
−0.50
1:2
1:1
0.0
1.0
2.0
3.0
4.0
m (mol/kg)

Fig. 7.1. (*cont.*)

followed by an increase after typically around 0.2 M. (Extension of the theory via more sophisticated statistical mechanics like the hypernetted chain approximation including ion size, or more refined models that allow interpenetration of hydration shells or charge-induced dipole effects does not do any better than simple extended Debye–Hückel theory.) The agreement is more or less quantitative for part of the alkali halide series, in the sense that fitted hydration distances of closest approach of cation and anion are additive.

But the agreement is illusory. Fitted radii are less than the sum of bare crystal ion radii, e.g. for Cs^+ or Li^+ halides, or for nitrates, for sulfates, phosphates and so on. This is so whether these ions are spherical or not.

The ranking order of the ions is not always the same. For example, for chlorides and divalent sulfates the sequence goes from the kosmotropic to the chaotropic side, but for monovalent sulfates it is reversed (these terms are defined below). For Li^+ and Na^+ electrolytes the sequence goes from chaotropic to kosmotropic, for Cs^+ it is reversed. Acetate usually belongs to the kosmotropic side, but not for the sodium series.

This is further confirmation that specific ion effects cannot be explained by electrostatic theories alone. Dispersion forces (which basically depend on the ionization potentials and polarizability) must be considered. Ion pairing, charge transfer, hydrogen bonding and formation of complexes or protonated species in solution – that reflect the different hydration properties of the ions – cannot simply be explained in terms of charge and ion size only.

In what follows we discuss some physicochemical properties of electrolyte aqueous solutions showing marked specific ion effects that show parallel trends. The mechanism through which ion specificity takes place will be analysed later.

Fig. 7.1. Activity coefficients of simple electrolytes for cations at fixed anion, and anions at fixed cation. From left to right and top to bottom. Monovalent chlorides (a): HCl (●), LiCl (■), NaCl (◆), KCl (×), NH_4Cl (○), RbCl (▷), CsCl (□). Divalent chlorides (b): $MgCl_2$ (×), $CaCl_2$ (□), $SrCl_2$ (◆), $BaCl_2$ (▲), $MnCl_2$ (▼), $CoCl_2$ (■), $NiCl_2$ (●), $CuCl_2$ (◁), $ZnCl_2$ (△), $CdCl_2$ (▽), UO_2Cl_2 (▷). Trivalent chlorides (c): $AlCl_3$ (×), $ScCl_3$ (□), $CrCl_3$ (◆), YCl_3 (▲), $LaCl_3$ (○), $CeCl_3$ (◣). Sulfates (d): Na_2SO_4 (×), K_2SO_4 (■), $(NH_4)_2SO_4$ (◆), Rb_2SO_4 (□), Cs_2SO_4 (△), $BeSO_4$ (◣), MgSO4 (○), $MnSO_4$ (◇), $NiSO_4$ (●), $CuSO_4$ (▲), $ZnSO_4$ (▼), UO_2SO_4 (▽). Potassium salts (e): KOH (×), KF (□), KCl (◆), KBr (△), KI (▽), $KClO_3$ (◣), $KBrO_3$ (◢), KNO_3 (◇), KOAc (●), KSCN (▲), KH_2PO_4 (▼). Caesium and silver salts (f): CsOH (●), CsCl (■), CsBr (◆), CsI (▲), $CsNO_3$ (▼), CsOAc (△), $AgNO_3$ (○). The lines are the fitting curves according to Equation (3.11). Data taken from Ref. 55. Lithium salts (g): LiOH (○), LiCl (●), LiBr (■), LiI (◆), LiClO4 (▲), $LiNO_3$ (▼), LiOAc (◣), $LiSO_4$ (□). The lines are the fitting curves for 1:1 and 1:2 electrolytes, according to Equation (3.11). Data taken from Ref. 55.

7.1.6 Viscosities

The effect of different electrolytes on the viscosity of water was observed for the first time by Poiseuille as long ago as in 1847 [56]. To explain the viscosity effects Jones and Dole [57] categorized ions as either kosmotropes or chaotropes. The former make salt solutions more viscous than pure water, the latter lower the viscosity of water. Over a wide range of concentrations (between 5 mM and 0.1 M), the viscosity η of a water solution can be fitted to a form:

$$\eta = \eta_0 \left(1 + A\sqrt{c} + B_{\mathrm{JD}}c\right) \quad (7.1)$$

where c is the concentration of the salt and η_0 the viscosity of pure water at the same temperature [57]. For very dilute solutions the term in c vanishes and that in $c^{1/2}$ dominates. The parameter A reflects the viscous drag due to the ionic atmosphere. This is supposed to retard the motion of an ion and increase the viscosity of the solution. Falkenhagen calculated and measured the values of A, which depend to good approximation only on the stoichiometry of the electrolyte (as expected for a dilute solution where the Debye–Hückel model applies). A is about 0.006 $L^{3/2}{\cdot}mol^{-1/2}$ for NaCl, 0.005 for KBr, 0.013 for K_2SO_4 and 0.03 for $CeCl_3$ [19]. But the term in c dominates in more concentrated solutions. The ion-specific Jones–Dole coefficient B can be negative (−0.014 L/mol for KCl and −0.090 for $CsNO_3$) or positive for salts containing multivalent ions.

A Hofmeister series exists, but the parameters deduced that describe hydrated ionic radii do not agree with those for activity coefficients.

The sequence, traceable from Fig. 7.2, is for sodium electrolytes (at 0.4 m):

$$\begin{aligned} PO_4^{3-} &\geq \text{citrate}^{3-} > HPO_4^{2-} > CO_3^{2-} > SO_4^{2-} > H_2PO_4^{-} > OAc^{-} > OH^{-} \\ &\geq HCO_3^{-} > HCOO^{-} > BrO_3^{-} > Cl^{-} > NO_3^{-} \approx Br^{-} > ClO_4^{-} > I^{-} \end{aligned}$$

and from Fig. 7.3 for potassium electrolytes (at 0.4 m):

$$HPO_4^{2-} > CO_3^{2-} > H_2PO_4^{-} > SO_4^{2-} > HCO_3^{-} > OH^{-} > Cl^{-} > NO_3^{-} \approx Br^{-} > I^{-}$$

Both trends follow the regular Hofmeister series from strong kosmotropes ('structure makers', solutions of which are more viscous than pure water) through chaotropes ('structure breakers', which give solutions that are more fluid than water at the same temperature).

For much higher concentrations, up to 5 M and more, more complicated formulae are necessary to fit the data. It is apparent that the extended formulae again reflect specificity [58]. For example:

$$\eta = \eta_0 \exp\left(A\sqrt{c} + Bc + Dc^2\right) \quad (7.2)$$

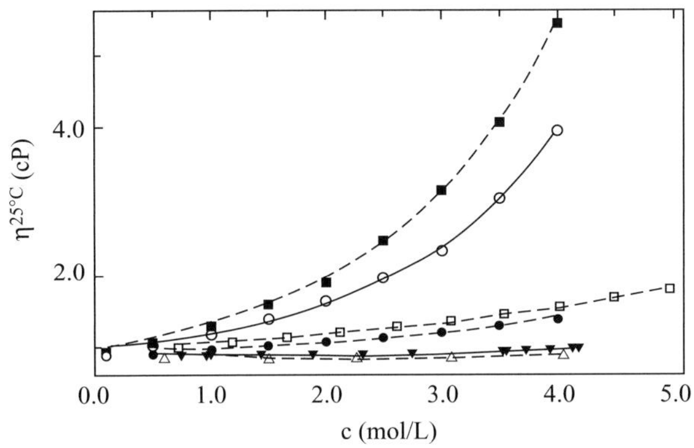

Fig. 7.2. Viscosities (η, in cP) of the electrolyte solutions as a function of concentration (in molar units) at 25 °C. The lines are the fitting curves according to Equation (7.2). Chlorides: LiCl (□), NaCl (•), KCl (▼), CsCl (△), $MgCl_2$ (■) and $CaCl_2$ (○). Data taken from Refs. 59 and 60 with permission. Copyright 1961 The Chemical Society of Japan and copyright 1989 American Chemical Society, respectively.

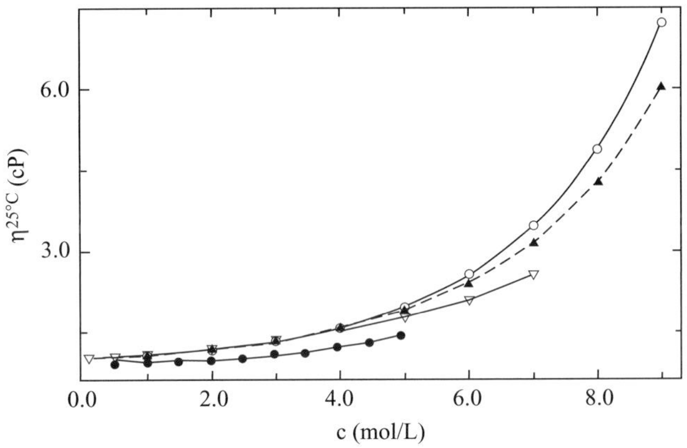

Fig. 7.3. Viscosities (η, in cP) of electrolyte solutions as a function of concentration (in molar units) at 25 °C. The lines are the fitting curves according to Equation (7.2). Sodium salts: NaI (•), $NaNO_3$ (▽), $NaClO_4$ (○) and NaSCN (▲). Data taken from Ref. 61 with permission. Copyright 1970 American Chemical Society.

7.1.7 Conductivity and self diffusion

Another property that is obviously significantly affected by the nature of the ions is conductivity [62,63]. Figure 7.4 shows the equivalent conductivity Λ for KCl, KBr, KI and KH_2PO_4 in water as a function of the electrolyte concentration at 25 °C. The order is now $I^- > Br^- > Cl^- \gg H_2PO_4^-$. This is the same trend that is shown by the activity coefficient of sodium salts.

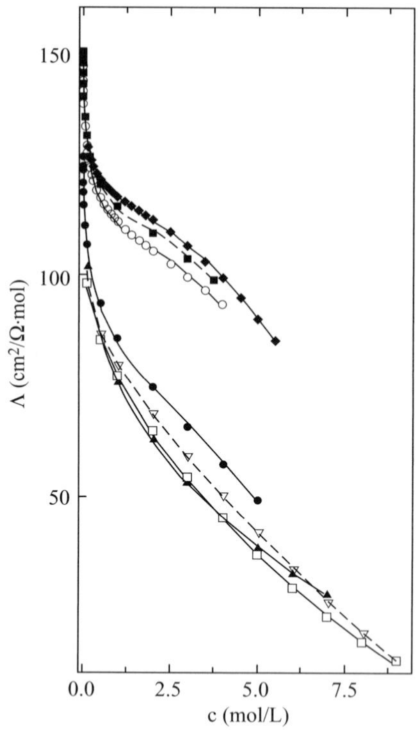

Fig. 7.4. Equivalent conductivity of KCl (○), KBr (■), KI (♦), NaCl (•), $NaNO_3$ (▲), $NaClO_4$ (□) and NaSCN (◁) at 25 °C as a function of *c* (in molar units). Lines are the fitting curves according to Equation (7.3). Data taken from Refs. 55, 61, 64 and 65 with permission. Copyright 1934, 1958 and 1970 American Chemical Society.

The equivalent conductivity data are fitted with the equation [64]:

$$\Lambda = \Lambda_0 - (\alpha\Lambda_0 + \beta)c^{1/2} + Bc + \alpha Bc^{3/2} \tag{7.3}$$

where Λ_0 is the limit equivalent conductivity (for $c \to 0$), α and β are explicit functions of the dielectric constant (ε_w) and viscosity of the solvent (η), and depend on temperature T and on the charge q_i of the ions as:

$$\alpha = \frac{\sqrt{2\pi N_A}}{3\sqrt{1000}(\varepsilon_w k_B T)^{3/2}} q_i^2 |q_i| \omega \tag{7.4}$$

$$\beta = \frac{\sqrt{2\pi N_A}}{3\pi\eta\sqrt{1000\varepsilon_w k_B T}} q_i^2 F \tag{7.5}$$

N_A is Avogadro's number, k_B the Boltzmann constant, and ω is a factor (for a symmetric binary electrolyte it is $2 - \sqrt{2}$) [66]. The first two terms of the right-hand side of Equation (7.3) derive from the Debye–Hückel–Onsager limiting law,

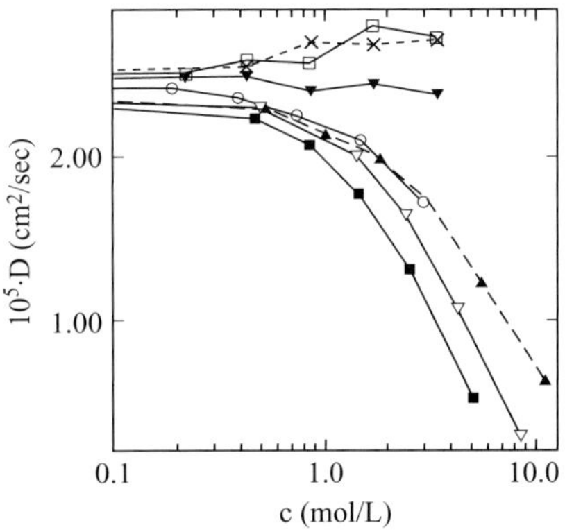

Fig. 7.5. Self diffusion coefficient of water D in salt aqueous solutions at 23 °C as a function of the electrolyte concentration c (in molar units), for KOH (○), KF (▲), KCl (▼), KOAc (▽), KBr (✕), K_2CO_3 (■) and KI (□). Data taken from Ref. 20 with permission. Copyright 1965 American Chemical Society.

which is valid only at infinite dilution, where electrostatics dominates and inter-ion interactions are negligible.

B is an empirical parameter that depends on the electrolyte.

The self diffusion coefficient of water (D) is largely dependent on the kind of electrolyte in solution. Figure 7.5 shows the value of D plotted as a function of the salt concentration (c) for different potassium salts. The trend clearly discriminates kosmotropic species (e.g. carbonate and fluoride) from the chaotropes (iodide and bromide). This evidence matches perfectly with the picture of water molecules being more restrained when they strongly interact with a kosmotrope, or more free to move in the presence of a less disturbing chaotrope. KCl occupies an intermediate position with almost no variation of D with c [67].

7.1.8 Refractive index, heat capacity and freezing point

The refractive index reflects the way a material responds to an electromagnetic wave in thc UV-vis frequency range of the spectrum: $n^2 \approx \varepsilon$, so that it is a measure of the dielectric properties of the sample. The addition of electrolytes to pure water alters the refractive index, depending on the nature of the salt and on its concentration [68]. The variation of n (at 20 °C and at the sodium D line) for several electrolytes as a function of their molar concentration is for Na^+ salts (at 0.5 M):

$$NO_3^- < Cl^- < ClO_3^- < OH^-, HCO_3^- < OAc^- < N_3^- < Br^- < H_2PO_4^- \\ < SCN^- < SO_4^{2-}, I^- < CO_3^{2-} < HPO_4^{2-} < PO_4^{3-} < \text{citrate}^{3-}$$

and for K^+ salts (at 0.5 M):

$$NO_3^- < Cl^- < OH^-, HCO_3^- < Br^- < H_2PO_4^- < SO_4^{2-}, I^- < CO_3^{2-} < HPO_4^{2-}$$

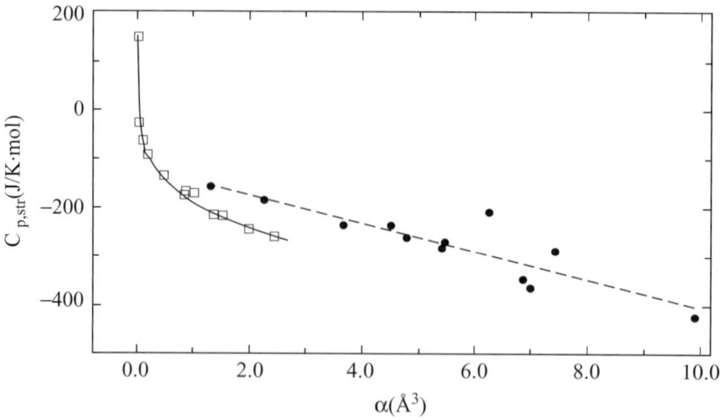

Fig. 7.6. Values of $C_{p,str}$ (in J/mol·K) as a function of the ion polarizability in solution a (in $Å^3$) for cations (□) and anions (•). The data for $C_{p,str}$ were taken from Ref. 70 with kind permission from Springer Science+Business Media.

Here the sequence does not follow the classic Hofmeister series, as strong kosmotropes (PO_4^{3-}) and strong chaotropes (I^-) both produce the highest n values, while intermediate species (NO_3^- or Cl^-) correspond to the lowest values. However, it must be noted that, while activity coefficients and adsorption at interfaces reflect and depend on ion–ion and ion–solvent interactions, the refractive index is the result of the interaction between an individual ion or molecule with the electromagnetic radiation field. Such interactions are reflected in the polarizability of the ion/molecule. Thus, refractive indices are additive, and – although they depend on the nature of the specific solute and solvent – do not follow the same trend as for the activity coefficients.

Changes in thermodynamic functions such as the hydration enthalpy, entropy, isobaric heat capacity and Gibbs free energy of aqueous solutions of electrolytes have been shown to depend on the specific salt added. For example, Fig. 7.6 illustrates the water structure contribution to the heat capacity of single cations and anions as discussed by Marcus [69]. The trends show that the ions significantly reduce the structural contribution to the heat capacity of the solvent, and that such modification is related to the electronic properties of the ion.

The lowering of the freezing point (ΔT) of a solution with respect to the pure solvent depends on the effective number density of solute particles and not on their nature, so it is sensitive to ion pair formation or other kinds of association processes. For dilute solutions ΔT is proportional to the molality m of the solution. Again, different electrolytes at the same concentration produce different ΔT values. Of course this is expected when the salts have different stoichiometry (e.g. Na_3PO_4, Na_2CO_3 and NaCl), and when the anion further dissociates (for example $H_2PO_4^- \rightleftharpoons HPO_4^{2-} + H^+$), producing more ions in solution. However, for 1:1

electrolytes – in principle – the number of ion species should be the same, but the measured ΔT are very different.

Increments for Na^+ (at $c = 0.5$ M):

$$H_2PO_4^-, HCO_3^- < NO_3^- < SCN^- < Cl^- < OH^- < Br^- < CO_3^{2-} < OAc^- < citrate^{3-}$$

Increments for K^+ (at $c = 0.5$ M):

$$NO_3^- < H_2PO_4^- < HCO_3^- < Cl^- < Br^- < I^- OH^- < HPO_4^{2-} < CO_3^{2-}$$

7.1.9 Other problems, with pH and buffers

Practically all methods of determining hydrated ion radii give as many different hydrated ion radii for a given ion as there are methods of measurement.

An additional complication is that we know from Chapter 6 that the accepted methods for pH measurements give results that are ion and electrolyte concentration specific (cf. Figs. 6.1 and 6.2). There is a certain circularity in this as the interpretation of pH measurements depends on a theory that does not admit specific ion effects. The interpretation assumes both the validity of the extended Debye–Hückel theory, and the equivalent double-layer theory of charged interfaces and their interactions at the core of colloid science. Indeed we remind the reader of the unhelpful advice of the International Union of Pure and Applied Chemistry that it is better to not trust pH measurements above 0.1 M salt, the extreme limit for the extended Debye–Hückel theory (see Section 6.3).

The problem is even worse with buffers, as they contain a weak acid or base and the corresponding conjugate species (e.g. $H_2PO_4^-/HPO_4^{2-}$ or NH_3/NH_4^+), used to set pH.

The same theory is used again, with a fitting parameter for buffer anion radius that takes on absurd values, e.g. 30 Å for a phosphate ion, a 'radius' that varies with electrolyte type and concentration and mixture. Experimentally, Evens and Niedz make a convincing demonstration that pH becomes a dependent variable in real biological systems [10].

The work of Salis et al. [13] shows that the Hofmeister series effects on apparent (i.e. measured by a glass electrode) pH are reversed for phosphate and cacodylate buffers supposedly fixed at pH 7. Even more surprisingly, with phosphate buffer the series reverses with replacement of Na^+ by K^+. These effects cannot be due to changes in bulk activities and have to reside in specific ion surface adsorption effects that are not electrostatic in origin. The same is true if one tries to measure pH in surfactant solutions, where the reproducibly irreproducible dynamic effects that have to be attributed to surface adsorption on the glass electrode make measurements impossible.

The problem is not resolved by use of a H_2/Pt electrode, because surface specific ion adsorption will be different for the platinum, H_2/water, or glass surface. Nor is it resolved by use of indicators, which also display Hofmeister effects [71].

7.1.10 *Classification of and non-universality of Hofmeister series*

The typical lyotropic series, as determined by Hofmeister in his pioneering works on the salt-induced precipitation of egg albumin, is [72]:

$$H_2PO_4^- > SO_4^{2-} > F^- > Cl^- > Br^- > NO_3^- > I^- > ClO_4^- > SCN^-$$

This means that, in this particular case, dihydrogen phosphate induces the largest effect, and thiocyanate the least. Intermediate ions (for example chloride) will produce intermediate effects.

However, several phenomena show the opposite trend; some others indicate a partial displacement between some ions along the series (for example nitrate can precede or follow bromide).

Other systems indicate a non-monotonic trend: for example, the ions' efficiency increases up to chloride and then it decreases again. In this case we may indicate this behaviour as a 'V-shaped' curve with a minimum or a maximum in the trend. The Hofmeister series can reverse direction for a given buffer with change of cation from sodium to potassium [13]. It can reverse direction when two different buffers supposed to give the same pH are used.

7.1.11 *Specific ion effects in direct force measurements*

For more than 50 years a central focus of colloid science has been on direct measurement of DLVO forces in electrolytes. The invention of the surface forces apparatus (SFA) by Tabor [73] and its development by Israelachvili and colleagues [74] to measure forces in liquids between molecularly smooth mica opened up a new dimension. The development of osmotic pressure techniques by Parsegian and colleagues discussed in Chapter 4 allowed further quantification of short-range hydration forces, and at high electrolyte concentrations. Both require very hard work and skill, in preparation, and purification. For the adaptation of atomic force microscopy to force measurement Ducker, Senden and Pashley provided an easier popular option [75], but one fraught with innumerable artefacts.

The almost universally claimed 'agreement' between DLVO theory and measurement was illusory.

Typically a measurement is fitted to the sum of double-layer and van der Waals forces, with extra parameters such as 'surface ion binding' constants of unknown origin, or the Stern layers of classical colloid science. The differences between 'theory' and experiment were then assigned to 'hydration' or 'extra-DLVO' forces.

However, specific ion effects in the forces showed up repeatedly, with differences that can be of an *order of magnitude* as we have seen repeatedly; even for 1:1 electrolytes at low concentrations (between 1 and 10 mM), e.g. NaBr vs. NaOAc, or $LiNO_3$ vs. NaCl, or NaCl vs. LiCl.

For interactions of molecularly smooth mica with alkali metal electrolytes (Li^+, Na^+, K^+, Rb^+ and Cs^+) 'secondary' hydration forces appear at low separations. These can be fitted to some extent by postulating a site binding model (charge regulation). The term secondary hydration encompasses the notion that adsorbed hydrated ions are compressed due to the presence of the electric potential due to the other surface as the surfaces come together. Work has to be done to release that hydrated water associated with the ions. This shows up in a short-range repulsive hydration, in effect a 'dehydration' force [76–79].

However, the inferred surface potentials in the presence of the different salts show a peculiar non-monotonic behaviour with changing concentration. That can be accommodated in part by invoking another effect missing, a surface entropy due to the finite size of supposedly adsorbed hydrated ions [80].

Such effects and more complicated phenomena that make no sense from an electrostatics- and hydration-only theory on different substrates like silica are being increasingly discovered as techniques of measurement improve [2,81–88].

But the early attempts to accommodate strange effects were pushing classical theory too far. Something is still missing: an explanation of the source of site binding, an ion-specific constant for the counterions. Naturally enough by now, the neglect of specific dispersion forces of adsorption is the prime suspect.

For a series of 1:1, 2:1 and 3:1 dilute aqueous electrolytes between mica, Pashley [76–78] found 'secondary' or 'extra-DLVO hydration' forces with fitted exponential decay lengths of 10, 20 and 30 Å respectively!

The phenomenological observation can be exploited in controlling flocculation, but explanation remains elusive [89].

Valiant extensions of the double-layer theory due to S. Marcelja, J. Quirk and co-workers by inclusion of 'ion fluctuation' forces applied to the (discontinuous stepwise) swelling of very high counterion content (up to 6 M) clays seem to have simply failed. (The term ion fluctuation forces used in this connection refers to reduced repulsion effects that occur with asymmetric (here divalent) counterions in electrolytes. They reflect the non-linear changes in electrolyte screening length seen in asymmetric electrolytes in Chapter 3.) Extensive simulation work by Bo Jonsson and colleagues captures those effects, but it cannot account for specificity, the different swelling of clays with substitution of one divalent counterion by another, for the same reasons.

Measured forces between mica surfaces in electrolytes do show up oscillations at very small distances due to the size of hydrated cations [90].

The problem, as we have rehearsed, is that the theory is deficient. So the assignment of differences between theory and experiment to 'hydration' is equally wrong in general.

The failure is independent of situations where real surface chemistry occurs. The dissolution of interfaces of silica at high pH produces polysilicic acid, which can bridge between interfaces, as for the real chemistry of cements.

The problem is more complicated still when dissolved atmospheric gas is removed, a matter that induces a further change in specificity and magnitude of forces for 'hydrophobic' surfaces. This will be explored in Chapter 8.

Nonetheless the illusion persists that DLVO provides the best theory, with compounding catastrophic results when applied to the interpretation of, for example, neutron scattering data on microstructured fluids.

7.1.12 Interfacial tensions and computer simulations

Faced with these problems, an increasingly large literature has been devoted to simulation of liquids and, of interest to us, of electrolytes at interfaces. But the best 'effective' two-body water potential does not explain the density maximum at 4 °C, or why ice floats on water. Whatever hydrogen bonding means in bulk water, it has some meaning that is not captured by such effective potentials. At 90 °C, water is not hydrogen-bonded at all and behaves in all respects thermodynamically almost identically to hydrazine. Simulations of the interfacial tension of water give results that differ from the measured value by at best 30%. Even for simulation of the phase diagram of argon, the simplest van der Waals hard-core liquid, the explanation of interfacial tension requires the inclusion of three-body interactions: see also remarks on the ground-state energies of noble gas solids in Chapter 2.

A further difficulty for the simulation approach is the fact that the cohesive energy or tensile strength of real liquid water is hundreds of times less than it ought to be theoretically. This can be traced to the defects imposed by dissolved atmospheric gas molecules, which, as we shall see in Chapter 8, seem to play a major role in hydrophobic interactions. The simulations model 'water models', not real water that contains these gas molecules.

At an air–electrolyte interface, extensive simulations appear to show positive adsorption of 'hydrophobic' ions such as bromide or iodide. The real interfacial region, two or three water layers (6–9 Å) thick, contains atmospheric gas ranging from 1 M to 10^{-2} M concentration across this region, at much higher concentration than, and necessarily associating with, any such physisorbed bromide ions. The simulation approach also has a difficulty, so far, in predicting negative adsorption for hydroxide ions, an awkward prediction apparently contrary to fact. (The electrical potential at the air–water interface is supposed to be negative and 25 mV, and due to hydroxide ions, although that itself could be an artefact of the theory used to interpret the measurement.)

We do not comment on the simulation industry that mimics protein structure with 37 000 temperature-dependent molecular interaction parameters that do not include water at all. However the simulation business develops, its major sin is that it is incredibly boring.

7.1.13 Dramatic Hofmeister effects in self assembly

Whatever the state of theory, very large differences in forces due to specific ion effects are real enough [91]. Changes in headgroup area of self assembled aggregates, and of interaggregate interactions, and of wetting induced by specific ion effects allow huge flexibility in a wealth of applications, formulation, colloid stability, microstructured fluid design, protein crystallization and nanotechnology that have hardly been recognized. These will be dealt with in Part II of this book.

7.1.14 Correlations and the approach of Collins

The lack of a self-consistent and general theory of molecular forces that could explain the experimental results and predict behaviour has prompted several researchers to relate the observed Hofmeister phenomena to some physicochemical properties of the intervening ions, since the end of the nineteenth century [92].

Various attempts have been made to explain the mechanism that governs specific ion effects. Although conceptually simple, specific ion effects imply the interplay of several different interactions. Even in the absence of more complicated self-assembled and macromolecular systems (i.e. with polymers, proteins and colloidal systems in general), a genuine solution of a simple salt in water involves cation–anion, cation–water, anion–water, cation–cation, anion–anion and water–water interactions. So, the global effect of the electrolyte on the properties of its solution will depend on the interplay between these different interactions, that is whether the interactions between the ionic species and water are stronger or weaker than those between water molecules in the bulk. Note that this interplay will result in a value of only the average activity coefficient ($\gamma_{\pm}$), of a particular salt in water (see Chapter 3).

The existence of a correlation between a particular quantity in a specific ion effect study (for example viscosity, pH or conductivity or cmc) with physicochemical parameters is important for a number of reasons:

1 It indicates the trace of 'Hofmeister fingerprints' in phenomena where specific ion effects occur, and suggests possible mechanisms that involve different kinds of intermolecular interactions (in particular of dispersion forces).
2 It is a way to order phenomena – in the absence of a definite theory – at the same level as Mendeleev's classification of the then-known elements in the periodic table.
3 It indicates the presence of a 'direct' or 'inverse' Hofmeister sequence.
4 It can be used to predict the behaviour of other electrolytes, based on their kosmotropic or chaotropic nature.

In attempting to account for specific ion effects one approach is that due to Collins. A brief account of his schema is as follows. He called attention to the hydration enthalpies of ions. In considering the interactions of solvating water molecules with an ion, the distance of a water molecule from the centre of a spherical ion increases with the size of the ion, until the water–ion interaction is weaker than the water–water molecule interaction. This is assumed to occur with a chaotropic ion (large size, small charge). Further, by considering the regular dependence of heats of solution of alkali halides (MX) on the Gibbs free energy of hydration, the dependence of the MX solubility on the ion size, and other experimental evidence, Collins derived a rule for the tendency of ions of opposite charge to form inner-sphere ion pairs, the so-called 'Law of Matching Water Affinities': the less soluble ion pairs are produced by ions that are closest to each other in their hydration energy, i.e. how strongly they bind to water [15]. In particular, salts with a positive enthalpy of solution (q) will produce cold solutions, and those with a negative enthalpy of solution will result in hot solutions.

Collins considers only electrostatic forces. There is a problem of course with 'ion size'. Quantum mechanical, dispersion and perhaps ion-induced dipole forces responsible for the specificity of ionic hydration are excluded.

When q is plotted as a function of the Gibbs free energy of hydration for the electrolyte (the sum of ΔG_{hydr} of the cation and of the anion), a typical 'volcano' or 'V-shaped' plot shown in Fig. 7.7 is obtained.

This plot indicates that when the cation and the anion have similar free energies of hydration – i.e. they are kosmotropic–kosmotropic (Li^+F^-) or chaotropic–chaotropic (Cs^+I^-) – then the salt has a positive q – that is, heat is taken up on solubilization, no strong interactions are established with the solvent, and the ions are likely to remain 'close' in ion–ion pairs. On the other hand, the combination of a chaotropic with a kosmotropic ion (Cs^+F^- or Li^+I^-) will result in a large value of ΔG_{hydr} (in absolute value), either on the far left- or far right-hand side at the foot of the volcano, and corresponds to a negative q (heat is released), indicating the formation of strong ion–water interactions. In summary, ion pairs are formed when the cation and the anion have similar ΔG_{hydr}. Interestingly, it is the global

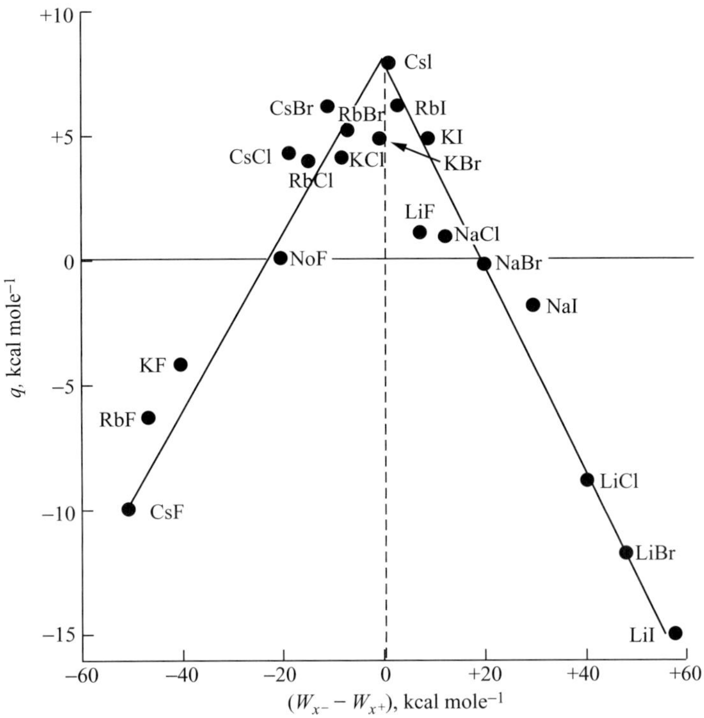

Fig. 7.7. Volcano plot after Collins: heat of solution q versus the Gibbs free energy of hydration W(anion)$-W$(cation). Reprinted from Ref. 15, copyright 2004, with permission from Elsevier.

cation–anion pair that affects the microscopic structure of water, rather than each single ion [93].

Furthermore, it is clear that water will be more or less strongly affected by the ions it interacts with. In fact H_2O is asymmetric, and therefore interacts in different ways: water will tend to face a cation with the oxygen, while it will point out from an anion with its hydrogens. This difference will result in a detectable change of water polarizability, as discussed by Leberman and Soper [94].

The elegant qualitative characterization of specific ion effects due to Collins gives a set of 'effective' working rules that often appear to be consistent and useful [95–98].

Insights into Hofmeister effects through parallels between ion peptide and ion–air–water interfacial partitioning assigned to a competition between electrostatic and hydration are misleading, since they exclude the extra, specific dispersion forces missing from the theory on which the conclusions are based.

If oil–water and other interfaces are admitted into the comparisons the parallelism breaks down.

But they cannot be quantitative rules, and remain ‘effective’ rules being based on primitive model linear electrostatic theories and ‘hydration’ alone [15]. They cannot include effects of temperature vital to biological situations.

The same deficiencies apply to computer simulation of specific ion effects such as interfacial tension with ‘effective’ water molecule potentials. The potentials necessarily are fitting parameters that have to be changed with temperature, and in any event give bulk and surface properties of water that, for example, give bulk densities erroneous by 20–30%.

The foundations of the characterization derive then from a theory that we have seen is frustratingly deficient and a bit of a hodge-podge. As already remarked, it can be proved that without the inclusion of dispersion forces, and inclusion at the same level as non-linear electrostatics, the theory is thermodynamically inconsistent [51]. The situation was as if we were attempting to account for properties of molecules from the periodic table without quantum mechanics, just like the Lewis or Kékulé picture of valence and bonds.

Other correlations have been established in the past and in more recent studies. The first attempt was made Setschenow in 1889, who studied the solubility of CO_2 in blood and in aqueous solutions of electrolytes [99]. He introduced a salting-out coefficient, K_s, that progressively changes in the sequence $F^- > Cl^- > Br^- > I^-$ [100]. This parameter has been related to other specific ion phenomena, for example the solubility of benzene in aqueous solutions [101], the salting-out of haemoglobin [102] and the adsorption of atmospheric gases at the air–water interface [103].

Later in the 1930s Voet introduced an empirical lyotropic number (*N*) that accounts for salt effects in the precipitation of starch [104].

Other parameters that reflect the different interactions of the ions with the solvent and with interfaces are the surface tension increment $\sigma = (\partial\Delta\gamma/\partial c)$, polarizability ($\alpha$), Gibbs free energy or entropy change of hydration (ΔG_{hydr} or ΔS_{hydr}, respectively), partial molal volume, molar refractivity (R'_2),[7] ion radius, solvation heats, Jones–Dole viscosity *B* coefficient and entropy change of water (ΔS_{II}).

The former – which quantifies the increment of water surface tension upon addition of salt – will be treated in Chapter 8, with bubble–bubble coalescence [46].

The entropy change of water [16] strongly depends on the specific ions, and reflects two opposing factors: one is the disorder induced by the ion in the structure

[7] The molar refractivity (in cm^3/mol) is derived from the Lorentz–Lorenz equation, and is calculated as $R'_2 = \frac{1000}{c}\left[\frac{n^2-1}{n^2+2} - \frac{1}{\rho_w}\left(\rho - \frac{cM}{1000}\right)\left(\frac{n_w^2-1}{n_w^2+2}\right)\right]$ where ρ_w and ρ are the density of pure water and of the salt solution at concentration c, M is the molar mass of the electrolyte, n_w and n are the refractive index of pure water and of the salt solution (c) at the same temperature and wavelength [105].

of the solvent molecules, and the other is the re-ordering effect due to the electric field of the ion that interacts with the permanent dipoles of water. So $\Delta S_{II} < 0$ (positive hydration) reflects a strong ordering of water around the ion, which is what occurs with Li^+, Mg^{2+}, Al^{3+}, F^-, SO_4^{2-} and PO_4^{3-} (kosmotropes). Instead the existence of negative hydration ($\Delta S_{II} > 0$), as for chaotropes such as Cs^+ or ClO_4^-, means that the disordering effect prevails. These arguments were supported also by previous results obtained by Samoilov [106,107]. Collins noted that the trend in entropy change of water with the anion is evidence of a chemical interaction, namely a charge transfer between the anion and the water molecule. This hypothesis is supported by the fact that anions (richer in electrons) are more effective than cations [108].

The standard molar Gibbs free energy change of hydration, ΔG_{hydr}, for an ion M^z is related to the process $M^z_{(g)} \rightarrow M^z_{(aq)}$, that is to the transfer of the ion from the gas phase to solution (and this goes back to the problem of the calculation of Born energy). These values are well presented and discussed by Marcus [109]: kosmotropes present the highest values of ΔG_{hydr}, as a result of their strong interactions with water.

The Jones–Dole viscosity B coefficient has already been discussed in Section 7.1.6 above.

The volumetric properties of ions depend on their hydration. For small ions (Li^+ or F^-) with high density charge and strongly hydrated the partial molar volume is negative due to electrostriction, i.e. to the volume contraction of the hydration shell because of the ion's electric field. The apparent molal volume of ions has been investigated since the nineteenth century and, like the activity and osmotic coefficients, depends on the concentration of the ionic species. This parameter reflects the different interactions established in a solution: ion–ion, ion–solvent and solvent–solvent.

Table 7.1 offers as an example the values of these physicochemical parameters for halides. It is easy to verify (see Fig. 7.8) that α, σ, ΔG_{hydr}, ΔS_{II}, N, B and K_s change monotonically from F^- to I^-. These quantities are all, more or less intuitively, related to the dispersion forces that the ion experiences in solution.

(More refined values of α are now available from recent *ab initio* quantum mechanical calculations to be described later. An essential determinant of specificity is the details of the *frequency* dependence of the polarization response of an ion, which reflects its many electron quantum states – just as the energy levels of atoms determine their capacity to form molecules.)

That the inclusion of dispersion forces in Hofmeister phenomena is essential is confirmed by correlating the observed effects to these physicochemical parameters and revealing 'Hofmeister fingerprints' [112]. This methodological approach has been verified in several works [29,43,45,48,49], and more recently Cremer has

Table 7.1. *Anion static polarizability (α in Å^3), surface tension increment (σ in mN·L/m·mol), Gibbs free energy change of hydration (ΔG_{hydr} in kJ/mol), entropy change of water (ΔS_{II} in J/mol·K), lyotropic number (N) [26], Jones–Dole viscosity B coefficient (in L/mol) [21], Setschenow constant (K_s, in L/mol) [92] and cloud point temperature for aqueous dispersions of diC$_8$-PC (T_c in K) [29] for sodium halides.*

Anion	α	σ	ΔG_{hydr}	ΔS_{II}	N	B	K	T_c
F^-	1.22	3	−465	70.3	4.8	0.107	0.23	318.5
Cl^-	4.22	1.64	−340	6.3	10.0	−0.005	0.046	315.2
Br^-	6.03	1.32	−315	−27.6	11.3	−0.033	−0.023	312.4
I^-	8.97	1.02	−275	−54.8	12.5	−0.073	−0.021	298.6

shown that in fact it reflects the different effects due to different ions: basically kosmotropic ions act by modifying the polarization of interfacial water molecules, and chaotropes work by affecting surface tension [113–115].

As an example, Fig. 7.8 shows the regular variation of some physicochemical properties with the anion static polarizability for sodium halides.

The papers cited and an extensive recent theoretical literature [4,6,11,22,91, 116–124], devoted to inclusion of dispersion forces into theory consistently, establish proof of concept only. Comparison with experiment was based on polarizability data for ions and ionic radii that are best guesses. They omit electrostatic ion-induced dipolar forces that are not usually important. However, they do show conclusively that these missing, specific (quantum mechanical) dispersion forces are as large as, or dominate, electrostatic forces.

More extended calculations of polarizabilities of ions and their complete frequency response based on *ab initio* quantum mechanics and proper definitions of ionic radii will be shown below to be on the way to doing rather better in getting a better handle on both forces and hydration.

What clearly emerges is that the largest surface tension gradient, free energy of hydration, viscosity *B* coefficient, entropy change of water, lyotropic number and the minimum cloud point temperature are produced by the kosmotropic

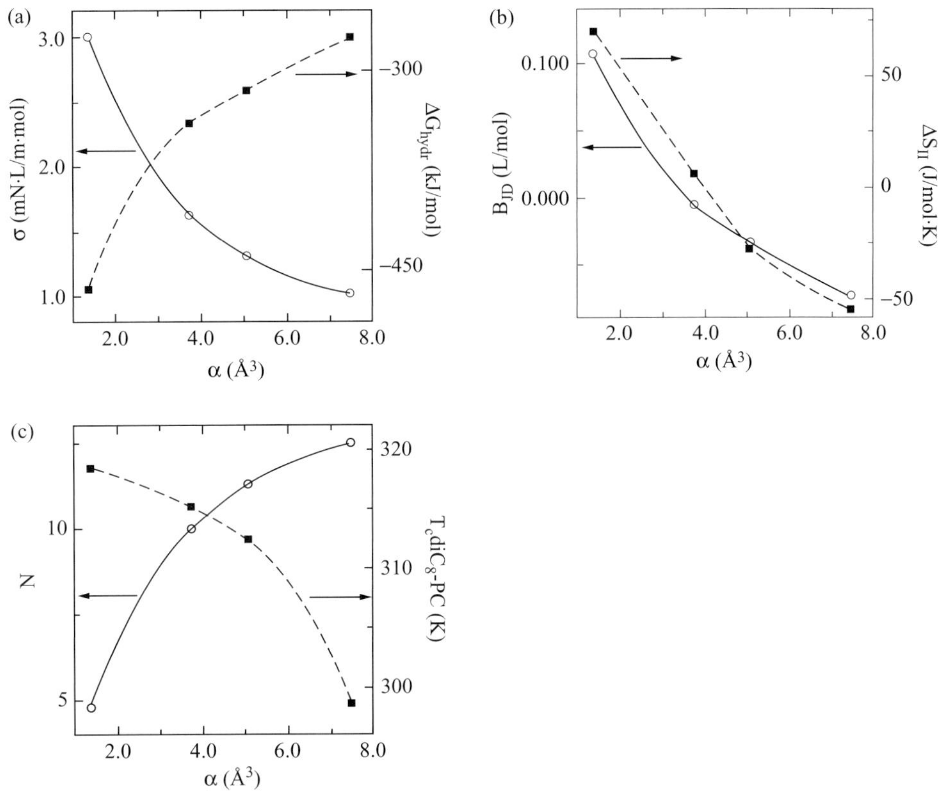

Fig. 7.8. Illustrative plots of different physicochemical properties as a function of the static anion polarizability in solution (α). (a) Surface tension increment (σ, ○) and Gibbs free energy change of hydration (ΔG_{hydr}, ■); (b) Jones–Dole viscosity B coefficient (B, ○) and entropy change of water (ΔS_{II}, ■); (c) lyotropic number (N, ○) and cloud point temperature of diC$_8$-PC (T_{c}, ■) in the presence of sodium fluoride, chloride, bromide and iodide. The lines are guides for the eye. The data are shown in Table 7.1. See the table for permissions.

fluoride, the halide with the strongest hydration properties and the smallest polarizability.

Trends get more complicated when other polyatomic species such as $H_2PO_4^-$ or NO_3^- or divalent species (CO_3^{2-} or SO_4^{2-}) are considered [112]. This indicates that the charge, distribution of the bond and lone pair electrons (VSEPR model),[8] acid/base properties and the geometry of the ions are important features that affect their behaviour in solution and at interfaces. (In a proper theory of dispersion forces these are all included.)

[8] Valence shell electron pair repulsion [125].

It is interesting to note that ion specificity emerges phenomenologically in several equations that capture the concentration dependence of some characteristic electrolyte parameters.

The first is activity coefficient. We know and we have seen in this book that according to the extended Debye–Hückel equation the average ionic activity coefficient is expected to obey this formula:

$$Log\gamma_\pm = -\frac{A\left|z_+z_-\right|\sqrt{c}}{1+Ba\sqrt{c}} + bc \approx -A\left|z_+z_-\right|\sqrt{c} + \left(A\left|z_+z_-\right|Ba + b\right)c \qquad (7.6)$$

where c is the concentration of the aqueous solution of the salt, the first term – in $c^{1/2}$ – accounts for purely electrostatic interactions ('A'), while the second term in c reflects ion specificity ('b').

Similar quadratic expressions in $c^{1/2}$ are found to fit the experimental data well also for other quantities: for example osmotic coefficient ϕ (see Fig. 3.3), viscosity and conductivities (see above), cmc of diC$_8$-PC dispersions in water [29], solubility of non-polar substances in water as described by the Setschenow constant (see Equation (3.16)) and the apparent molal volume of an electrolyte in water that is well described by [17]:

$$\phi_V = \phi_{V^0} + S_V\sqrt{c} + b_V c \qquad (7.7)$$

where ϕ_{V^0} is the apparent molal volume at infinite dilution, while S_V and b_V were defined as 'deviation' (i.e. specificity) constants.

What is interesting in this argument is that the concentration dependence carries two terms: the one in $c^{1/2}$ accounts for electrostatic ion–ion interaction and prevails when the solution is very dilute (Debye–Hückel regime). In principle it does not depend on the nature of the electrolyte but only on its stoichiometry. The second term, in c, reflects non-electrostatic (dispersion) forces, ion–ion and ion–solvent interactions, emerges only for moderate or high concentration, and is ion-specific.

7.2 Manifestations of Hofmeister effects in biology and biochemistry

Bearing all this in mind we digress now to bring together and consider some examples that underline how far the classical theory of physical chemistry is removed from real-world biology and biochemistry. Hofmeister himself worked first in a pharmacology department, and his initial papers on specific ion effects dealt with the relative efficiencies of different salts, at the same concentrations, in precipitating suspensions of egg white lecithin [72,112]. This early work had motivations in biochemistry and physiology. The series of precipitating effectiveness for cations at fixed anion and more strongly for anions at fixed cation was the same as that for suspensions of several inorganic colloids. So the idea grew up that the Hofmeister series was fixed and universal. His work was done before ideas on pH, activities, osmotic and other properties of electrolyte solution properties began to be quantified after the work of van t'Hoff, and through the following half century represented by the monumental work of Robinson and Stokes, and of Harned and Owen. (Even pH or buffers were concepts still on the horizon in Hofmeister's time.)

Activity coefficients exhibited the same kinds of Hofmeister series. This appeared to confirm the idea that the effects should be associated with the bulk properties of aqueous salt solutions, a matter apparently reinforced a fortiori by the efficacy of magnesium sulfate on constipated chemists. In the evolution of the intuition about such matters, bulk hydration induced by ion–water interactions, adduced from heats and enthalpies of solution, and electrostatic forces appeared to handle the job of explanation, as further quantified by the Collins approach above.

In electrolyte theory proper the ad hoc extension of linear Debye–Hückel theory to include ion 'size' or more sophisticated statistical mechanical theories beyond the primitive (continuum solvent) model up to the present worked in accommodating phenomena. They worked in the sense that to explain activities, the models had to allow a civilized model, i.e. hydration and interpenetration of ionic hydration shells described by phenomenological ion pair specific 'Gurney' potentials. So it seemed as if the intuition could be applied to 'binding' of counterions to specific protein ionizable sites.

In the development of these theoretical ideas only electrostatic interactions and hydration appeared. Interactions between ions of like charge were ignored. The theory behind measurements of pH, of buffers (cf. Section 6.3) and other weakly associated electrolytes, of membrane potentials and zeta potentials all rested on this theoretical edifice. In the above situations, where theory failed, it was swept to one side, or accommodated by even more parameters. (An example is what happens when a net neutral phosphatidylcholine vesicle is exposed to a dilute phosphate buffer. Phosphate ions adsorb, contrary to intuition, and the electrically 'neutral' vesicle has a zeta potential of 140 mV!)

Nowhere did the quantum mechanics essential to the specificity of chemistry appear explicitly. It surfaced first, and ominously, in the DLVO theory of colloidal interactions. Here a specific quantum mechanical dispersion force (Lifshitz) had to be recognized to account for flocculation of lyophilic particles in electrolyte solutions. Just to confuse matters further, the DLVO theory did not involve bulk activities. The theoretical edifice of electrolyte theory was taken over partially and results of direct force measurements were accommodated, as we have seen, by invoking hydrated ion sizes and hydration (inner and outer Helmholtz planes of different dielectric constants) at interfaces, or site binding. But the specificity of forces and the differences in magnitude by factors of 10 or more with interchange of counterion or co-ion that we have seen were simply not explained. Further, as we shall see in the following chapters, there is a fundamental irreconcilability between theories of self assembly and these classical theories of forces. Another major inconsistency was identified as described in Chapter 6, in the inadmissible ansatz that treated quantum mechanical forces separately from electrostatic forces.

We make no apology for retracing this confusing morass of 'laws' and procedures which the generations of textbooks of the adepts of the freemasonry of physical chemistry – including ourselves – had set in stone.

Concepts like buffers, pH, pK_as, zeta and membrane potentials, ion-binding specific ion effects and the interpretation of measurements based on non-predictive theory are so integral to biochemistry and molecular biology that if they are, as we claim, so flawed, some repairs to the edifice are in order.

From the beginning Hofmeister was bemused on the question of whether surface or bulk electrolyte effects were dominant in his observations. Subsequent to his work a problem emerged that engendered confusion for the next century. With some protein systems, the Hofmeister series reversed in order above and below the pK_a of the protein. See for example, Refs. 126 and 127 for an excellent treatment of the situation that existed in the 1920s.

Loeb's work was particularly interesting [127]. His claim, working with a series of acids and bases and gelatin and some other proteins, was that if pH was controlled, that is in our terminology, above and below the isoelectric point of the proteins, or pK_a, the Hofmeister series did not exist! Swelling and precipitation were all consistent and with bulk viscosities of the acids or bases used – it all fell into place, apparently, and was consistent with 'chemical' binding of anion or of cation to ionizable sites on the polymer. But this did not explain the reversal of the Hofmeister series for neutral salts above and below the isoelectric point of the protein. Gustavson [126] pointed out in an extensive work that his argument did not apply to the Hofmeister series or its reversal above and below the pK_a in neutral salts (strong electrolytes). The debate at that time was whether it was due to chemical binding or the new colloid science ideas about physisorption, not about hydration and bulk effects.

Biology probably offers the broadest set of examples where specific ion effects take place. Biological systems are almost always based on water and comprise a wide collection of ions. Typically these occupy the kosmotropic side or intermediate position in the direct Hofmeister ranking: phosphates, sulfate, carbonate, chloride, nitrate, sodium, potassium, calcium, magnesium and a few others. Instead, chaotropes are less present in biological samples, or even toxic (iodide, thiocyanate, cyanide, caesium, barium and so forth).

We have investigated Hofmeister effects, in self energies, activities and interfacial tensions of electrolyte solutions, and on pH and buffers. We now proceed to explore the manifestation of Hofmeister effects in successively more complex molecular systems.

7.2.1 Optical rotation of chiral molecules

The first are the effects on optical activity of α-amino acids and saccharides [94,128].

Optical activity (α) is produced by the difference in refractive index ($n_L - n_D$) of a chiral chemical for circularly left- and right-polarized light [129]:

$$\alpha = \pi d \frac{n_L - n_D}{\lambda} \tag{7.8}$$

where λ is the wavelength and d (in mm) the distance travelled by light in the sample. The specific rotation (or optical activity $[\alpha]$) depends on temperature T, λ, and concentration of the solution c (in g/mL) [130]:

$$[\alpha]_\lambda^T = 100 \frac{\alpha}{cd} \tag{7.9}$$

The effect of different anions on the optical activity of several α-amino acids and glucose has been investigated [130]. It was found that this specific ion effect follows the regular Hofmeister series, and correlates well with the anion polarizability in solution for the halides. Anisotropic and polyatomic anions (SCN^-, ClO_3^-, ClO_4^-, NO_3^- and $H_2PO_4^-$) produce a more complex result, presumably (in conventional language) due to the capacity of these species to participate in hydrogen bonding and coordination. The mechanism is probably related to the direct and specific interactions that the ions establish with the chiral solute, and to their different hydration properties, that perturb the solute–water interactions, and therefore lead to a modification of their conformation in solution.

7.2.2 Polymers

Next in the hierarchy are polymers. Conformational changes occur with changes in counterion. Ion exchange Cl^- to OAc^- with chitosan, the basic ingredient of crustacian shells, unravels the polymer, which is then solubilized. The standard electrostatic, Manning theory – ion condensation – to characterize this conformational change does not encompass specific ion effects. Inclusion of dispersion forces does [117]. See also Chapter 12.

Structural features (helix-to-coil transitions) of natural polymers such as carrageenans and agarose [131,132] and the swelling properties of polymers (e.g. poly (styrene sulfonic acid)) strongly depend on electrolytes [133].

The same applies to polyelectrolytes such as DNA. It seems that conformational change induced by specific ion adsorption associated with drugs or the local physicochemical environment must affect gene expression.

7.2.3 Proteins

The Hofmeister effects in precipitation of proteins were revisited and their reversal above and below the isoelectric point or pK_as[9] was studied in careful experiments by Tardieu and colleagues recently [115,134].

This work and its analysis is typical of a continuing recurring debate: the inclusion of specific ion adsorption due to dispersion forces appears to explain the effects quantitatively. However, while this and a large series of other papers associated with Boström and Ninham and colleagues seem to establish at least a first-order proof-of-concept in accounting for Hofmeister effects, they can be criticized. This is because the analyses were based on a *continuum solvent* approximation, do not include hydration and are based on a set of data for polarizabilities and ionic radii that are too crude.

We have reached an impasse. On the one hand:

1 The Collins approach takes electrostatics at the level of extended Debye–Hückel theory, and hydration as deduced from heats of solution, with an additional postulate of Gurney potential – phenomenological rules for interpenetration and compatibility of hydrated ions. The specificity is taken as given experimentally, from a particular set of data on the thermodynamics of electrolyte solutions.

2 The inclusion of dispersion forces approach is more fundamental. In principle it includes the first approach. It attempts to predict the specificity, but uses a primitive model. It attempts to handle hydration through extensions of the Born theory, which ignores solvent structure at a molecular level. Our problem is, by obtaining better data on ionic polarizabilities, to effect a reconciliation.

7.2.4 Rhodopsin and cytochrome c

Much the same Hofmeister effects have been explored in the elegant work of Vogel [135] and of Boström et al. [136] on rhodopsin.

Others have worked on effects on cytochrome c [119,137], and many other proteins and their precipitation [138]. See particularly the work of Cremer [139].

[9] In the electrostatic Poisson–Boltzmann description, the pH at the surface of a protein is set by a competition for the model charged protein interface of counterions and hydronium ions. This regulates the degree of association of the active ionizable group of particular pK_a and hence the surface charge, as in the discussion of Chapter 3. (Proteins normally have just two ionizable groups, carboxylates (pK_a around 4.7 for the equilibrium $R\text{-}COOH \rightleftharpoons R\text{-}COO^- + H^+$) and ammonium ($pK_a$ 4 for the equilibrium $R\text{-}NH_3{}^+ \rightleftharpoons R\text{-}NH_2 + H^+$). This cannot explain the effect).

7.2.5 Enzyme action

(A) Restriction enzyme

The Hofmeister phenomena associated with restriction enzyme activity and the energetic mechanisms that drive the catalysis were discussed in detail in Section 6.4.

This shows up more dramatically than any other example the limitations and irrelevance of classical ideas on electrolytes, electrostatic interactions, hydration, buffers and pH. The mechanism seems to involve a cooperative mini phase transition, with 'hydrophobic' interactions coupled to the interplay between dissolved gas and electrolyte, and cavitation and specific free radical production known from sonochemistry [14,38,40,41,96].

(B) Horseradish peroxidase

This mechanism of restriction enzyme activity is not universal. Horseradish peroxidase, perhaps the most investigated of all enzymes, shows Hofmeister effects that have much potential for exploitation that has not yet been realized. The mechanism appears to involve surface-mediated catalysis between hydration shells of adsorbed reacting species [96].

(C) Non-aqueous media

Enzymatic action in non-aqueous media shows up specific ion effects, but it is probable that these are due to trace amounts of virtually unremoveable water sequestered in active sites. The mechanism is probably much as for the restriction enzyme problem [41].

7.2.6 Bacteria

Hofmeister phenomena emerge also with living organisms. In fact *Staphylococcus aureus* and *Pseudomonas aeruginosa* strains were found to be sensitive to the presence of different electrolytes in the growth medium (see Fig. 7.9) [43], and the anion efficiency in affecting the bacterial growth rate is:

S. aureus: Cl^-, CH_3COO^- > succinate > NO_3^- > $HCOO^-$ > Br^- > ClO_3^- > citrate, maleate > BrO_3^- > SCN^-, SO_4^{2-} > I^- > (H_2O) > SeO_4^{2-} > $H_2PO_4^-$ > F^- > BF_4^-, CNO^-, CO_3^{2-}, PO_4^{3-}, IO_3^-.[10]

P. aeruginosa: (H_2O) > Cl^- > NO_3^- > OAc^- > Br^- > I^-, SCN^-.

The results seem to indicate that the presence of salts at physiological concentration (0.9 M) greatly affects the growth rates of these two strains, presumably because of the ion-specific dependence of a rate-limiting enzyme and/or specific

[10] Succinate: $^-OOC-CH_2-CH_2-COO^-$; citrate: $HO-C(CH_2-COO^-)_2-COO^-$; maleate: $^-OOC-CH=CH-COO^-$(cis).

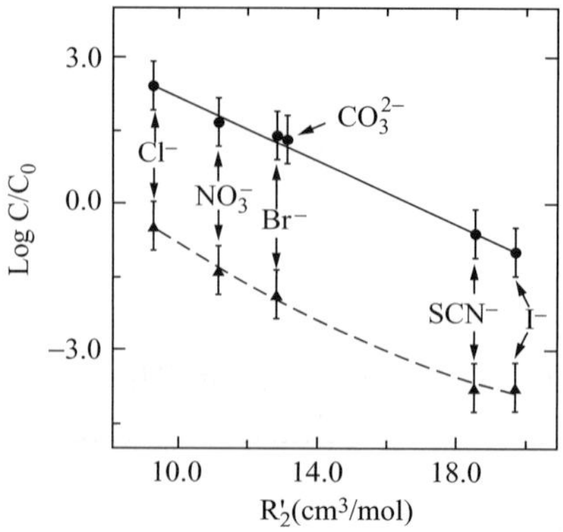

Fig. 7.9. Growth rate (expressed as log of the salt/water counts) as a function of the molar refractivity R_2' of the anion. Data taken from Ref. 43 with permission. Copyright 2005, IOP Publ. Ltd.

interactions of ions with the bacterial outer surface. The potential for exploitation of such effects seems obvious.

7.2.7. Wool and leather

At an even higher level of organization, there are very large specific effects on retention of water on fibres such as wool. Natural wool fibres absorb large amounts of water, about 30% w/w in normal temperature and relative humidity conditions. This has relevant industrial textile applications. The presence of different electrolytes on water absorption by wool has been investigated [45]. It was found that the direct Hofmeister series for anions and cations apply: F^-, $HCOO^-$ > WO_4^{2-} > SO_4^{2-} > VO_3^- > Cl^- > SiF_6^{2-} > $H_2PO_4^-$, IO_3^- > NO_3^- > SeO_4^{2-} > ClO_3^- > Br^- > NO_2^- > OAc^- > BrO_3^- > I^- > ClO_4^-, SCN^- > N_3^- (sodium salts), and Al^{3+} > Cr^{3+} > Mg^{2+} > Ca^{2+} > K^+, NH_4^+ > Rb^+, Na^+, Sr^{2+} > Cs^+ > Ba^{2+} (for chlorides). But the reverse sequence is found for NH_4^+ > K^+ > Na^+ >Ba^{2+}, Ca^{2+} > Mg^{2+}, La^{3+} (nitrates).

Similar results were found for leather, where the effects and their reversal have been exploited [126].

7.2.8 Hofmeister effects in medicine

There are many examples in medicine and biology that reflect specific ion effects. These are explicit in the examples of restriction enzymes and growth of bacteria. The effects are ultimately due to Hofmeister effects, i.e. to physical chemistry. Specific binding to enzymes affects conformation and folding, and at secondary level hydrophobicity provides the stage for the source of energetics of key rate-limiting steps.

If we limit our brief to simple electrolytes, perhaps the most familiar and most effective Hofmeister effect in medicine is that associated with the use of lithium acetate or chloride in the treatment of manic depression [140]. This was discovered in Western Australia around 1940. It replaced the then in vogue alternative of lobotomy, for which a Nobel prize in medicine was awarded! At 1 gram per day lithium chloride is a very effective treatment. The amount seems quite small to induce such a huge change. No biochemistry seems to be involved. If we consider that an average human has 5 litres of blood, mainly sodium chloride at 0.15 M concentration, or about 0.75 moles of sodium, a gram of LiCl is about 0.024 moles of lithium, a ratio of 30 sodium ions to 1 lithium ion, not so dramatically large a ratio as seems at first apparent. Specific ion exchange of sodium for lithium at the surface of an axon evidently affects the binding and release of molecules such as dopamine and serotonin that control mood.

NO_2^- ions have a vasodilatation effect [141].

Cataracts and neurodegenerative diseases are related to phase separation and aggregation of peptide macromolecules [142], and the presence of salts produces significant shifts in the thermodynamics that regulates these processes [143].

The absence of trace amounts of iodine in salt that is specifically bound as iodate (IO_3^-) in the thyroid gland leads to cretinism [144], especially endemic in Tibet, where it affected a huge proportion of the population. It has been cured by addition of small amounts of sodium iodide, a 40-year project self-funded by a lone volunteer working over 40 years. It is of course routinely added to table salt in Western countries for the same reasons. No explanation of such effects is possible within the confines of classical physical chemistry that excludes specific ion effects.

Western diets of bread, butter, canned food and table salt are made using pure sodium chloride. This is a small by-product of the industry that extracts salt from the ocean, the aluminium and hydrochloric acid industries. In the process all multivalent salts are precipitated out, to yield pure sodium chloride. Even without any additional salt a person on a normal Western diet takes in about 10 grams per day of sodium chloride. A standard human has about 50 grams of salt circulating in the blood. The blood, predominantly sodium chloride, contains also potassium, calcium, magnesium, zinc, copper, selenium, vanadium, manganese and other trace elements in a mixture that is the same as the ocean electrolyte mix in the Permian ocean from which we emerged 200 million years ago. The result is that kidneys are placed under huge stress to restore the optimal physiological balance. This is due to the perturbation and imbalance produced by the extra sodium chloride. That is obvious if animals are fed with salt lacking these elements. Pets invariably die. People, as in parts of Sweden and New Zealand, and animals exposed to natural drinking water deficient in trace elements such as selenium suffer appalling diseases, remedied by restoration of the balance.

Japanese miso, an essential dietary supplement, used to be made exclusively with sea salt. Modern variants, many of which cut corners by using pure NaCl, negate the entire purpose of the supplement.

How kidneys work to restore the balance is a matter of much interest for desalination processes. Dugongs, known also as manatees, are mammals that live on an exclusive diet of sea grass. This contains salt, albeit balanced, but, at the concentration of the present ocean, at least twice as high as the Permian and blood concentration. The excess salt is successfully eliminated by unknown processes of physical chemistry, as it is by seagulls and salt-tolerant plants.

The same story on the manifestation of Hofmeister effects occurs in plants. The addition of trace amounts of vanadium, copper and zinc oxides to superphosphate fertilizer renders the growth of cereals possible in vast farming areas such as South Western Australia containing poor soils. The use of such agricultural methods coupled to nitrogen fixing by clover is ultimately destructive – growth of the preexisting natural multiple species is inhibited by ploughing, which destroys microflora and fungi that extract such elements by specific ion binding in organic ring compounds.

We discuss the problem of ion pumps in Chapter 12. The process whereby biological cells, e.g. red blood cells, maintain a concentration of potassium inside the cell much higher than outside, vice versa for sodium and similarly for chloride, is universally assigned to active biochemical ion pumps. This central postulate was arrived at by the failure of classical physical chemistry to account for the observations. But the theory depended on an electrostatics-only theory of 'Donnan equilibrium'.

The inclusion of dispersion forces and specific ion binding to cell proteins opens up the possibility of at least a contribution to pumps for purely physical chemistry mechanisms.

Finally, in this litany of biological examples of Hofmeister effects that are now coming into sight as part of the armoury of tools available to physical chemists, we mention the use of a mixture of 19 g of Na_2HPO_4, 7 g of NaH_2PO_4 made up to 118 mL of water as a virtually immediate most effective enema. This can hardly be a bulk effect, and may be due to the consequences of specific ion adsorption and pH changes at the surface of the controlling smooth muscle nerve cell surfaces.

The meaning of pH in a physiological mixture of colloids like the stomach remains completely obscure, as is explicit in the work of Evens and Niedz (cf. Chapter 6) [10].

7.3 Inorganic and other systems

In inorganic chemistry, the precipitation and formation of nanoparticles, their size and shape and perhaps crystal symmetry can be controlled at will with concentrated sugar solutions and by varying sugar isomer and/or temperature. An example is the synthesis of magnesium hydroxide nanoparticles [145].

So too the addition of background indifferent salts shows up Hofmeister effects in the nanoparticle structure, a matter that has scarcely been explored except in the microemulsions of Chapter 12.

Of course there are a vast number of papers dealing with specific ion effects in inorganic chemistry. For example, the adsorption of ions at Hg electrodes was investigated by Conway, who showed that the phenomenon follows a Hofmeister series [146].

Regarding environmental chemistry, the positive or negative adsorption of anions at the air–water interface plays a relevant role in determining the concentration of some electrolytes in the atmospheric aerosol over the oceans [147,148], the solubility of oxygen in salt solutions [149], in the adsorption of atmospheric gases at the air–water interface [150] and in the formation of clouds [151].

7.4 Towards a resolution by inclusion of dispersion forces

What follows is substantive new work that goes some way to resolving the major problem of specific ion effects which is a central issue of this book.

It is due substantially to Drew F. Parsons.

The discussion is necessarily technical in many parts. But the advances are presented in language understandable to the average twenty-first-century English-speaking high-school graduate who is familiar with mathematics. Or should be.

Essentially what emerges is that the proper inclusion of dispersion forces even within the primitive model of electrolytes does indeed capture the essentials of Hofmeister effects quantitatively. Further extension that includes 'hydration' consistently within the civilized model now becomes possible and is under way.

In this and preceding chapters, not to put too fine a point on it, we have a mess on our hands.

Within the continuum solvent approximation we know how to remedy the situation. At that first level we know what to do. We have shown that in colloid science the DLVO theory fell down in its fundamental ansatz that treats electrostatic double layer interactions separately from van der Waals or Lifshitz dispersion interactions. That statement is a *proof*, not a hypothesis [51]. The Lifshitz theory in principle handles all many-body dispersion interactions through its reliance on dielectric susceptibilities as a function of the electromagnetic frequency of the interacting media. But in so incorporating Lebedev's vision into the DLVO theory, with the treatment of electrostatic forces in a non-linear theory and the specific quantum forces in a linear theory, we have violated two fundamental physical principles, the Gibbs adsorption isotherm, the very definition of an interface, and the gauge condition on the electromagnetic field, i.e. charge continuity. Ion specificity is lost.

The many-body forces acting in ion–solvent, ion–ion and ion–interface interactions can be extracted from extensions of Lifshitz theory that include molecular size and shape. These forces can then be incorporated with electrostatic forces at the same non-linear level. Then specific ion effects make their appearance. Another problem is that of a definition of finite ion 'size'. These challenges are presented by the whole hierarchy of forces that concern us: self free energies, activity coefficient, interfacial interactions and colloidal particle interactions. We first of all have to remove these defects in theory. When this is done, we can then refine the theory further, if necessary, to include solvent structure *consistently*.

7.4.1 Ion polarizabilities and their frequency dependence

An inhibition to so extending the theory to include dispersion forces correctly has been an almost complete lack of data on ionic polarizabilities and their frequency dependence.

The modelling of activity coefficients, modifications to Born energies, surface adsorption and interactions has been done using two different approaches. The first, the 'bottom-up' molecular approach discussed in Chapter 2, uses somewhat arbitrary effective Lennard-Jones potentials for simulation studies, or estimated 'effective' potentials of mean force with water molecules in statistical mechanical theories [153–165].

The second which concerns us is the 'top-down' approach that builds in all many-body dispersion forces consistently from a knowledge of polarizabilities of ions and of dielectric susceptibilities of the *bulk* solvent as a function of frequency. It can then be refined and extended to include ion and surface specific solvation and hydration.

This first approximation, using experimental bulk solvent properties, is equivalent to the Laplace interface approximation to interfaces of Chapter 2. There is always a profile of solvent order at a real interface, and around an ion, imposed by ion–solvent interactions, so that eventually we expect to have to include ionic hydration.

Both top-down and bottom-up approaches have an additional complication to deal with. An interface, such as air–water or mica or silica–water, is always going to have some inhomogeneity, fuzziness, surface specific hydration at the molecular level that extends to a few molecular layers. Further, even with exact ionic polarizabilities for the ions there are always going to be some uncertainties in dielectric data even for an ideal macroscopic medium. Further, most media have anisotropic dielectric properties.

The first attempts to include dispersion forces used unreliable data and inconsistent ionic radii. These first steps towards the resolution took data from available refractive index information on ionic solutions and guessed single-frequency relaxation frequencies, and an arbitrary choice of ion sizes [1,3,91,116,117,120,121, 166–169].

Many papers established proof of concept, but quantitative prediction eluded them. Hofmeister series emerged, sometimes in the observed order, but sometimes not.

7.4.2 Consistent definitions of ion size

There are several points to consider in relation to the question of ion size. First, existing estimates based on ionic crystal dimensions or other criteria tend to treat an ion as a hard sphere. The ionic charge is then found strictly inside that sphere, and nowhere outside it. The real electron distribution of an ion is found in a spatially diffuse electron cloud, described by quantum mechanics. Furthermore models of the Lifshitz dispersion energy between ions which include finite ion size are based on a diffuse spatial spread for the ion. This means there is in principle a discrepancy between electrostatic properties calculated using a hard sphere ion radius and dispersion energies using a softer Gaussian ion radius.

The volume of a hard sphere with radius a_s is $4\pi a_s^3/3$, while the volume of a Gaussian sphere with soft Gaussian radius a_g is $\pi\sqrt{\pi}a_g^3$. A conversion could be made between the hard sphere radius and the soft Gaussian radius by conserving the ion volume and taking these two volumes to be equal. But the simple electrostatic Born energy calculated using the hard sphere model will differ from the equivalent value calculated by a Gaussian model by more than 10%, even after taking equivalent radii (conserving the ion volume) [6].

Another practical point in favour of a diffuse Gaussian description is that many important ions are nonspherical. Nitrate, formate and chlorate ions are planar, thiocyanate is linear. A Gaussian model may be naturally extended to the nonspherical case, whereas a 'hard ellipsoid' is a much less convenient generalization of the hard sphere [6].

More important than the details of the model used to describe the ion volume is the question of where that volume is obtained from in the first place. Ion radii may be estimated from crystal data, but, irrespective of any criticisms in interpreting these data, they are not available for every ion of interest. Instead we may formulate a consistent approach to the determination of ion size by turning to *ab initio* quantum

Table 7.2. *Ion sizes of spherical ions, determined from* ab initio *quantum chemistry.* a_g *(in Å) is the soft Gaussian radius and* V_g *(in* cm^3/mol*) is the equivalent ionic volume.* a_s *(in Å) is the equivalent hard sphere radius (values in brackets include the empty outer shell of metal cations).* a_{crys} *(in Å) refers to experimental crystalline radii.* Me_4N^+ *refers to tetramethylammonium. Reprinted from Ref. 6 with permission. Copyright 2009, American Chemical Society.*

Ion	a_g	V_g	a_s	a_{crys}
F^-	1.02	3.5	1.12	1.33
Cl^-	1.69	16.3	1.86	1.81
Br^-	1.97	25.6	2.16	1.96
I^-	2.12	32.0	2.33	2.20
Li^+	0.38	0.2	0.42 (0.95)	0.59
Na^+	0.61	0.8	0.67 (1.33)	0.99
K^+	0.96	3.0	1.06 (1.77)	1.37
Rb^+	1.12	4.7	1.23 (1.92)	1.52
Cs^+	1.47	10.7	1.62 (2.14)	1.67
Fr^+	1.44	10.1	1.59 (2.23)	1.80
Be^2+	0.28	0.1	0.31 (0.75)	0.27
Mg^2+	0.50	0.4	0.55 (1.16)	0.57
Ca^2+	0.80	1.7	0.88 (1.59)	1.00
Sr^2+	0.96	3.0	1.05 (1.77)	1.18
Ba^2+	1.21	5.9	1.33 (1.99)	1.35
Ra^2+	1.35	8.3	1.49 (2.08)	1.48
O^{2-}	1.95	24.7	2.14	1.21
ClO_4^-	2.17	34.4	2.39	
SO_4^{2-}	2.29	40.5	2.52	
PO_4^{3-}	2.40	46.2	2.64	
AsO_4^{3-}	2.44	48.7	2.68	
NH_4^+	1.09	4.4	1.20	
Me_4N^+	2.47	50/8	2.72	

chemistry. Quantum chemical calculations have the dual advantage of explicitly dealing with the diffuse electron cloud and being applicable, in principle, to any ion. A quantum calculation is able to generate an estimate of ion volumes by determining the volume bounded by a surface at which the electron density has dropped down to a small level close to zero. By mapping this boundary surface onto a Gaussian model, the soft Gaussian radius (and equivalent hard sphere radius) may be obtained. The results of this procedure are shown in Table 7.2. The equivalent hard sphere radii for halides and alkali metal cations are close to crystalline radii found previously in the literature, which means we can be confident in applying the procedure to other ions whose sizes are unknown.

7.4.3 The deployment of ab initio quantum mechanics

The polarizability of an ion is determined from the number of different electron transitions available to that ion, each with a different strength. Each electron transition has a distinct energy with a corresponding frequency, which means that the strength of the polarizability depends on the frequency of the radiation that the ion is interacting with. That is, the dynamic polarizability may be described as a sum over many modes, each representing a different electron transition:

$$\alpha(i\omega) = \alpha_0 \sum_i \frac{f_i}{1 - (\omega/\omega_i)^2} \tag{7.10}$$

The weights f_i must sum to unity. Although static polarizabilities α_0 may be found in chemical data tables for some ions, the modal frequencies ω_i and their weights f_i are not known and tabulated. Recent work gives this information (D. F. Parsons and B. W. Ninham, *Langmuir* (2009), in press).

An approximate model has been suggested using the single, most extreme 'electron transition', that is ionization. In this single-mode model the frequency ω_I is determined from the ionization energy of the ion, with the dynamic polarizability given as:

$$\alpha(i\omega) = \frac{\alpha_0}{1 - (\omega/\omega_i)^2} \tag{7.11}$$

Ab initio quantum chemical methods may be employed to get the full polarizability response of the ion, over all modes. We will compare the success, or otherwise, of the single-mode ionization model with *ab initio* results.

When applying quantum chemical results, it is important to bear in mind the level of numerical approximation used. Broadly speaking, quantum chemical methods may be placed into three categories. The first level of approximation is known as the Hartree–Fock (HF) method. This method starts with single-electron orbitals for each atom in the molecule or ion being calculated. These single electron orbitals are comparable with the s-, p-, d- orbitals of the lone hydrogen atom. The Hartree–Fock method multiplies the single-electron orbitals together in all possible permutations to form multi-electron molecular orbitals. Multi-electron orbitals are added together to form the final electron cloud of the particle. The computational business of the calculation lies in determining the weights accorded to each possible molecular orbital. The key point concerning the Hartree–Fock method is that the contribution of each single electron to the overall electron cloud is considered separately. Electron correlations refer to the notion that the probability of finding one electron at a given point in space will be diminished if a second electron is already likely to be found there. Electron correlation is neglected in the Hartree–Fock method.

Certains ions and molecules have highly delocalized electrons, where, instead of being fixed in a specific region of space around a specific atom, the electron is equally found spread across several atoms, for instance via π-bonds between atoms. This occurs in the conjugated double or triple bonds of benzene rings and polyunsaturated fatty chains. It also occurs in carboxyl groups, found in formate, acetate, citrate and phosphate or arsenate. The high delocalization of the electrons in these ions, with the electron cloud spread over many atoms, means electron correlation will be high. For this reason the Hartree–Fock quantum chemical method provides a poor description of these ions.

The second level of quantum chemical methods, then, accounts for electron correlation. They start from Hartree–Fock molecular orbitals, mix and combine them further, accounting for spatial overlap between orbitals. Various electron-correlation methods are available, providing ever larger orders of correction. The oldest are the Møller–Plesset methods, with second-order (MP2) and fourth-order (MP4) corrections being most popular. More recently, quantum chemists have favoured the coupled-cluster methods, of which there are multiple variants such as CCSD, CCSD(T) and QCISD (they differ in the order of excitation or order of perturbation used to mix the Hartree–Fock orbitals). These methods provide the highest accuracy known to quantum chemistry, but are relatively computationally expensive, restricting the size of ion (as determined by the total number of electrons) which may be calculated.

The third class of quantum chemical methods is density functional theory (DFT). Rather than building up from individual single-electron orbitals, DFT seeks to calculate the total electron density for the whole ion or molecule right from the start. Electron correlation is explicitly included in the calculation from the beginning via the given DFT method's 'functional'. The functional describes how overlap between molecular orbitals is to be treated, and is the chief property that distinguishes one DFT method from another. A range of different models for the functional are available; the one which provides best agreement with experimental static polarizabilities is the PBE0 (or PBE1PBE) functional [170–172], not to be confused with the PBE (or PBE0PBE) functional. Aside from explicitly incorporating electron correlation, the other advantage of DFT methods is that they are computationally faster than calculations using Hartree–Fock + higher-level electron correlation, while yielding results of comparable quality.

The results of an *ab initio* quantum calculation of the dynamic polarizability of the chloride ion are shown in Fig. 7.10. Electron correlation is provided via the CCSD method. The *ab initio* polarizability is compared against the single-mode ionization model. The most significant difference is that the single-mode ionization model drops to zero at a frequency far too early compared with the *ab initio* result. This is incredibly important since the higher-frequency behaviour in the visible-UV

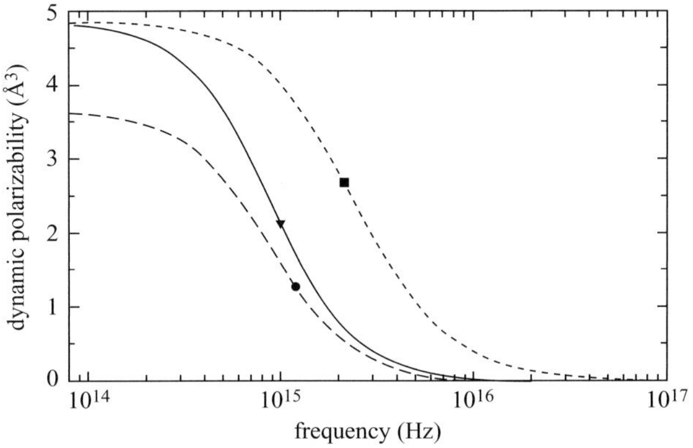

Fig. 7.10. Dynamic polarizability $\alpha(i\omega)$ of Cl^- comparing full *ab initio* data (■) against the single-mode ionization IP model. Two alternatives are given for the latter, taking the static polarizability α_0 from literature (•) and α_0 from the *ab initio* calculation (▼).

spectrum contributes the bulk of the dispersion energies. The dispersion self-energy of the ion, for instance, is, ignoring minor retardation effects due to the finite speed of light, given by

$$U_{\text{self–disp}} = \frac{4k_{\mathrm{B}}T}{\sqrt{\pi}\mathrm{a}^3} \sum_{n=0}^{\infty}{}' \frac{\alpha^*(i\omega_n)}{\varepsilon(i\omega_n)} \tag{7.12}$$

where $\omega_n = 2\pi n k_{\mathrm{B}}T/\hbar$. α^* refers to the so-called 'excess polarizability', a solvent modification of the ionic polarizability which will be explained later. The first non-zero frequency for $n = 1$ at room temperature, is $2.5{\cdot}10^{13}$ Hz, already in the infrared. The single-mode ionization model for the polarizability drops to zero by about $6{\cdot}10^{15}$ Hz, in the near UV spectrum. This corresponds to the 265th term in the dispersion energy sum. The *ab initio* polarizability however, does not drop to zero until around $4{\cdot}10^{16}$ Hz (8 nm), at $n = 1590$, pushing into the spectrum of soft X-rays. Hence a further 1300 or so non-zero terms contribute to the sum making up the dispersion energy. Consequently the strength of the actual dispersion forces will be much stronger than predicted by the single-mode ionization model of the polarizability. The high-frequency components are crucial! What this means is that for short-range interactions between ions the distinction usually made between bound and free electronic states, chemical bonds and physical interactions, disappears.

The importance of the high-frequency spectrum (up to soft X-rays) is relevant not only to ionic polarizabilities, but also to the medium in which they are found (water, and other materials in the case of surface interactions), that is, to the dielectric function $\varepsilon(i\omega)$. In Fig. 7.11 we compare the dielectric spectrum of a 1981

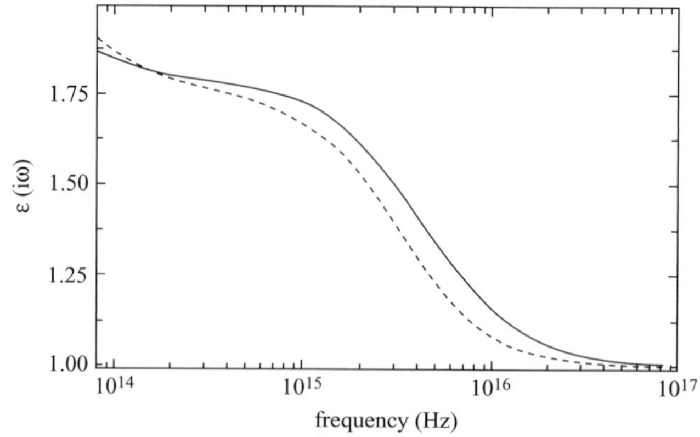

Fig. 7.11. Dielectric spectrum of water (in UV-vis range), comparing the Dagastine–Prieve–White model (full line) [173] to the Parsegian–Weiss model (dotted line) [174].

model of water by Parsegian and Weiss, derived from experimental optical spectra, to a more recent (2000) re-evaluation of the optical data by Dagastine, Prieve and White [173].

Just as the *ab initio* polarizability data extend out to soft X-rays, so too does the dielectric model of Dagastine–Prieve–White compared to the Parsegian–Weiss model [174].

This point is particularly poignant for the calculation of surface dispersion interactions, where the polarizability of the ion is modulated against the difference in the dielectric responses of the two media on either side of the interface. The strength of the ion–surface dispersion interaction may be characterized by a coefficient B:

$$B = -\frac{k_B T}{2} \sum_{n=0}^{\infty}{}' \frac{\alpha^* (i\omega_n) \Delta (i\omega_n)}{\varepsilon_w (i\omega_n)} \tag{7.13}$$

$$\Delta (i\omega_n) = \frac{\varepsilon_s (i\omega) - \varepsilon_w (i\omega)}{\varepsilon_s (i\omega) + \varepsilon_w (i\omega)} \tag{7.14}$$

The prime ($'$) next to the summation sign indicates that the zero-frequency component at $n = 0$ should be taken with a factor $1/2$. $\Delta(i\omega)$ refers to the difference between the dielectric spectra of the two media, $\varepsilon_w(i\omega)$ is the dielectric spectrum of the solvent, water, and $\varepsilon_s(i\omega)$ is the dielectric spectrum of the surface material (see Equation (7.14)). While water has a far larger dielectric constant at zero frequency, that zero-frequency component is only one point among several thousand. At the remaining high frequencies relevant to dispersion interactions, the dielectric response of water is comparable to that of other materials. It might be slightly higher or slightly lower, or indeed the two dielectric spectra may even cross several

Table 7.3. *Ion–surface dispersion coefficients B (10^{-50} J·m^3) at the oil–water interface. Oil is represented by the Parsegian–Weiss model for tetradecane (C_{14}). Two estimates of B values are shown, derived from the Parsegian–Weiss (PW) and Dagastine–Prieve–White (DPW) models for water.*

		B	
	Ion	PW model	DPW model
Nonhydrated ions:			
	F^-	−0.32	0.21
	Cl^-	−0.67	−0.15
	Br^-	−0.77	−0.41
	I^-	−1.10	−0.59
	$HCOO^-$	−0.67	0.05
	AcO^-	−0.94	0.11
	Li^+	−0.01	0.03
	Na^+	−0.04	0.13
	K^+	−0.20	0.30
	Rb^+	−0.32	0.35
	Cs^+	−0.48	0.29
	NH_4^+	−0.30	0.32
Hydrated ions:			
	Li^+	−0.96	0.54
	Na^+	−0.91	0.89

times. This means different components of the overall dispersion interaction may vary between repulsive and attractive, depending on the frequency.

Consider the dispersion interaction of ions at the oil–water interface. For the dielectric model of 'oil', we accept the Parsegian–Weiss model for tetradecane (C_{14}). Table 7.3 shows the surface *B* coefficients for a range of simple ions, comparing the Parsegian–Weiss model of water against the Dagastine–Prieve–White model. Ionic polarizabilities are taken from *ab initio* calculations with CCSD electron correlation. Negative *B* values indicate an attractive ion–surface interaction, positive values give a repulsive interaction. The difference is starkest for the cations, both alkali metals and ammonium. The Parsegian–White model predicts an attractive interaction between cation and oil surface while the Dagastine–Prieve–White model predicts a repulsive cation–surface interaction. This more refined model also predicts a distinct difference in behaviour between the bromide ion and the chloride or acetate ion, a point important to Hofmeister effects. As mentioned already, bromide induces very different behaviour in lipid bilayers from chloride or acetate, a point to which we shall return later.

Table 7.4. Ab initio *static polarizabilities (in Å^3), calculated without and with electron correlation (DFT/PBE0). The fourth column gives the proportion of the electron correlation correction as a percentage of the total polarizability.*

Ion	Uncorr.	Corr.	% corr.
Cl^-	4.22	4.82	12.4
Br^-	4.35	4.75	8.4
I^-	8.97	9.97	10.0
Na^+	0.13	0.14	9.1
K^+	0.80	0.83	3.8
H^+	0.88	0.93	5.2
OH^-	2.75	4.22	34.9
CO_3^{2-}	6.04	9.84	38.6
SO_4^{2-}	6.13	7.84	21.8
$HCOO^-$	3.94	5.02	21.6
AcO^-	5.61	6.87	18.4
PO_4^{3-}	9.89	18.18	45.6
AsO_4^{3-}	11.43	21.12	45.9

We close this section with a quantification of the significance of electron correlation on ionic polarizabilities. Table 7.4 shows the static polarizabilities of a range of ions calculated first without electron correlation (Hartree–Fock values), and then with electron correlation via DFT calculations under the PBE0 functional. For simple ions – alkali metal cations and halides – the correction due to electron correlation is relatively mild, less than 15%. For ions with delocalized electrons the correction due to electron correlation is phenomenal, comprising 46% of the total polarizability of phosphate and arsenate, 39% of carbonate and 20% of formate, acetate and sulfate. Electron correlation is also powerful in the hydroxide ion, contributing 34% of the total polarizability. Although only static polarizabilities are shown in Table 7.4, it is worth commenting that electron correlation also impacts the modal frequencies in the dynamic polarizability, shifting the frequency at which the polarizability drops to zero.

It is noteworthy that the greatest electron correlation, at 46%, occurs with phosphate (and its toxic cousin arsenate). This is not merely a consequence of the triple charge of the ion; citrate is also triply charged but has electron correlation commensurate with the singly charged carboxylates, formate and acetate. This has profound implications, given the central role that phosphate plays in biology, forming the backbone of nucleic acids, the energy transfer molecules ADP and ATP and the headgroups of phospholipids which form cell membranes. In biochemical analysis too, phosphate is frequently used as a buffer solution to maintain pH. The significance of the large effect of electron correlation in phosphate is that the dispersion interactions of the ion will be greater than we would otherwise anticipate. This is particularly important in biological systems where the overall salt concentration is sufficiently high (0.1 M) that electrostatic interactions are screened, to the point where dispersion interactions may compete equally.

The electron correlations associated with phosphate make its behaviour qualitatively different from other ions and radicals. In particular, when adsorbed on the surface of enzymes there must be a strong coupling of modes associated with phosphate and those of the enzyme which drives the catalysis of ADP–ATP reactions.

7.4.4 Excess polarizabilities

The polarizability of an ion refers to the strength of the induced dipole formed when an external electric field distorts the electron cloud of the ion. In a medium such as water, however, it is not only the ion which gets polarized by an external electric field; the medium itself is also polarized. The response of the medium is described by its dielectric susceptibility as a function of the entire electromagnetic spectrum.

This means that an ion in solution does not simply experience the full external field, instead it experiences a modulated electric field formed from the original external electric field, plus a 'depolarization field' due to the polarized medium. By treating an ion as a simple dielectric sphere with volume V_i and dielectric function $\varepsilon_\iota(i\omega)$, Landau, Pitaevskii and Lifshitz [175] deduce the resulting induced dipole, with the corresponding effective ionic polarizability being:

$$\alpha^*(i\omega) = \frac{3V_i}{4\pi} \frac{[\varepsilon_i(i\omega) - \varepsilon_{\rm m}(i\omega)]}{[\varepsilon_i(i\omega) + 2\varepsilon_{\rm m}(i\omega)]} \tag{7.15}$$

$\varepsilon_{\rm m}(i\omega)$ is the dielectric spectrum of the medium (e.g. water).

The dielectric function of the ion is derived from its polarizability (its intrinsic polarizability, that is its polarizability in vacuum). The most straightforward approach is to apply the simple model $\varepsilon = 1 + 4\pi n\alpha$, where n is the number density of the ion, that is $n = 1/V$. In other words,

$$\varepsilon_i(i\omega) = 1 + 4\pi \frac{\alpha_i(i\omega)}{V_i} \tag{7.16}$$

A more sophisticated model may be considered [9,176], based on the Clausius–Mossotti relation between polarizability and dielectric function. However, this model is highly sensitive to the ratio of ion polarizability to volume; if the ion volume has been slightly underestimated then a negative dielectric value for the

Table 7.5. *Effective (excess) static polarizabilities (in Å^3) of ions in water.*

	Static polarizability	
Ion	Intrinsic (in vacuum)	Effective (in water)
Li^+	0.03	−0.04
Na^+	0.14	−0.14
K^+	0.83	−0.55
Rb^+	1.38	−0.88
Cs^+	2.37	−2.01
NH_4^+	1.31	−0.81
H^+	0.93	−0.58
OH^-	4.21	−1.19
F^-	1.57	−0.63
Cl^-	4.82	−3.03
Br^-	4.75	−4.79
I^-	9.97	−5.95
$HCOO^-$	5.02	−3.44
AcO^-	6.87	−1.35
ClO_4^-	5.49	−6.53
PO_4^{3-}	18.18	−8.47
Citrate^{3-}	21.20	−17.31
CO_3^{2-}	9.84	−6.06
SO_4^{2-}	7.84	−7.66

ion will be generated, which is physically impossible (violating causality). The simple formula offered above is more robust to measurement errors in the ion volume.

The effective polarizability $\alpha^*(i\omega)$ of the ion in solution has come to be known as 'excess polarizability' since, because of the $\varepsilon_i - \varepsilon_m$ term, it describes the polarization response of the ion after subtracting the polarization response of the medium. The term 'excess polarizability' is in a sense misleading, since the difference in the actual polarizabilities of ion and solvent is not taken directly. The term 'effective polarizability' seems more descriptive.

For reference, the effective (excess) static polarizabilities of a range of ions in water are given in Table 7.5. Because of the very large dielectric constant of water at zero frequency, the ionic effective polarizabilities are in general negative, meaning that the induced dipole of an ion in solution due to a constant external electric field will point in the opposite direction to that field.

We should remember, however, that the zero-frequency term is not particularly significant in the determination of dispersion energies. Contributions at UV-vis frequencies are more important. The effective polarizability spectrum of chloride

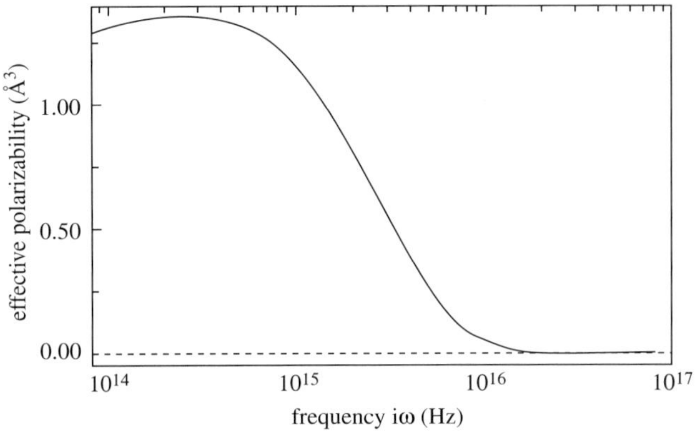

Fig. 7.12. Effective (excess) polarizability of a chloride ion in water, as a function of frequency. The dielectric spectrum of water is taken as that given by the Dagastine–Prieve–White model.

in water is shown in Fig. 7.12. The effective polarizability is positive at most non-zero frequencies for this ion, but becomes negative again (albeit with small magnitude) at frequencies above $2{\cdot}10^{16}$ Hz. In principle, depending on the details of the ion's intrinsic polarizability relative to the dielectric spectrum of water, the effective polarizability could cross between positive and negative several times as a function of frequency.

7.4.5 Born energies revisited: induction forces

Having established a consistent definition of ion size, for both the Born electrostatic (Chapter 3) and dispersion self energies (Chapter 4), and having evaluated frequency-dependent polarizabilities of ions we can now proceed to see how inclusion of dispersion and ion polarization effects affects matters.

In making the comparison we remark that the standard Born energy used to estimate literature values of free energies of transfer of an ion and partition coefficients of ions from one medium to another uses the continuum solvent approximation [177].

To be consistent we use the same approximation for dispersion effects (Table 7.6).

Recall the Born energy result from Chapter 3:

$$U_B = \frac{Q^2}{2\varepsilon_r a_g} \tag{7.17}$$

where a_g is the Gaussian radius of the ion (for an ion with a diffuse spatially spread charge), Q is the charge of the ion and ε_r is the dielectric constant of the medium ($\varepsilon_r = 78.3$ for water).

Table 7.6. *Born and dispersion contributions to self free energies (in kJ/mol) of typical spherical ions in oil and in water. Note that the specificity comes in most part from the dispersion component, which can be even greater than the electrostatic part! (D. F. Parsons and B. W. Ninham, Langmuir, in press.)*

	Dispersion self			Electrostatic self			Total self			Dispersion solvation		Electrostatic solvation		Total solvation	
Ion	In water	In oil	In vacuum	In water	In oil	In vacuum	In water	In oil	In vacuum	In water	In oil	In water	In oil	In water	In oil
F^-	515	621	1461	3.44	132	270	518	753	1731	−946	−840	−266	−138	−1213	−978
Cl^-	80	169	615	4.18	160	327	84	329	942	−535	−446	−323	−167	−858	−613
Br^-	26	113	498	3.59	138	282	29	251	780	−472	−385	−278	−144	−750	−529
I^-	35	121	530	3.33	128	261	38	249	791	−495	−408	−258	−133	−753	−541
$HCOO^-$	73	166	560	2.36	90	185	76	256	744	−486	−394	−182	−94.3	−668	−488
AcO^-	74	167	610	2.19	84	172	76	251	7812	−537	−444	−169	−87.5	−706	−531
Li^+	1060	1186	1830	4.97	191	390	1061	1377	2220	−774	−645	−385	−199	−1159	−844
Na^+	1118	1250	1973	4.30	165	337	1123	1415	2311	−855	−724	−333	−172	−1188	−896
K^+	580	695	1395	5.01	192	392	585	887	1788	−815	−701	−387	−200	−1202	−901
Rb^+	460	570	1237	4.61	177	361	465	747	1598	−776	−667	−357	−184	−1133	−852
Cs^+	204	301	783	4.14	159	324	208	460	1107	−579	−482	−320	−165	−899	−647
NH_4^+	466	575	1248	6.47	248	507	473	823	1755	−782	−673	−500	−259	−1282	−932

The corresponding dispersion self energy at finite temperature T is a sum over frequencies $\omega_n = 2\pi n k_B T/\hbar$:

$$U_{\text{self-disp}} = \frac{4k_B T}{\sqrt{\pi} a^3} \sum_{n=0}{}' \frac{\alpha^*(i\omega_n)}{\varepsilon(i\omega_n)} \tag{7.18}$$

We ignore the finite velocity of light here. We choose for water the representation of Dagastine, Prieve and White. In the case of self-energies the alternative data of Parsegian and Weiss give results which are not too different. For oil we use the Parsegian–Weiss model for tetradecane [174].

At finite temperature the integral becomes a sum over frequencies ($\omega_n = 2\pi n k_B T/\hbar$).

We ignore the finite velocity of light here.

Note that the results are importantly quite sensitive to correct polarizabilities. The first such estimates with crude polarizability data are significantly different [4].

7.4.6 *Dispersion and induction forces*

The polarizabilities of each ion lead to two kinds of ion–ion interactions: induction forces (permanent charge-induced dipole interaction), where the permanent electrostatic field due to the charge of one ion induces a dipole in the other, polarizable, ion; the second kind is the dispersion interaction (induced dipole–induced dipole) due to the polarizabilities of both ions at once.

Since the induction force is electrostatically derived, it depends only on the static polarizability of the ion. The interaction between a single ion 1 with charge q_1 (and electrostatic potential $\psi_1(r)$) and ion 2 with static excess polarizability $\alpha_2{}^*$ is

$$U^{\text{ind}}_{1\text{on}2}(r) = -\frac{q_1 \alpha_2^* \psi_i'(r)}{8\pi\varepsilon_0\varepsilon_r r^2} \tag{7.19}$$

(Note the use of excess polarizabilities of the ions in solution, rather than intrinsic polarizabilities.) Just as the electrostatic field of charge 1 induces a dipole in charge 2, so too does charge 2 induce a dipole in charge 1. The total induction force between the two ions will be the sum:

$$U_{\text{ind}}(r) = U^{\text{ind}}_{1\text{on}2}(r) + U^{\text{ind}}_{2\text{on}1}(r) \tag{7.20}$$

In the limit of infinite dilution where the surrounding electrolyte provides no screening, the electrostatic potential is Coulomb, $\psi_i(r) = q_i/4\pi\varepsilon_0\varepsilon_r r$, and the total induction is:

$$U_{\text{ind}}(r) = -\frac{1}{8\pi\varepsilon_0\varepsilon_r r^4}\left(q_1^2\alpha_2^* + q_2^2\alpha_1^*\right) \tag{7.21}$$

In an electrolyte with finite concentration, expressed via the inverse Debye length κ_D, the electrostatic potential is screened, $\psi_i(r) = (q_i/4\pi\varepsilon_0\varepsilon_r r)\exp(-\kappa_D r)$ (within the linear Debye–Hückel approximation), decaying in line with the Debye length. Consequently the induction force is also screened:

$$U_{\text{ind}}(r) = -\frac{e^{-\kappa r}(1+\kappa r)}{8\pi\varepsilon_0\varepsilon r^4}\left(q_1^2\alpha_2^* + q_2^2\alpha_1^*\right) \tag{7.22}$$

The dispersion interaction between the ions (with Gaussian radii a_1 and a_2 respectively) is:

$$U_{\text{disp}}(r) = -\frac{C(r)}{r^6} \tag{7.23}$$

where:

$$C(r) = 6k_B T F(r) \sum{}' \frac{\alpha_1^*(i\omega_n)\alpha_2^*(i\omega_n)}{\varepsilon_r(i\omega_n)} \tag{7.24}$$

with, neglecting the finite speed of light:

$$F(r) = \frac{1}{3}(P_1 P_2 + 2Q_1 Q_2) \tag{7.25}$$

$$P_i = erf\left(\frac{R}{a_i}\right) - \frac{2}{\sqrt{\pi}}\frac{R}{a_i}\exp\left(-\frac{R^2}{a_i^2}\right) \tag{7.26}$$

$$Q_i = erf\left(\frac{R}{a_i}\right) - \frac{2}{\sqrt{\pi}}\frac{R}{a_i}\left(1+\frac{R^2}{a_i^2}\right)\exp\left(-\frac{R^2}{a_i^2}\right) \tag{7.27}$$

At long distances where $r > a_i$, $F(r) \approx 1$ and $C(r)$ becomes constant.

The ion–ion dispersion energy at contact (where $r = a$) will show a Hofmeister series and may be relevant, for instance, to interpretations of ion binding at surfaces. Contact energies are tabulated in Table 7.7, for interactions between two ions of the same type in aqueous solution, at a separation equal to the Gaussian radius of that ion. The Hofmeister series obtained is $F^- < Cl^- < I^- < Br^- < HCOO^- < OAc^-$ and $Na^+ < Li^+ < K^+ < Rb^+ = NH_4^+ < Cs^+$. Contact energies (that is, energies of adsorption) may differ by factors of 4–10. In general, the smaller magnitude of the radius of small ions means they attain higher contact energies. This does not account for the tightly held hydration later of small, kosmotropic ions (F^-, OAc^-, Li^+, Na^+) which is addressed in the following section in the context of bulk activity coefficients.

The formulae used above come from the Mahanty–Ninham formalism that applies to the self polarization of an ion treated as an electrically neutral species [7,178–180].

Table 7.7. *Contact dispersion energies between ions in aqueous solution at a separation of one radius.*

Ion	Contact energy (kJ/mol)
F^-	−10.7
Cl^-	−1.24
Br^-	−0.59
I^-	−0.83
$HCOO^-$	−0.45
OAc^-	−0.47
Li^+	−21.1
Na^+	−23.7
K^+	−11.0
Rb^+	−8.27
Cs^+	−2.36
NH_4^+	−8.22

It can be extended as in Chapter 7 of Ref. 7, to include the effects of charge-induced dipole polarization fluctuation contributions. These temperature- and frequency-dependent contributions are also ion-specific. (They are significant as remarked and shown in the chapter on self assembly of ionic micelles, where they can be shown to contribute about 10% of the hydrophobic free energy of transfer of a surfactant molecule to a micelle.)

The inclusion of ion-induced dipole effects is the subject of an active debate in the literature at the present time. They occurred earlier in the 1950s under the title Kirkwood–Shumaker fluctuation forces. Since they are screened by the Debye length in electrolytes they are generally not so important compared with high-frequency dispersion forces proper. The important point is that they are included in the same general formalism that deals with ion and polarization dispersion fluctuation effects (see Chapter 7 in Ref. 7, and Refs. 181 and 182).

7.4.7 Applications to activities and inclusion of water structure

In their simplest form, the self energies described above refer to the energy of a single ion in solution, that is at the limit of infinite dilution. In a real solution, interactions between ions provide additional effects, for instance adding electrolytic corrections to the self energies, which depend on the concentration of the electrolyte.

Interactions between ions will shift the effective concentration of the electrolyte, thereby shifting the osmotic pressure of the solution, or shifting the equilibrium position, for instance the equilibrium position between hydrogen ions and phosphate, hydrogen phosphate and dihydrogen phosphate. In the latter example it is apparent that dispersion interactions between ions will have an impact on the pH of the solution.

The 'effective concentration' of ion i is known as the 'activity' a_i of the solution, and is related to the actual concentration via an ion activity coefficient γ_i, $a_i = \gamma_i c_i$.

When interactions between ions are attractive, the effective concentration (the activity) is less than the actual concentration. Each ion occupies a smaller volume within the solution than it would if there were no interactions between ions so the amount of available solvent volume is effectively greater, and therefore the effective salt concentration is less. The activity coefficient is then less than unity. Conversely when interactions are repulsive, the activity is greater than the actual concentration (each ion takes up a greater volume of space, so the available solvent volume is less) and the activity coefficient is greater than unity.

There are a variety of ionic interactions to consider. The most obvious are the electrostatic interactions due to the permanent charges of the ions. This case has been considered in Chapter 3, leading to the Debye–Hückel limiting law (first term in Equation (7.6)). The Debye–Hückel law contains no ion specificity, predicting the same activity coefficients for all electrolytes formed from the same charges.

The second ion–ion interaction is an inner core interaction due to the finite size of the ions. Real ions are not point charges, but have a high electron density at their core which prevents one ion from coalescing onto another, centre onto centre, even if the overall charge of one ion has the opposite sign to the other. As the two ions press close to one another, the overlap of their electron clouds drives a repulsive interaction, acting to push them apart. The simplest model for this phenomenon is to treat each ion as a hard sphere (we might take the equivalent hard sphere radius derived from a Gaussian spatial spread of the ion, Section 7.4.2). We obtain a 'distance of closest approach', equal to the sum of the two hard sphere radii. The distance of closest approach becomes increasingly important as the electrolyte concentration increases, providing a limit to how closely the ions in a highly concentrated solution may pack together. The repulsive core interaction increases activities, and corresponds to the second term in Equation (7.6).

In dilute solution the induction interaction decays with r^{-4}, whereas the dispersion interaction decays more quickly, going approximately as r^{-6}. Therefore at low concentrations we may expect induction forces to provide a greater correction than dispersion interactions. Due to screening of the induction force at higher concentrations, however, dispersion forces (and inner core repulsion) will dominate.

The various interaction energies are combined to give an average interaction energy E^*, which is the ion–ion interaction energy averaged over the possible positions of each ion. The total interaction energy between two individual ions is

$$u_{ij}(r) = u_{ij}^{\text{el}}(r) + u_{ij}^{\text{disp}}(r) + u_{ij}^{\text{ind}}(r) + u_{ij}^{\text{core}}(r) \tag{7.28}$$

$u_{ij}^{\text{el}}(r)$ is the electrostatic interaction, $u_{ij}^{\text{disp}}(r)$ the dispersion interaction, and $u_{ij}^{\text{ind}}(r)$ is the induction interaction; $u_{ij}^{\text{core}}(r)$ refers to core repulsion. The probability of finding ion j at a distance r from ion i is given by the Boltzmann distribution:

$$g_{ij}(r) = \exp\left[-\frac{u_{ij}^{\text{el}}(r) + u_{ij}^{\text{disp}}(r) + u_{ij}^{\text{ind}}(r) + u_{ij}^{\text{core}}(r)}{k_{\text{B}}T}\right] \tag{7.29}$$

The average energy per mole of electrolyte is then [183]:

$$E = \frac{c}{2}\sum_{ij} \nu_i \nu_j \int u_{ij}(r)\, g_{ij}(r)\, d^3r \tag{7.30}$$

where c is the concentration of the electrolyte and ν_i are the stoichiometric coefficients of each ion in the electrolyte (e.g. $\nu_i = 1$ in a 1:1 electrolyte). The general calculation of the activity coefficient involves an integration of the average interaction energy over temperature. This process may be simplified by using the relationship between temperature and concentration expressed in the Debye length, which is strictly valid only for dilute electrolytes, but leads to the simple expression:

$$\ln \gamma_{\pm} = \frac{E}{\nu k_{\text{B}}T} \tag{7.31}$$

where $\nu = \sum_i \nu_i$.

In this way, with the various ion–ion interaction energies, we may calculate activity coefficients as a function of salt concentration.

Once ion specificity is introduced via intrinsic ion size and ion polarizabilities (in both dispersion and induction forces), a Hofmeister series appears. For example in the potassium halides (and acetate), the anion series is calculated as $F^- < Cl^- < Br^- < I^- < OAc^-$. This follows elements of the experimental series, which goes $Cl^- < Br^- < I^- < F^- < OAc^-$. That is chloride, bromide and iodide are correctly ordered, but the calculated position for F is incorrect.

The fluoride ion differs from the halides by being ‘kosmotropic’, where the others are ‘chaotropic’. At this point we recall the Collins concept explaining these qualities [12,184]. ‘Kosmotropic’ ions have a tightly ‘bound’ hydration shell, while the hydration shell of chaotropes is only loosely ‘bound’. The degree of binding is itself driven by dispersion and electrostatic forces together. We apply this idea by saying that the tightly bound hydration shell of kosmotropes affects the distance of closest approach between these ions. The layer of water effectively increases the

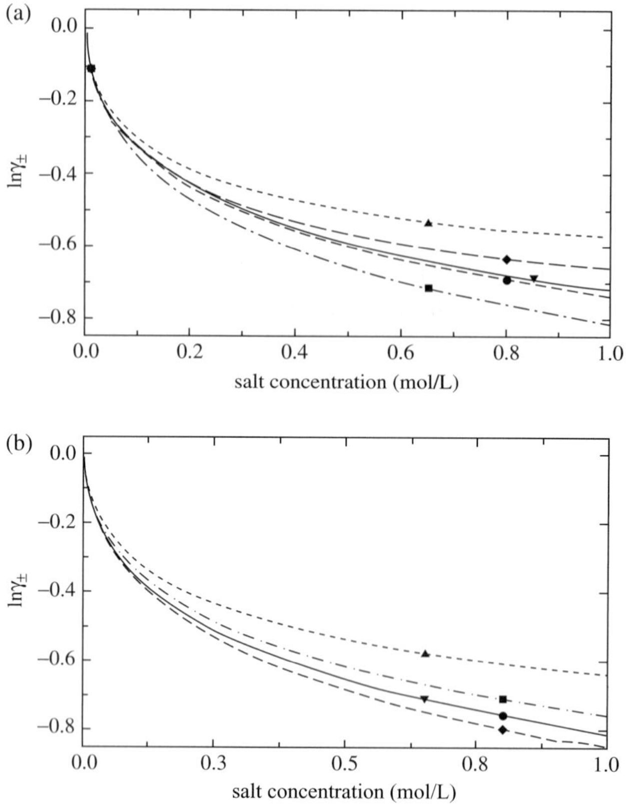

Fig. 7.13. Theoretical activity coefficients $\ln\gamma_\pm$ of: (a) potassium halides and acetate, KF (♦), KCl (■), KBr (•), KI (▼) and KOAc (▲); (b) alkali chlorides, LiCl (▲), NaCl (■), KCl (•), RbCl (▼) and CsCl (♦).

inner core size of the kosmotropic ion. This idea is modelled most simply by adding the radius of a water molecule to the hard sphere radius of a kosmotropic ion (or adding two water radii to the smallest kosmotropes Li^+ and F^-), to determine the distance of closest approach.

With this adjustment to the size of kosmotropic fluoride and acetate, the activity coefficients, shown in Fig. 7.13, then do appear in the correct experimental order

$$Cl^- < Br^- < I^- < F^- < AcO^-$$

In the alkali metal series, lithium and sodium are kosmotropic while rubidium and caesium are chaotropic. We use outer ionic radii for metal ions Li^+–Rb^+, which includes an empty outer electron orbital (ns°). For caesium, the inner ionic radius is used. The presence of f-orbitals make Cs^+ different (the notion of empty outer electron orbital does not apply). The activity coefficients for the alkali chlorides given these assumptions are shown in Fig. 7.13. With the tightly bound water layer accounted for in Li^+ (double water layer), Na^+ (single water layer), the series goes $Cs^+ < Rb^+ = K^+ < Na^+ < Li^+$, in line with experiment. Without the kosmotropic water layer, the incorrect series $Li^+ < Na^+ < Cs^+ < Rb^+ = K^+$ would be obtained.

We see that ion specificity in the activity coefficients is driven largely by ion size, with hydration contributing to the ion size of kosmotropic ions. The same phenomenon has been seen by Vrbka and Parsons with regard to osmotic coefficients, related to activity coefficients (L. Vrbka and D. F. Parsons, work in progress). Using these ion sizes to define distances of closest approach, the dispersion component (including induction) is relatively small, 5–10% of the total. It is not, however, negligible. For instance if electrostatic interactions alone are used, without dispersion, then Rb^+ appears in the wrong order, $Cs^+ < K^+ = Na^+ < Rb^+$ instead of $Cs^+ < Rb^+ = K^+ < Na^+$.

However, as for the large dispersion contribution to (Born) self energies of ions, the dispersion interactions have a large effect in determining effective hydrated ion size.

It is worth commenting further on the nature of the ionic radii used to determine distances of closest approach. The underlying reason why a distance of closest approach needs to be applied is due to quantum mechanics. When a cation approaches an anion, both electrostatic and dispersion interactions are attractive. As the two ions come into contact, their electron clouds start to overlap, which generates a repulsive force. This core repulsion prevents the two ions from coalescing on top of each other, and is the underlying mechanism behind the concept of 'distance of closest approach'. Neither electrostatic nor dispersion interaction models describe this kind of core repulsion; we have treated it with a simple hard core (hard sphere) model, saying that one ion may not approach closer than a given fixed distance from the other ion.

We derived the hard sphere radii for core repulsion from the Gaussian radii used to calculate dispersion interactions. This choice may be evaluated with the help of quantum mechanical calculations of the energy of two ions as a function of the distance between them. The core repulsion may be determined by subtracting the electrostatic, dispersion and induction interactions. A sample calculation, for instance for Cl^-–Cl^-, indicates the distances of closest approach used above correspond to a hard core repulsion of around $4k_BT$. In real systems it is not too difficult for thermal fluctuations to achieve this energy, causing ions to compress closer than the 'distance of closest approach' which we have used. Even $10k_BT$ events may routinely observed experimentally. A core repulsive barrier of $100k_BT$ could serve as the criterion for establishing the radius for hard core repulsion. In the Cl^-–Cl^- example, repulsion of $100k_BT$ is found at 0.6 d_0, where d_0 is our original distance of closest approach.

The interesting point about reducing the distance of closest approach in this manner is that it shows how the short-range dispersion interactions play a larger role in determining ion specificity than the simpler argument above suggests.

In fact the choice is diabolical. We can apparently ignore dispersion interactions essentially by an appropriate adjustment of radii. Then apparently we could get away with Debye–Hückel theory as in the classical theory of pH and double layer, with some lip service to 'hydration'. But the bargain is Faustian. Hydration so introduced varies from one experimental quantity to another and with temperature.

At the same time developing a more rigorous model of the ion suggests we have also to develop a more sophisticated model of the solvent. In the current model, hydration is handled simply by its effect on the distance of closest approach for kosmotropic ions. A more complete model would include the polarizability of the entire hydrated cluster, including polarization of the water molecules in the hydration layer. The ion will experience a localized water environment, which may alter its effective or excess polarizability. In this way, a more detailed description of the hydration layer is expected to have strong implications for ion specificity. We describe a further solvent effect, that of solvent–solvent correlations, in the next section.

7.4.8 Solvent structure: solvent–solvent correlations

In the previous section, the effect of ion–ion interactions on the overall effective concentration of an electrolyte solution was explored, leading to Hofmeister series differentiating the effect of an electrolyte depending on the specific ions involved.

We saw also that the correct series could not be obtained unless the hydration shell of kosmotropic ions was included. In other words, ion–solvent interactions are important. These were modelled in a simple manner, by changing the distance of closest approach between ions. More sophisticated ion–solvent interactions may be considered, to account for changes in the dielectric response of the hydration layer (or, equivalently, changes in the total polarizability of the hydrated ion).

At the same time, a question not yet raised is the question of solvent–solvent interactions, or solvent–solvent spatial correlations. In classical electrolyte theories the solvent is modelled as a structureless medium (or continuum) characterized solely via its dielectric constant ε (or dielectric susceptibility $\varepsilon(i\omega)$).

In fact the solvent does have molecular structure, or spatial structure at a molecular scale. The molecular nature may be modelled in computer simulations including explicit water molecules, but this approach is computationally costly. This limits the size of the system which may be studied (and consequently constrains the range of electrolyte concentrations which may be simulated).

The simple continuum model may be retained, but expanded to include spatial structure and solvent–solvent spatial correlation by replacing the dielectric constant with a spatially dependent dielectric function $\varepsilon(r, r')$. The dielectric constant normally determines the polarization of the medium (or more correctly, the electric displacement D) due to an electric field, $D = \varepsilon E$. Solvent spatial correlations are included by replacing the constant ε with:

$$D(r) = \int \varepsilon\left(r, r'\right) E\left(r'\right) d^3 r' \tag{7.32}$$

The idea of the integral over all space is that the polarization of the solvent at point r depends on the polarization of all nearby points (r'). The function $\varepsilon(r, r')$ describes the correlation between solvent molecules at different points in space.

The simplest model of $\varepsilon(r, r')$ will be a monotonically decaying function. That is, the further away a solvent molecule at point r' is, the less of an effect it will have on the polarization of a solvent molecule at point r. The molecular scale of the solvent in this model is represented by the spatial correlation decay length [185].

A more sophisticated model will give $\varepsilon(r, r')$ an oscillating decaying form [186]. This describes the alternate layering of water dipoles, from positive end to negative to positive, and so on.

The effect of solvent–solvent correlations is illustrated in Fig. 7.13, using a monotonic model. The correlation length is taken to match the size of a water molecule [185]. Solvent spatial correlation introduces a correction of around 15% to the magnitude of activity coefficients.

7.4.9 Applications to adsorption and interfacial tensions

The dispersion interaction of an ion with an interface is (Ref. 7 and Chapter 4 in this book):

$$U_{\text{surf}}(x) \approx -\frac{B}{x^3} \tag{7.33}$$

where x is the distance of the ion from the interface and B is derived from the excess polarizability, as shown in Equation (7.13) above. This formula is valid when the centre of the ion is a diameter or two away from the interface; as the ion makes contact with the surface, the full expression does not diverge to infinite energy, but flattens out to a finite value at $x = 0$.

When the B coefficient is positive, the interaction is attractive, in general leading to positive adsorption of the ion at the interface. Conversely a negative B value indicates a repulsive interaction, generally leading to depletion of the ion at the surface. The B coefficients for ions in aqueous solution at a variety of surfaces is

Table 7.8. *Ion–surface dispersion coefficients B (10^{-50} J·m^3) at the air–water, oil(C_{14})–water, silica–water, alumina–water, mica–water and quartz–water interfaces.*

Ion	Air	Oil (C_{14})	Silica	Alumina	Mica	Quartz
Nonhydrated ions:						
Cl^-	3.11	−0.15	−0.76	−2.71	−1.21	−1.01
Br^-	2.82	−0.41	−0.86	−2.79	−1.30	−1.13
I^-	4.19	−0.59	−1.25	−4.11	−1.92	−1.66
$HCOO^-$	3.15	0.05	−0.62	−2.47	−1.07	−0.85
OAc^-	4.53	0.11	−0.89	−3.52	−1.51	−1.20
Li^+	0.09	0.03	0.01	−0.03	−0.01	0
Na^+	0.43	0.13	0.01	−0.16	−0.04	0
K^+	1.62	0.3	−0.14	−0.92	−0.34	−0.21
Rb^+	2.37	0.35	−0.26	−1.46	−0.56	−0.38
Cs^+	2.99	0.29	−0.45	−2.07	−0.84	−0.62
NH_4^+	2.21	0.32	−0.25	−1.38	−0.53	−0.36
Hydrated ions:						
Li^+	4.91	0.54	−0.68	−3.27	−1.31	−0.95
Na^+	6.48	0.89	−0.74	−4.06	−1.56	−1.07

shown in Table 7.8. The dielectric response of water is given by the Dagastine–Prieve–White model [173].

Let us first consider the surface of an uncharged oil droplet in water. In the classical theory of electrolytes, described by the standard Poisson–Boltzmann, the fact that the surface is uncharged means there is no electric field between oil droplets and therefore no preferential adsorption of either cation or anion at the surface. When dispersion interactions are introduced, we are given a surface interaction which is independent of electric charge. Since the strength of the interaction is specific to each ion, as seen in the B values, we will achieve preferential adsorption of one ion over the other.

Let us add sodium chloride to the oil–water mixture. At 1 mM concentration, for instance, ion-specific dispersion interactions lead to the concentration profile of Na^+ and Cl^- ions shown in Fig. 7.14. Chloride is preferentially adsorbed, leading to a negative effective surface charge on the surface on the neutral oil droplet. The separation of charges leads to a non-zero electrostatic potential leading out from the oil surface, as shown in Fig. 7.15. A negative surface potential (zeta potential) is induced at the surface of the uncharged droplets.

The dispersion-induced charge separation becomes stronger as the concentration increases, measurable via a concentration-dependent zeta potential, shown in

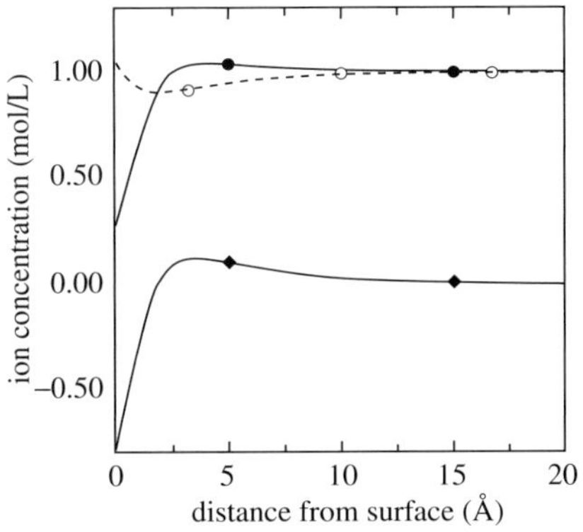

Fig. 7.14. Total charge concentration of 1 M NaCl near an uncharged oil surface. Sodium (•), chloride (◦) and total charge (♦).

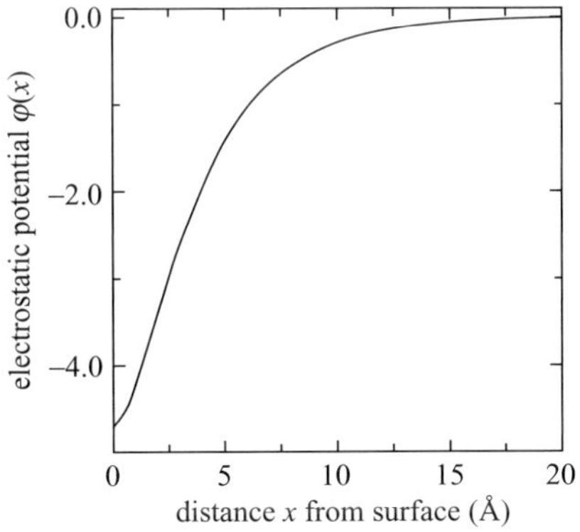

Fig. 7.15. Electrostatic potential at uncharged oil surface with 1 M NaCl.

Fig. 7.16 for both sodium halides and alkali chlorides. Hofmeister series are evident: $F^- < Cl^- < Br^- < I^-$ and $Li^+ < Na^+ < K^+ < Rb^+$.

Positive adsorption of the chloride ion and surface depletion of sodium can be expressed via the total surface excess Γ, defined as the difference of the ion concentration near the surface from the bulk concentration:

$$\Gamma_i = \int_0^\infty [c_i(x) - c_{i0}]dx \tag{7.34}$$

The surface tension increment, that is the change in surface tension γ as salt concentration increases, is directly proportional to the surface excess of the ions,

$$\frac{d\gamma}{dc} = -\frac{k_B T}{c} \sum_i \Gamma_i \tag{7.35}$$

Table 7.9. *Surface tension increments $d\Delta\gamma/dc$ (in $mJ{\cdot}L/m^2{\cdot}mol$) for an oil–water interface with added salt.*

Salt	$d\Delta\gamma/dc$
NaCl	0.467
NaBr	0.334
NaI	0.265
NaCl	0.467
KCl	0.359
RbCl	0.345

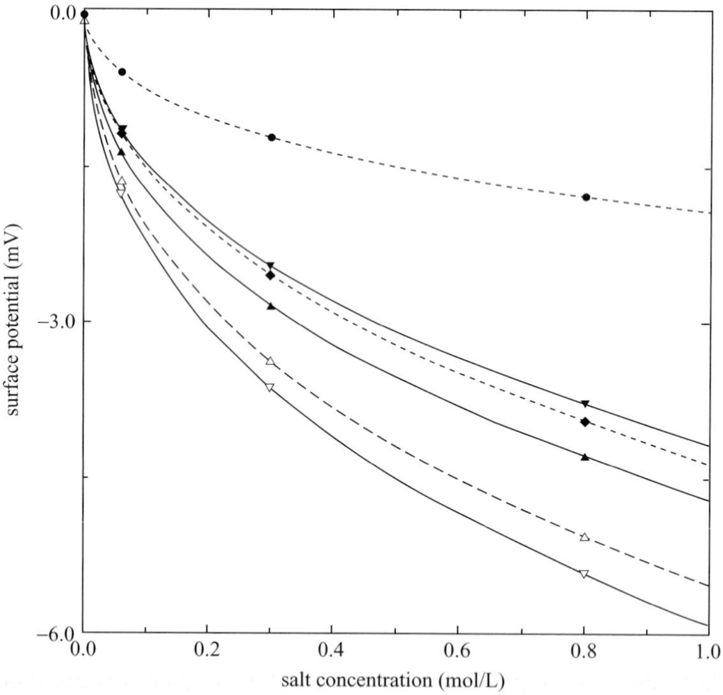

Fig. 7.16. Surface zeta potential at an uncharged oil interface with NaCl (▲), NaBr (△), NaI (▽), LiCl (•), KCl (♦) and RbCl (▼).

The surface tension increment at the oil–water interface is given for a range of sodium and chloride salts in Table 7.9. The Hofmeister series in the surface tension increments is $I^- < Br^- < Cl^-$ and $Rb^+ < K^+ < Na^+$. This matches the series observed experimentally at the oil–water interface [181], although magnitudes are different.

The surface tension increment at the air–water interface is more readily measured experimentally than for the oil–water interface. The experimental Hofmeister series at the air–water interface goes in the same direction as calculated above for oil–water. Interestingly, the dispersion *B* coefficients at the air–water interface suggest that the series would go in the *opposite* direction, with $I^- > Br^- > Cl^- > F^-$, and indeed that is what the theoretical calculation predicts. The difference between theory and experiment in this case is two-fold. On the one hand gas molecules from the air phase penetrate into the aqueous surface region. The dissolved gas changes the nature of the solvent environment around each ion in the surface region. We may speak of aerated, rather than hydrated, ions. However, the effect is not strong enough alone to reverse the Hofmeister ordering. On the other hand water molecules will be oriented at the air–water interface, with the ordered water layers introducing an anisotropic dielectric response at the interface, changing the value (and direction) of the ion–surface interactions. These effects are not observed at the oil–water interface since the solubility of oil in water is much lower than the solubility of gas molecules in water (bulk aqueous concentration of dissolved gas is around 1–5 mM, much higher at the interface). Likewise anisotropic ordering at the oil–water interface is also less than found at the air–water interface.

The impact of ion-specific surface interactions has been measured macroscopically via pressure measurements of neutral phospholipid layers. Petrache and coworkers measured pressure curves as a function of the distance between layers [25].

They found the pressure in bromide solution to be nearly an order of magnitude stronger than in chloride solution (both with K^+ as counterion). We may reproduce these results approximately by modelling the membrane tail as 'oil' and treating the headgroup to be relatively thin enough that we may neglect it. As in the NaCl example above, we again find positive adsorption of chloride and bromide ions, however to different degrees. The total ion concentration profile (over cation and anion) is shown in Fig. 7.17. Note the appearance of an *electrolytic* double layer, with a layer of negative charge close to the surface, surrounding by an outer positively charged layer (mostly clearly seen here in the charge profile for KCl). This does not generally appear in conventional DLVO electrostatic theory, where the secondary charge simply diminishes the magnitude of the overall charge, rather than reversing the sign of the overall charge as seen here. The 'double layer' in conventional theory refers to a single layer of bound surface charge (such as negative oxide groups in silica or mica, or positive amine groups in proteins), surrounded by a single layer of opposite sign due to the electrolytic counterion in solution. In the example shown here there is no surface charge; the 'double layer' appearing here (seen also in Fig. 7.14 above) is due to both cation and anion in solution. If we were to add a surface charge, we would find a 'triple layer', formed

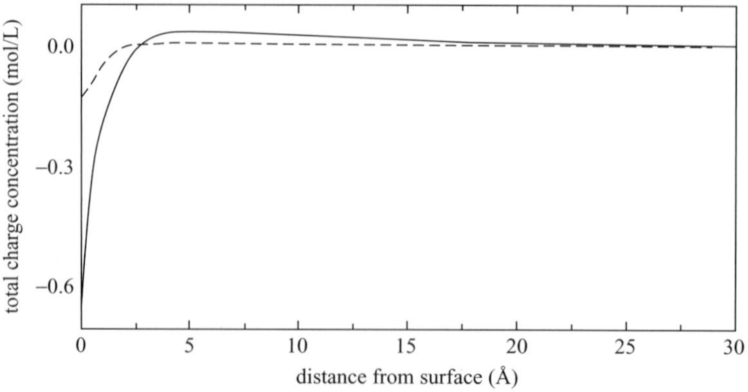

Fig. 7.17. Total charge concentration of 0.1 M KCl (full line) and KBr (dotted line) near an uncharged oil surface.

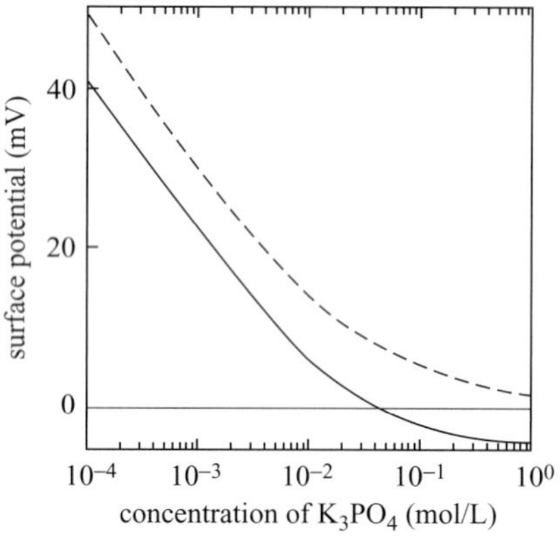

Fig. 7.18. Surface potential (solid line) at a model protein surface in a solution of potassium phosphate, as a function of electrolyte concentration. The dotted line shows the theoretical surface potential with electrostatic interactions only (DLVO, no dispersion interactions).

from a layer of surface charge, followed by a layer of counterions in solution, surrounded by a layer of coions in solution.

Charging of neutral vesicles occurs in phosphate buffers, giving rise to a zeta potential of up to −140 mV [187]. So too can surface dispersion interactions reverse the charge of a surface, leading to a corresponding change in the sign of the zeta potential. Take the example of a protein surface, with pH placed above the isoelectric point such that the surface charge is positive. We use the simple model for the protein dielectric response, of Tavares et al. [188].

We follow the zeta potential for potassium dihydrogen phosphate as a function of salt concentration, shown in Fig. 7.18. At sufficiently high concentration, the surface adsorption of dihydrogen phosphate ions leads to charge reversal from

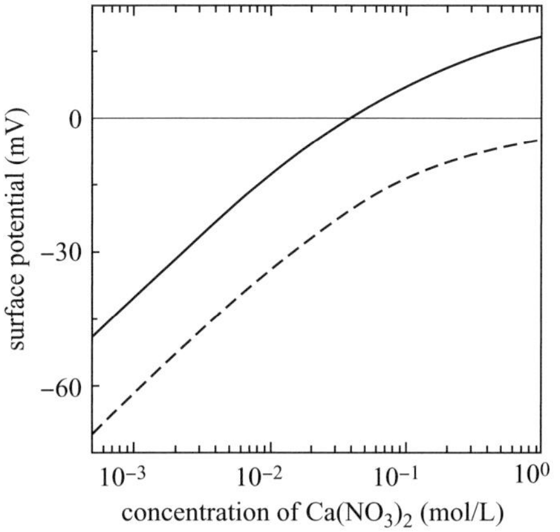

Fig. 7.19. Surface potential (solid line) at mica surface in a solution of calcium nitrate, as a function of electrolyte concentration. The dotted line shows the theoretical surface potential with electrostatic interactions only (DLVO, no dispersion interactions).

positive to negative. (The calculation in this case used potassium phosphate; in the real solution there would be additional effects due to equilibrium with hydrogen phosphate and phosphate.) The charge reversal seen here is due entirely to the ion–surface dispersion interaction. It enables the phosphate anion to adsorb positively to the surface in excess of the adsorption expected from electrostatics alone. If only electrostatic interactions were in play (the traditional DLVO model) then there would only be enough adsorption of the counterion to negate the original surface charge, bringing the surface potential to zero, shown in Fig. 7.18.

The reversed polarity at high concentration is driven solely by the large polarizability of the phosphate ion. It is not an electrostatic effect in the sense that it is independent of the original surface charge: the same high concentration surface potential, around −5 mV, appears regardless of whether the protein surface has 0.01 or 0.001 positive charges per square nanometre.

Charge reversal has been observed experimentally, for instance with calcium nitrate at a mica surface [189]. This was explained by Kékicheff, Marčelja and coworkers by invoking a site-binding model, creating a surface equilibrium between surface charge sites and Ca^{2+} ions. In fact the experimental data may be reproduced theoretically without any such surface equilibrium simply by including ion–surface dispersion interactions. The resulting graph of surface potential against concentration is shown in Fig. 7.19. The site-binding model may be interpreted as a special case of the full model described here, where only the ion–surface interaction of ions directly on the surface is considered. By contrast, the extended model presented above also includes the surface interactions of ions at some distance from the surface, giving the model greater predictive power.

The theoretical surface potential displayed in Fig. 7.19 has nearly exact quantitative agreement with the experimental data of Kékicheff et al. [189,190]. It is worth commenting on a more subtle aspect of the theoretical model required to achieve this level of accuracy. Experimental data show a surface potential around +10 mV at 0.1 M concentration of $Ca(NO_3)_2$. The theoretical calculation reproduces this point and extends further to very high concentration (10 M), where the surface potential asymptotes at around +25 mV. This was achieved by invoking a hydrated ion model for the kosmotropic ion Ca^{2+}, on the same lines as the hydration model introduced above in the discussion of bulk activity coefficients. The hydration model modifies the ion in two ways: first the polarizability of the hydrated ion is increased by adding the polarizability of the water molecules in the hydration shell to the intrinsic polarizability of the bare ion. The number of water molecules in the hydration shell is taken as the hydration number. Using the Marcus values [191], the hydration number of Ca^{2+} is 7. Secondly the size of the hydrated ion is increased by adding the width of the hydration layer, taken to be the radius of one water molecule.

If hydrated kosmotropic ions are not invoked then agreement with experiment is lost. For instance with calcium nitrate at mica, the application of unhydrated Ca^{2+} ions means that zero surface potential will be reached at 0.5 M instead of 0.03 M, the surface potential at 0.1 M will be −7 mV rather than +10 mV, and the high concentration asymptotic surface potential will be only +9 mV instead of +25 mV.

7.4.10 Comment on frequency contributions

The sum over all frequency components involves the frequency-dependent dielectric properties of substrate and the solvent as well as the ionic polarizabilities.

The potentials can be and are either attractive or repulsive. Some contributions to the sum over frequencies can be positive, others negative.

The complete formula is a mess. But it contains a lot of information. We shall write it out below in the non-retarded case (for an oil–water interface), with some trepidation, and advise the reader to attend only to our comments.

$$U_{\pm}(x) = \frac{k_B T}{2x^3} \sum_{n=0}^{\infty}{}' \frac{\alpha^{\pm}(i\omega_n)\left[\varepsilon_w(i\omega_n) - \varepsilon_{oil}(i\omega_n)\right]}{\varepsilon_w(i\omega_n)\left[\varepsilon_w(i\omega_n) + \varepsilon_{oil}(i\omega_n)\right]} \tag{7.36}$$

One comment is that the complete formula, as for original Lifshitz theory, includes retardation (Chapter 6). That is the ion, or any other adsorbant, is subject to a sum of forces from different Fourier frequencies, in the microwave, infrared and visible and far ultraviolet regions. The higher-frequency components are damped out successively, exponentially as distance of the ion from the interface increases.

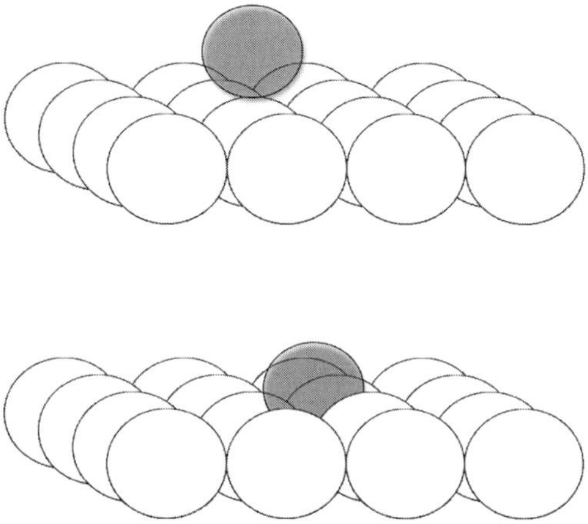

Fig. 7.20. Illustration of different types of 'contact' between an adsorbing ion (grey ball) and the molecules of the substrate (white spheres). Above: the ions sit on top of the substrate surface. Below: the ions nestle into the surface; adsorbed molecules interact with more nearest neighbours, and experience a much higher energy of interaction.

The simple form Bx^{-3}, equivalent to pairwise summation, can therefore be much too crude (it is also temperature-dependent as described in Chapter 6).

The distribution of ions near a charged interface is determined by a complex non-linear competition between electrostatic and dispersion forces acting on all the ions:

$$N_{\pm}(x) = N_0 \exp\left[-\frac{U_{\pm} \pm e\phi}{k_B T}\right] \tag{7.37}$$

Particularly for long-range interactions between colloidal particles in the presence of an electrolyte, the interactions can change quite dramatically even in sign because of this effect.

Another important remark is that at short distances, 'contact' depends on the molecular structure of the surface and size of the adsorbing ion (see Section 7.4.6).

With several adsorbed molecular species, the lowering of self energies is a major factor in catalysis, especially on metals where the coupling of electronic modes of the substrate reduces energy barriers to molecular formation even further. At the 'surface' of large enzymes the coupling of substrates is via cooperative infrared, vibrational modes of the macromolecular enzyme.

We do not consider here changes in interfacial tensions at the air–water interface with dissolved electrolyte. The changes predicted are about right in magnitude, but a planar model for such an interface is too crude. The interface is highly anisotropic, a matter that could be accounted for by admitting anisotropic dielectric properties that can estimated. The problem is more complicated than that because of the presence of atmospheric gas in large quantities – it goes from 1 to $5{\cdot}10^{-3}$ M – in the first few molecular layers. Studies via simulation using effective water models of interfacial tension changes induced by electrolyte ignore this complex interfacial layer, which seems not permissible.

7.4.11 Applications to forces in colloid science: a work in progress

The previous section addressed the question of the concentration profile of ions near a surface. Due to the ion–surface dispersion forces, the interaction between ion and surface becomes ion-specific. Depending on the details, the frequency dependence of the polarizability of an ion and the dielectric response of the media on either side of the interface, the ion–surface interaction may become attractive or repulsive, with magnitudes depending on the ion, and on the complex competitive interplay between electrostatic and dispersion forces. Hence the ion concentration profile becomes ion-specific.

The ion concentration profile lies at the heart of classical double-layer theory, which describes the pressure or force that arises between two surfaces as they approach one another. As the ion concentration profiles at each surface start to overlap, this induces a shift in those concentrations, affecting both the enthalpic component (the electrostatic interaction energy) of the total surface–surface interaction energy as well as the entropic component due to the ion distribution. These two components combine in classical double-layer theory to give the double-layer pressure between the two surfaces due to the ion distribution:

$$P_{\mathrm{DL}} = k_B T \sum_i \left[c_i\left(x_{\mathrm{m}}\right) - c_{i0}\right] \qquad (7.38)$$

where x_{m} is the position in the middle between the two surfaces where the electrostatic potential reaches a minimum magnitude (this will be the midpoint, when the surfaces are identical). That is, the double-layer pressure is simply given by the difference in the concentration of the ions at the midpoint from their bulk concentration. If the ion concentrations are enhanced at the midpoint, the double-layer pressure will be positive (surfaces repel); if there is ion depletion at the midpoint, the pressure is negative (surfaces attract).

In classical double-layer theory where dispersion interactions are left out, there is no ion specificity in the concentration profiles and therefore no ion specificity in the double-layer pressure.

When dispersion forces are added, ion-specific surface forces are obtained. A dispersion component is added to the double-layer pressure; its magnitude is determined by the change in the total dispersion interaction of the ions between the two surfaces with respect to a change in the distance between surfaces. That is:

$$P = P_{\mathrm{DL}} - 2\sum_i \int_0^{L/2} c_i\,(x)\,\frac{dU_i^{\mathrm{disp}}(x)}{dL}dx \tag{7.39}$$

This formula assumes the two surfaces are the same, so that $x_m = L/2$.

If, as discussed above, we take the dispersion interaction between an ion and one surface to be $u_{\mathrm{d}}(x) = B/x^3$, then the dispersion interaction between both surfaces is:

$$U_i^{\mathrm{disp}}(x) = \frac{B}{x^3} + \frac{B}{(L-x)^3} \tag{7.40}$$

and so the change in dispersion energy with respect to surface separation is:

$$\frac{dU_i^{\mathrm{disp}}\,(x)}{dL} = -\frac{3B}{(L-x)^4} \tag{7.41}$$

When ion–surface interactions are attractive ($B < 0$), then the dispersion component to the pressure between two surfaces will be positive. Vice versa, if ion–surface interactions are repulsive ($B > 0$) then there will be a negative correction to the pressure. In other words, when ions bind to the surface as a result of ion–surface interactions, this will have the effect of repelling the two surfaces away from each other. When there is ion depletion at the surface, the consequence will be an attractive force between surfaces.

These comments refer to the direct affect of ionic dispersion forces. There is also an indirect affect insofar as those same dispersion forces shift the electrostatic potential and shift the concentration profile of the ions between the surfaces. The impact of this indirect dispersion effect is not straightforward to predict, being confounded by the requirement to maintain overall electroneutrality between the surfaces.

7.4.12 An explicit example

By way of example, consider the 2006 experiment of Petrache, Zemb, Belloni and Parsegian [25].

They set up uncharged neutral lipid lamellar multilayers in solutions of Cl^- and Br^- (K^+ as counterion). Measuring the external osmotic pressure of the system, which is equivalent to measuring the pressure between surfaces, they found a stark difference in pressure between KCl and KBr solutions: the pressure for KBr at surface separations greater than 20 Å was an order of magnitude stronger than for KCl. We may attempt to model this system using a Poisson–Boltzmann description

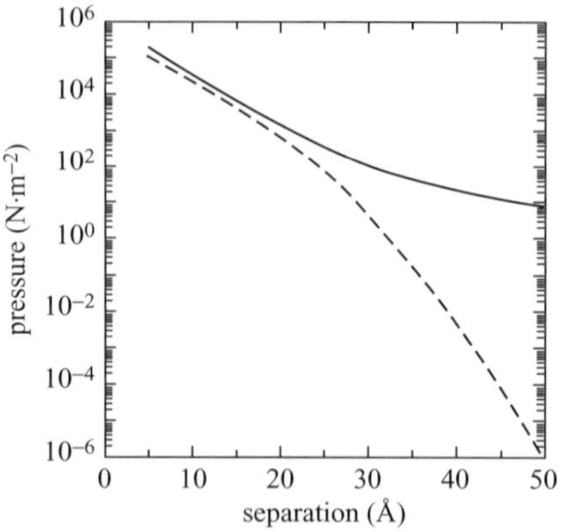

Fig. 7.21. Pressure–distance curves for 1 M solutions of KCl (dotted line) and KBr (full line) between oil (C_{14}) surfaces (a model for DLPC).

of the electrostatic potential, incorporating ion–surface dispersion interaction into the ion concentration profile, used in the Poisson–Boltzmann calculation.

The key question for the ion–surface dispersion interaction is the dielectric model for the surface. To get reliable results, in general we need a reliable description of the dielectric function of the surface, in this case of the lipid layers, in particular at high frequencies, UV-vis and up to soft X-rays. Classical experiments in colloid science typically used, and still use today, hard surfaces such as mica (an alumina), silica or quartz. The dielectric spectra of these materials are well described. There is a much greater variety in the surfaces of biology and related applications, ranging from phospholipids to proteins to quaternary ammonium surfactants. Where a lipid headgroup is small relative to the tail, such that its dielectric response is dominated by the tail, we may find the possibility of neglecting the headgroup and modelling the surface as a simple oil.

Suppose then we accept the dielectric function of tetradecane as a model for the surface of the Petrarche lipid (the actual lipid is the 12-carbon chain DLPC, 1,2-dilauroyl-*sn*-glycero-3-phosphocholine). Calculating Poisson–Boltzmann concentration profiles including ion–surface dispersion interactions, we obtain the pressure–distance curves shown in Fig. 7.21. First, we find a strong ion specificity in the pressure curves: the pressure of the two salts is different by an order of magnitude. This is particularly notable since this calculation is performed with zero surface charge, which in classical double-layer theory would yield zero electrostatic potential, zero shift in concentration profiles from bulk concentrations, and zero double-layer pressure. Second, we see that the ion specificity does indeed match experiment: the stronger pressure is seen in KBr solution. The absolute magnitudes of our model match experimental pressures reasonably well; a better match would

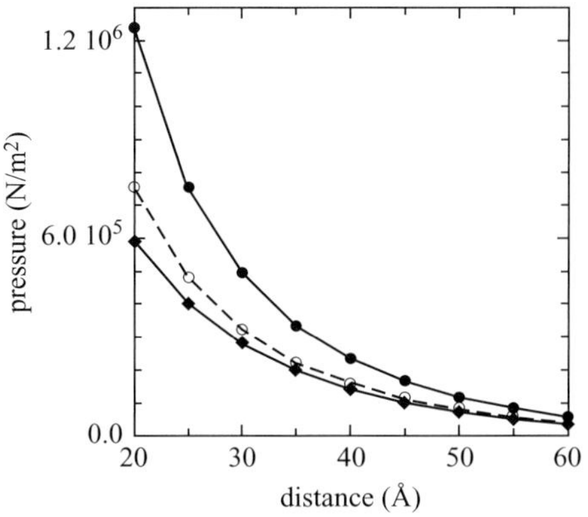

Fig. 7.22. Pressure between two silica surfaces for various chloride salts (50 mM) at pH 8. Kosmotropic ions (Li^+, Na^+) are hydrated. LiCl (•), NaCl (○) and KCl (♦).

be found by applying a more correct model of the (anisotropic) lipid dielectric function.

Another example is found in the Hofmeister series of chloride salts at silica and alumina surfaces. Theoretical pressure curves between two silica surfaces calculated by Parsons, Böstrom and coworkers (D. F. Parsons, M. Boström, T. J. Maceina, A. Salis and B. W. Ninham, submitted for publication) are shown in Fig. 7.22. An ordinary Hofmeister sequence is found: Li $>$ Na $>$ K. This series was measured experimentally by Franks and coworkers [192]. What is interesting is that at alumina surfaces Franks observed the reverse sequence, K $>$ Na $>$ Li. We have already discussed the role of hydration of kosmotropic ions. Theoretical pressure curves between alumina surfaces, including hydration of Li^+ and Na^+, are shown in Fig 7.23. When hydration is neglected, the same series is found as seen between silica surface, which for alumina is the wrong series. The correct reversed series is obtained once kosmotropic ions are hydrated (hydration was also used in the silica calculation, Fig. 7.23, where it affects magnitudes but not the overall order).

It seems likely that hydration, or more correctly dehydration, together with dispersion forces, may play a role in explaining long-standing apparently mysterious behaviour like 'secondary hydration forces' [76,77,90]. If ion–surface interactions are able to induce ion dehydration [193,194], stripping the hydration layer of kosmotropic ions close to the surface, then the energy cost required to achieve dehydration will result in additional repulsion.

The agreement claimed for DLVO and double-layer theory over 50 years with experiment seems to be illusory.

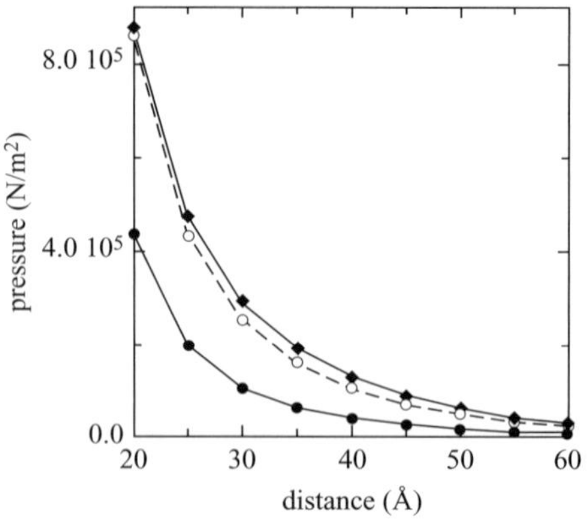

Fig. 7.23. Pressure between two alumina surfaces for various chloride salts (50 mM) at pH 12. Kosmotropic ions (Li^+, Na^+) are hydrated, enabling the system to achieve the experimentally observed reverse Hofmeister series. LiCl (•), NaCl (○) and KCl (♦).

We remark finally that while dispersion forces in activity coefficients can be swept under the carpet and hidden in sensitive ionic hydration radii, that is not so for surface forces. The reasons seem obvious. In the former case the interactions are short-range $V(R) \approx R^{-6}$. In the latter the adsorption potential is approximately R^{-3}.

7.4.13 Anisotropy and water ion clusters, hydroxide and other anisometric ions

Simple ions such as Na^+ or Cl^- may be satisfactorily treated as spherical. The interactions are isotropic, for instance their polarizabilities lead to induced dipoles with the same magnitude, irrespective of the direction of the external electric field. Ions such as ammonium, perchlorate or phosphate are tetrahedral rather than spherical. Nevertheless their polarizabilities are balanced in each direction such that they are still isotropic. Their interactions may therefore be treated in the same way as the truly spherical ions; effectively they *are* 'spherical'.

Other ions are by no means spherical at all; just as their geometry is not a sphere, so too their interactions are starkly anisotropic. Examples include linear ions such as thiocyanate and planar ions such as nitrate, chlorate and formate.

7.4.14 The anisotropic tensor

At a first level of approximation we can ignore these anisotropies and coerce a spherical approximation over these ions. Quantum chemical methods may provide

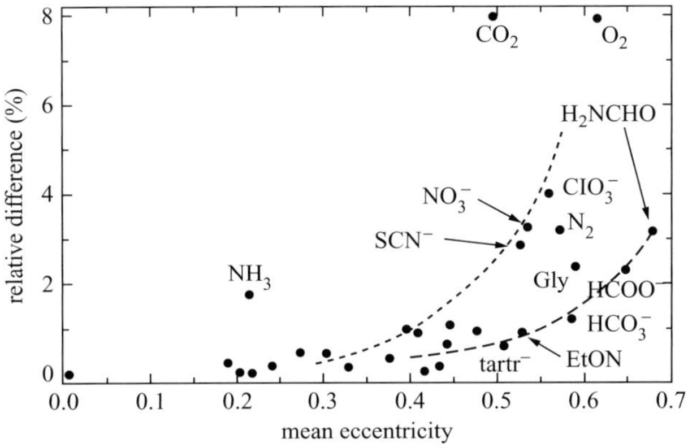

Fig. 7.24. Relative difference in self energy (in %) between nonspherical anisotropic model and spherical isotropic model. tartr$^-$ indicates tartrate, and EtON denotes ethanolamine-H$^+$ [183]. Reprinted from Ref. 183. Copyright 2009 with permission from Elsevier.

a reasonable estimate of the actual volume of the ion, accounting correctly for its nonsphericality. This nonspherical volume can then be converted to a mean spherical radius. Similarly a mean isotropic polarizability may derived by taking the average of the anisotropic polarizabilities along each axis:

$$\overline{\alpha} = \frac{\alpha_x + \alpha_y + \alpha_z}{3} \tag{7.42}$$

However, at some point we need to move beyond coarse spherical approximations and consider the anisotropies of the ions in full.

A first glimpse at the impact of anisotropy can be found in ion self energies. The relative difference between anisotropic self energy and mean spherical self energy as a function of eccentricity is illustrated in Fig. 7.24. The eccentricity describes how strongly the shape of the ion deviates from spherical; the higher the eccentricity the less spherical the ion.

Curiously, the sensitivity to anisotropy appears be larger when there is an axis of symmetry (that is when two of the three anisotropic polarizability values are equal). This is suggested by the two curves inscribed on the figure; the curve containing the linear thiocyanate ion (and oxygen or carbon dioxide gas molecules) and the planar nitrate and chlorate ions shows a slightly greater anisotropic effect.

The anisotropic effect is not large for the self energy, giving a correction of only 3–8%. This is because by definition the self energy is the interaction of an ion with its own electric field. The ion is always aligned with itself, therefore the anisotropic correction is minimized.

The anisotropic effect between two nonspherical ions on the other hand, or between a nonspherical ion and a surface, will be much greater. In this case there

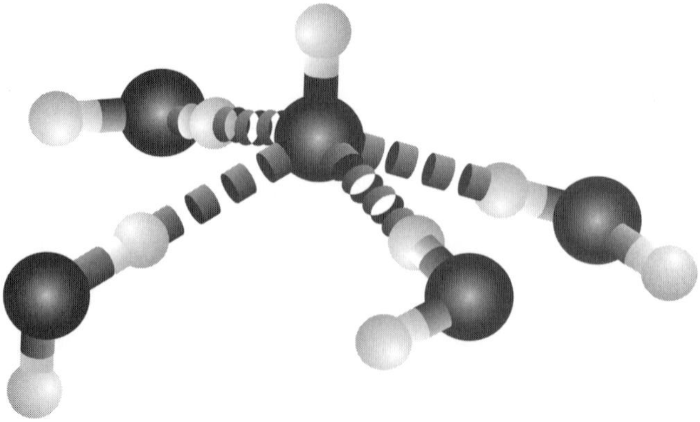

Fig. 7.25. Hydrated hydroxide $[OH{\cdot}4H_2O]^-$.

will be a full range of possible orientation angles between ion and ion or ion and surface. The interaction will be stronger at some angles, weaker at others (potentially even changing direction). These questions bring us to the frontiers of electrolyte theory.

The question of surface anisotropic effects is worth exploring further. The very existence of the surface itself introduces anisotropy into the system, which as just stated in the previous paragraph can via ion–surface interactions induce further ion-specific anisotropic effects. Interesting behaviour arising here will be particular striking for the hydroxide ion.

Natively the hydroxide ion OH^- is an anisotropic linear ion. Its hydrated geometry is, however, unusual. When thinking about a hydrated ion, we usually think of the water molecules in the hydration shell as forming a cage around the ion, with the water molecules spread more or less evenly around the ion, maintaining an overall spherical shape. This is the case, for instance, with a hydrated sodium ion.

By contrast, the hydration shell of the hydroxide ion, containing three or four water molecules, forms a plane. The hydroxide ion is situated in the centre of the hydration plane and oriented perpendicularly with H directed outwards, as illustrated in Fig. 7.25. It is evident that the planar (or pyramidal) form of hydrated hydroxide will have vastly different interactions at an interface than will the linear naked hydroxide ion (or a mean spherical model). This anisotropic hydroxide effect has implications for the controversy over hydroxide at the surface, where some experimental teams have reported positive adsorption of hydroxide at the air–water interface, while computer simulation teams, with some experimental support, have reported only mild concentrations of hydroxide at the surface, if not outright depletion.

7.4.15 Hints at specificity in complex matter and biology

We began this study of molecular forces with a search for an understanding of the origins of specificity of molecular forces between molecules and ions that are not included, or at least hidden, in classical descriptions of physical chemistry.

Once identified, the further implicit proposition was that that specificity must somehow conspire, collectively, with the size, shape and material properties of multimolecular aggregates in several ways. First, the interactions imposed by the surrounding medium of proteins and polymers affect and set their optimal conformations. Second, the interactions must drive the highly specific biochemical associations that we might subsume under the lock and key idea, apparently obvious geometrically. But the coming together of enzyme and substrate cannot be random. These are exquisitely repeatable machines. Third, we want to understand how such long-range, not just contact, forces provide the energetics of reactions consequent on association. Fourth, we want to understand how this hierarchy of forces gives rise to supraself assembled aggregates such as biological cells.

Some hints on these matters seem to have come into sight, in the Hofmeister effects, in the Lifshitz theory and its extensions, even in the example of restriction enzyme action. That involves another central neglected factor, the role of dissolved gas, to be discussed in the following chapter. Self assembly of surfactants and lipids is the subject of the second part of this book.

But Nature inevitably has many more tricks up its sleeves. We discuss here a few of these that we can identify in embryo form.

7.4.16 Material properties – effects of dielectric anisotropy in colloidal interactions: repulsive van der Waals interactions

We have considered above the effects on adsorption forces and interactions of anisotropy of the polarizability of ions such as hydroxide and nitrate. The considerations obviously apply to non-ionic molecules, e.g aromatic ring compounds from benzene to calyxarenes, and to their specific interactions with ions.

In all of our studies of interactions of colloidal particles so far we have restricted analysis to like materials and those of isotropic dielectric susceptibilities and uniform bulk medium properties.

Although we have not considered the problem explicitly, if the interacting particles have opposite surface charge or potential, the classical double-layer interaction becomes attractive not repulsive. This is a classical problem of colloid science, leading to little-understood heterocoagulation. With the inclusion of ionic dispersion forces it becomes a much more subtle issue.

By the same token, the force between two unlike uncharged materials across a liquid is not always attractive. This is explicit in the formulae of Lifshitz, and shows up most subtly in the studies of wetting that we mentioned in Chapter 5. There, for the case of oil films spreading on water near the saturated vapour pressure of oil [7,195], and of the different frequencies contributing to the force between air and water across oil, some give positive contributions, some negative.

A recent force measurement that demonstrates repulsive van der Waals forces is that of Ref. 196.

7.4.17 Anisotropic media

If one or more of the interacting media have anisotropic dielectric properties, as for liquid crystals, there can arise a strong torque on the systems as the surfaces approach. The interaction can be repulsive or attractive depending on those dielectric properties. Some theoretical analyses are those of Refs. 197–199.

Anisotropy in dielectric properties shows up almost everywhere, especially for an ideal mica surface, and its effects are almost invariably ignored.

Multiply layered structures of different and anisotropic dielectric properties show up in biology in multiwalled vesicles or in soap films. They display van der Waals forces different from those of single layers [197–199].

Such extensions of Lifshitz theory have some applications and implications for materials technology and adhesion [200].

Of more interest to us are biomembranes or biomimetic membranes or vesicles and lamellae of surfactant aggregates as discussed in Chapters 9–11. Here the double-chained hydrocarbon tails of lipids of the membranes (typically diC_{12}–diC_{18}) that form a bimolecular leaflet are lined up perpendicular to the membrane surface. Their dielectric properties are highly anisotropic. Further the headgroups of the lipids that face water (see Fig. 9.3) such as phosphatidylcholine or ethanolamine form an anisotropic layer. The entire membrane is a multiply layered structure:

water–headgroups (anisotropic)–hydrocarbon chains (anisotropic)–headgroups–water.

An ion or other species approaching the membrane 'sees' hydrocarbon at a large distance, and headgroups at a small distance. The force it 'feels' can be repulsive at one distance and attractive at another due to the multilayer and anisotropy effects.

If it is itself anisometric, it can also experience a torque of alignment as it approaches a specific target membrane-bound protein. The subtleties of such interactions are further extended by recognition of cooperative headgroup dipolar interactions and hydration that depend on heagdroup dipolar moment and size [201].

And membrane-bound proteins can diffuse appropriately in response to forces imposed by an approaching protein with which they wish to do business.

The point of such discussion is not that anyone would want to calculate these forces, unless they were seriously deranged, but to note these other sources of flexibility available for biorecognition.

As a matter of fact even the surface of a simple oil like an alkane is anisotropic, with a gradual liquid crystal type of alkane order, and consequent dielectric anisotropy, from the surface to the bulk liquid. It depends on chain length, and can be seen if one decomposes measured surface free energies into enthalpy and entropy as a function of temperature by numerical differentiation. The same applies self-evidently to the surface of water.

With magnetic and conducting materials the Lifshitz theory can easily be extended, but the consequences for interactions in magnetic fluids, for example, have hardly been recognized [202,203].

7.4.18 Shape and size

An extreme example of the influence of shape on interactions is that between long cylinders. There is a large torque that lines them up parallel to each other as they approach [204,205]. At smaller distances they see each other more like planes, not thin rods [7].

7.4.19 Conduction processes

With cylinders that are conducting as with conducting polymers or nanotubes of carbon or polyelectrolytes such as DNA with surrounding mobile adsorbed ions, shape becomes very important. Formulae in those cases are given in Chapter 5.

The forces between such cylinders are extremely long-range and in fact non-additive, a circumstance that may well be important in recognition [206].

That there is enormous flexibility available to molecular interactions of colloidal particles due to size and shape can be seen by considering the interactions between two spheres of different radii. The gravitational interaction between them is inversely proportional to the centre-to-centre distance $\sim R^{-1}$. But for molecular forces the interaction is very much more complicated even for spheres made up of simple pairwise additive van der Waals molecules (Chapter 2). This is apart from all the subtleties of molecular forces due to material properties that we have already elucidated. Size matters! A little-known fact too is that even for equal-sized spheres, the energetically preferred association is a line of spheres, not a clump.

But shape matters too. This can be seen if we consider inorganic colloids that will have crystal faces that are of different shape and differently charged. In immunology inorganic colloidal particles of alumina occur in Freund's adjuvant, an essential ingredient that used to be used in vaccines to adsorb proteins (antibodies) on (anisotropic) crystal faces so that they present to antigens in an optimal configuration.

How inorganic colloids agglomerate and grow to form non-Euclidean spiral structures characteristic of the shells of snails and other sea creatures that we are familiar with depends on the interactions between different faces of microcrystallites. We touch on 'microfossils' in Chapter 12. For the moment we remark on the packing of isosceles triangles, or of the two model shapes 'kites' and 'darts' of Penrose tiling, that underlies the subject of quasi-crystals and that can be explained analytically [207].

It was a problem that was explored by D'Arcy Thompson in *On Growth and Form* [208]. Why they should pack so regularly in ever-increasing perfect spirals is evident if we admit different forces between different faces.

7.5 Exploitation of specific ion effects

The old Christian prayer for forgiveness, 'because we have done those things we ought not to have done and have left undone those things we ought to have done', may well apply to us. We cannot claim to have a complete story. But we can claim to have a better insight into the immense variability of molecular forces. These we shall exploit systematically in the design of self-assembled microstructured fluids and some biomimetic systems in Part II of this book.

What has emerged is that the foundations of the subject have been shown to be flawed. From self energies of ions to activities to interfacial tensions and surface potentials and zeta potentials to colloidal particle interactions and the cornerstone DLVO theory, these flaws continue to propagate through the literature and cause vast confusion to those more interested in optimal experimental design.

The same applies as we have seen even to the meaning of quantities such as pH and buffers. When the flaws are corrected some things fall better into place.

Leaving aside the additional flexibility available to us through control of dissolved gas to be discussed in the following chapter – which controls some so-called 'hydrophobic' forces – we now know how to turn forces on or off with changes of salt type and concentration. Many of these things are known empirically. For example, a housewife will use vinegar rather than sodium chloride for removing difficult-to-remove proteins from surfaces. She knows that spilt red wine is inhibited from adhering to cloth by flocculating the dispersed proteins with sodium chloride! The Hanseatic League used sodium hydroxide to denature proteins and preserve cod fish.

We can solubilize a polymer such as chitosan, a totally soluble ionic cationic polysugar, by ion-exchanging the counterion from chloride to acetate; do the same for a polyanion, by ion-exchanging say Na^+ for Li^+ and deposit each successively on a surface. The multiply layered catanionic multilayer can then be fixed in place by titrating in excess NaCl. Ion-exchange columns and electrophoresis, which are essential tools in biology, routinely exploit and are based on specific ion adsorption without any fundamental understanding of how it works. That shows up in the apparently weird behaviour of 'buffers' and pH. At the same 'pH' one buffer will work, e.g. phosphate; another is poisonous, e.g cacodylate. Such dramatic effects are resolved if one understands why.

Crystallization of proteins for X-ray structural analyses presents large technical problems that can be solved by ion exchange to counterions that impose large repulsive forces, and degassing to reduce hydrophobic interactions (Chapter 8). The crystallization can then be effected by gradual reduction of those forces by titration of other, more strongly bound, counterions.

An archetypical example of the importance of and awareness of Hofmeister effects is in enhanced oil recovery. A vast amount of work over 50 years was accumulated on the efficacy of flooding reservoirs with brines to displace oil contained in capillaries, with disappointing results. The work was all done on a particular standard sample reservoir rock. But not all reservoirs are the same. Some will adsorb ions due to dispersion forces and change the reservoir surfaces from oil wet (hydrophobic) to water wet (hydrophilic) so that recovery changes from 10% to 90%! An example of changes in hydrophobicity of surfaces induced by specific ion adsorption is in Ref. 209.

Progress has been inhibited through reliance on DLVO theory and that of the electrical double layer, the truth of which was established by incantation. The fact that the theory did not work beyond $5 \cdot 10^{-2}$ M salt, as for Debye–Hückel theory, was repeatedly emphasized by Deryaguin to his students, even for the lyophilic systems for which it was devised (private communication). It was known to clay chemists and soil scientists. Most colloid and surface and surfactant chemists worked with only a few common alkali halide salts because of the conviction that classical theories were correct. So that the variability available in controlling forces by Hofmeister effects was hidden.

The vast existing literature on Hofmeister phenomena, mainly in biochemistry, shows that quite often it is impossible to compare results or deduce any systematic design rules from different experiments, because most investigations were restricted to incomplete sets of data.

This is mainly due to some methodological deficiencies:

1 Some papers report results obtained by adding different electrolytes that do not share either the cation or the anion. For example, the comparison is made of LiCl, NaBr and KNO_3. So a 'salt specificity' will emerge, but the discussion will not benefit from insights due to a systematic comparison between electrolytes with, for example, a fixed cation and varying anion. The most recent evidence that we have rehearsed shows that specific 'ion' effects are due to a complex interplay between the composition of the whole electrolyte, the nature of the surface and other factors and not just to the nature of the anion or that of the cation alone. There is no single Hofmeister series. Thus the series for refractive index increments and free energies of transfer of ions depends on single ion dynamic polarizabilities. The series for activities depends on interactions between ions involving such polarizabilities. The series for forces between particles depends on charge of the surfaces and their particular dielectric properties as well as the electrolyte type and concentration and mixture. The measurement of pH, or pH supposedly set by a buffer, follows a different specific salt-induced Hofmeister sequence for two buffers at the same prescribed pH. However complicated, these phenomena are now more or less accessible through the same formalism.

2 Awareness of the effects then suggests it is sensible to study a series of anions keeping the same cation, and vice versa. Possibly, fluoride, chloride, bromide, iodide, thiocyanate, nitrate, perchlorate and sulfate should be investigated for the anions, and lithium, sodium, potassium, caesium, magnesium, calcium and aluminium for the cations. Of course this is possible only when no specific 'chemical' processes are involved, e.g. redox reactions. The evidence that changing the cation may lead to a reversed Hofmeister series further supports this consideration.

3 The electrolyte concentration is another issue. We have pointed out that dispersion forces – which control specific ion effects – dominate over electrostatic interactions at moderate and high salt concentrations. Therefore it is important that the salt effects be investigated at both low (1 mM or below) and higher concentrations (between 10 mM and 1 M, where dispersion forces certainly dominate). It is a myth that dispersion forces are irrelevant at low concentrations. The examples discussed above show this explicitly, for forces with Br^- vs. OAc^-, wth $LiNO_3$ vs. NaCl, and with adsorption of very-low-concentration buffer anions such as phosphate which can have a surface ion adsorption exchange capacity of 100 to 1 compared with chloride.

4 Some ions show anomalous behaviour, due to their characteristic composition and nature. Species such as F^-, $H_2PO_4{}^-$ and OAc^- participate in hydrolysis equilibria, with changes in the pH of the solution and in the concentration of the intervening species (for example $OAc^- + H_2O \rightleftharpoons HOAc + OH^-$). Furthermore they are capable of establishing strong hydrogen bonds with water. Hydrogen bonding to surfaces can have dramatic effects on forces [210].

5 Shape and symmetry are other factors that may affect the behaviour of anions (see Chapter 3): halides are spherical, which means that the way they interact with water, interfaces and large molecules is isotropic (see also polarizability). In contrast, NO_3^- and SCN^-

are not spheres, and their interactions are anisotropic. Polyatomic ions introduce another complication because their geometry (and electron cloud distribution) may depend on the environment where they are dispersed, particularly for heavy elements where empty or partially empty d orbitals are available.

6 Especially in biological studies, a more or less significant amount of buffer components is used. This is necessary to stabilize or solubilize proteins and other macromolecules, but of course the presence of the buffer ions (for example $H_2PO_4^-$, HPO_4^{2-} and PO_4^{3-}) significantly alters the global ion effect. Furthermore, buffers act by modulating the pH and the surface charge of polymers. So it is worth paying special attention to the nature and concentration of the buffer used. A careful investigation of the effect of the composition of the buffer and of the total ionic strength invariably produces surprising results and insights.

7 It is not necessary to have precise knowledge of the quantitative specificity of forces, only an awareness of specificity so that it can be exploited.

References

[1] M. Boström, E. R. A. Lima, F. W. Tavares and B. W. Ninham, *J. Chem. Phys.* **128** (2008), 135104 1–4.

[2] M. Boström, V. Deniz, G. Franks and B. W. Ninham, *Adv. Coll. Interface Sci.* **123–126** (2006), 5–15.

[3] M. Boström, V. Deniz and B. W. Ninham, *J. Phys. Chem. B* **110** (2006), 9645–9649.

[4] M. Boström and B. W. Ninham, *J. Phys. Chem. B* **108** (2004), 12593–12595.

[5] M. Boström and B. W. Ninham, *Biophys. Chem.* **114** (2005), 95–101.

[6] D. F. Parsons and B. W. Ninham, *J. Phys. Chem. A* **113** (2009), 1141–1150.

[7] J. Mahanty and B. W. Ninham, *Dispersion Forces*. London: Academic Press (1976).

[8] R. Kjellander, S. Marcelja, R. M. Pashley and J. P. Quirk, *J. Chem. Phys.* **92** (1990), 4399–4407.

[9] R. R. Netz, *Curr. Op. Coll. Interface Sci.* **9** (2004), 192–197.

[10] T. J. Evens and R. P. Niedz, *Scholarly Res. Exch.* (2008), 818461.

[11] M. Diesendorf and B. W. Ninham, *J. Math. Phys.* **9** (1968), 745–752.

[12] K. D. Collins, *Biophys. Chem.* **119** (2006), 271–281.

[13] A. Salis, M. C. Pinna, D. Bilaničová, M. Monduzzi, P. Lo Nostro and B. W. Ninham, *J. Phys. Chem. B* **110** (2006), 2949–2956.

[14] A. Voinescu, P. Bauduin, C. Pinna, D. Touraud, W. Kunz and B. W. Ninham, *J. Phys. Chem. B* **110** (2006), 8870–8876.

[15] K. D. Collins, *Methods* **34** (2004), 300–311.

[16] G. A. Krestov, *Thermodynamics of Solvation*. New York: Ellis Horwood (1991).

[17] F. J. Millero, *Chem. Rev.* **71** (1971), 147–176.

[18] S. R. Raju and S. V. Talekar, *Physiol. Chem. Phys.* **10** (1978), 375–380.

[19] G. J. Safford, P. S. Leung, A. W. Naumann and P. C. Schaffer, *J. Chem. Phys.* **50** (1969), 4444–4467.

[20] D. W. McCall and D. C. Douglass, *J. Phys. Chem.* **69** (1965), 2001–2011.

[21] H. D. B. Jenkins and Y. Marcus, *Chem. Rev.* **95** (1995), 2695–2724.

[22] M. Boström, W. Kunz and B. W. Ninham, *Langmuir* **21** (2005), 2619–2623.

[23] L. Vrbka, M. Mucha, B. Minofar, P. Jungwirth, E. C. Brown and D. J. Tobias, *Curr. Op. Coll. Interface Sci.* **9** (2004), 2004, 67–73.

[24] P. B. Petersen and R. J. Saykally, *J. Phys. Chem. B* **110** (2006), 14060–14073.

[25] H. I. Petrache, T. Zemb, L. Belloni and V. A. Parsegian, *Proc. Natl. Acad. Sci. USA* **103** (2006), 7982–7987.

[26] K. P. Ananthapadmanabhan and E. D. Goddard, *Langmuir* **3** (1987), 25–31.

[27] H. Schott, *J. Coll. Interface Sci.* **173** (1995), 265–277.

[28] K. Weckström and A. C. Papageorgiou, *J. Coll. Interface Sci.* **310** (2007), 151–162.

[29] M. Lagi, P. Lo Nostro, E. Fratini, B. W. Ninham and P. Baglioni, *J. Phys. Chem. B* **111** (2007), 589–597.

[30] C. A. Ericsson, O. Söderman, V. M. Garamus, M. Bergström and S. Ulvenlund, *Langmuir* **20** (2004), 1401–1408.

[31] H. I. Petrache, S. Tristram-Nagle, D. Harries, N. Kucerka, J. F. Nagle and V. A. Parsegian, *J. Lipid Res.* **47** (2006), 302–309.

[32] L. Piculell and S. Nilsson, *J. Phys. Chem.* **93** (1989), 5596–5601.

[33] R. L. Baldwin, *Biophys. J.* **71** (1996), 2056–2063.

[34] M. W. Washabaugh and K. D. Collins, *J. Biol. Chem.* **261** (1986), 12477–12485.

[35] K. Saegusa and F. Ishii, *Langmuir* **18** (2002), 5984–5988.

[36] A. Kabalnov, U. Olsson and H. Wennerström, *J. Phys. Chem.* **99** (1995), 6220–6230.

[37] M. T. Ru, S. Y. Hirokane, A. S. Lo, J. S. Dordick, J. A. Reimer and D. S. Clark, *J. Am. Chem. Soc.* **122** (2000), 1565–1571.

[38] H.-K. Kim, E. Tuite, B. Nordén and B. W. Ninham, *Eur. Phys. J. E* **4** (2001), 411–417.

[39] P. Bauduin, F. Nohmie, D. Touraud, R. Neueder, W. Kunz and B. W. Ninham, *J. Mol. Liq.* **123** (2006), 14–19.

[40] M. C. Pinna, A. Salis, M. Monduzzi and B. W. Ninham, *J. Phys. Chem. B* **109** (2005), 5406–5454.

[41] D. Bilaničová, A. Salis, B. W. Ninham and M. Monduzzi, *J. Phys. Chem. B* **112** (2008), 12066–12072.

[42] A. Salis, D. Bilaničová, B. W. Ninham and M. Monduzzi, *J. Phys. Chem. B* **111** (2007), 1149–1156.

[43] P. Lo Nostro, B. W. Ninham, A. Lo Nostro, G. Pesavento, L. Fratoni and P. Baglioni, *Phys. Biol.* **2** (2005), 1–7.

[44] S. Rossi, P. Lo Nostro, B. W. Ninham and P. Baglioni, *J. Phys. Chem. B* **111** (2007), 10510–10519.

[45] P. Lo Nostro, L. Fratoni, B. W. Ninham and P. Baglioni, *Biomacromolecules* **3** (2002), 1217–1224.

[46] C. L. Henry, C. N. Dalton, L. Scruton and V. S. J. Craig, *J. Phys. Chem. C* **111** (2007), 1015–1023.

[47] B. Lonetti, P. Lo Nostro, B. W. Ninham and P. Baglioni, *Langmuir* **21** (2005), 2242–2249.

[48] P. Lo Nostro, J. R. Lopes, B. W. Ninham and P. Baglioni, *J. Phys. Chem. B* **106** (2002), 2166–2174.

[49] P. Lo Nostro, B. W. Ninham, M. Ambrosi, L. Fratoni, S. Palma, D. Allemandi and P. Baglioni, *Langmuir* **19** (2003), 9583–9591.

[50] G. S. Attard, S. Fuller and G. J. T. Tiddy, *J. Phys. Chem. B* **104** (2000), 10426–10436.

[51] B. W. Ninham and V. V. Yaminski, *Langmuir* **13** (1997), 2097–2108.

[52] A. A. Zavitsas, *J. Phys. Chem. B* **105** (2001), 7805–7817.

[53] M. Y. Kiriukhin and K. D. Collins, *Biophys. Chem.* **99** (2002), 155–168.

[54] V. Knecht, H. J. Risselada, A. E. Mark and S. J. Marrink, *J. Coll. Interface Sci.* **318** (2008), 477–486.
[55] R. A. Robinson and R. H. Stokes, *Electrolyte Solutions*. London: Butterworths (1959).
[56] J. M. L. Poiseuille, *Ann. Chim. Phys.* **21** (1847), 76–109.
[57] G. Jones and M. Dole, *J. Am. Chem. Soc.* **51** (1929), 2950–2964.
[58] G. Hefter, P. M. May, P. Sipos and A. Stanley, *J. Mol. Liq.* 103–**104** (2003), 261–273.
[59] T. Satoh and K. Hayashi, *Bull. Chem. Soc. Jpn* **34** (1961), 1260–1264.
[60] M. Afzal, M. Saleem and M. T. Mahmood, *J. Chem. Eng. Data* **34** (1989), 339–346.
[61] G. J. Janz, B. G. Oliver, G. R. Lakshminarayanan and G. E. Mayer, *J. Phys. Chem.* **74** (1970), 1285–1289.
[62] A. Anderko and M. M. Lencka, *Ind. Eng. Chem. Res.* **36** (1997), 1932–1943.
[63] J. F. Chambers, J. M. Stokes and R. H. Stokes, *J. Phys. Chem.* **60** (1956), 985–986.
[64] G. Jones and C. F. Bickford, *J. Am. Chem. Soc.* **56** (1934), 602–611.
[65] J. F. Chambers, *J. Phys. Chem.* **62** (1958), 1136–1138.
[66] A. Chandra and B. Bagchi, *J. Chem. Phys.* **110** (1999), 10024–10034.
[67] D. W. McCall and D. C. Douglass, *J. Phys. Chem.* **69** (1965), 2001–2011.
[68] J. V. Leyendekkers and R. J. Hunter, *J. Phys. Chem.* **81** (1977), 1657–1663.
[69] Y. Marcus, *J. Solution Chem.* **23** (1994), 831–848.
[70] Y. Marcus, *Chem. Rev.* **109** (2009), 1346–1370.
[71] Y. Avnir and Y. Barenholz, *Anal. Biochem.* **347** (2005), 34–41.
[72] W. Kunz, J. Henle and B. W. Ninham, *Curr. Op. Coll. Interface Sci.* **9** (2004), 19–37.
[73] D. Tabor and R. H. S. Winterton, *Proc. R. Soc. Lond. A* **312** (1969), 435–450.
[74] J. N. Israelachvili and D. Tabor, *Proc. R. Soc. Lond. A* **331** (1972), 19–38.
[75] W. A. Ducker, T. J. Senden and R. M. Pashley, *Langmuir* **8** (1992), 1831–1836.
[76] R. M. Pashley, *J. Coll. Interface Sci.* **83** (1982), 531–546.
[77] R. M. Pashley, *J. Coll. Interface Sci.* **80** (1981), 153–162.
[78] R. M. Pashley, *Adv. Coll. Interface Sci.* **16** (1982), 57–62.
[79] B. W. Ninham and V. A. Parsegian, *J. Theor. Biol.* **31** (1971), 405–428.
[80] S. J. Mikavic and B. W. Ninham, *J. Coll. Interface Sci.* **134** (1990), 305–311.
[81] M. Dishon, O. Zohar and U. Sivan, *Langmuir* **25** (2009), 2831–2836.
[82] M. Colic, G. V. Franks, M. L. Fisher and F. F. Lange, *Langmuir* **13** (1997), 3129–3135.
[83] G. V. Franks, S. B. Johnson, P. J. Scales, D. V. Boger and T. W. Healy, *Langmuir* **15** (1999), 4411–4420.
[84] S. B. Johnson, P. J. Scales and T. W. Healy, *Langmuir* **15** (1999), 2836–2843.
[85] S. B. Johnson, G. V. Franks, P. J. Scales and T. W. Healy, *Langmuir* **15** (1999), 2844–2853.
[86] G. V. Franks, *J. Coll. Interface Sci.* **249** (2002), 44–51.
[87] G. V. Franks and L. Meagher, *Coll. Surf. A* **214** (2003), 99–110.
[88] G. V. Franks and Y. Gan, *J. Am. Ceram. Soc.* **90** (2007), 3373–3388.
[89] R. M. Pashley and J. P. Quirk, *Coll. Surf.* **9** (1984), 1–17.
[90] R. M. Pashley and J. N. Israelachvili, *J. Coll. Interface Sci.* **101** (1984), 511–523.
[91] M. Boström, D. R. Williams and B. W. Ninham, *Langmuir* **18** (2002), 6010–6014.
[92] M. G. Cacace, E. M. Landau and J. J. Ramsden, *Q. Rev. Biophys.* **30** (1997), 241–277.
[93] R. Mancinelli, A. Botti, F. Bruni, M. A. Ricci and A. K. Soper, *J. Phys. Chem. B* **111** (2007), 13570–13577.

[94] R. Leberman and A. K. Soper, *Nature* **378** (1995), 364–366.
[95] E. Sedlák, L. Stagg and P. Wittung-Stafshede, *Arch. Biochem. Biophys.* **479** (2008), 69–73.
[96] M. C. Pinna, P. Bauduin, D. Touraud, M. Monduzzi, B. W. Ninham and W. Kunz, *J. Phys. Chem. B* **109** (2005), 16511–16514.
[97] L. M. Pegram and M. T. Record Jr., *Chem. Phys. Lett.* **467** (2008), 1–8.
[98] L. M. Pegram and M. T. Record Jr., *J. Phys. Chem. B* **111** (2007), 5411–5417.
[99] H. L. Clever, *J. Chem. Eng. Data* **28** (1983), 340–343.
[100] J. Setschenow, *Z. Phys. Chem.* **4** (1889), 117–125.
[101] W. F. McDevit and F. A. Long, *J. Am. Chem. Soc.* **74** (1952), 1773–1781.
[102] A. A. Green, *J. Biol. Chem.* **95** (1932), 47–66.
[103] E. Demou and D. J. Donaldson, *J. Phys. Chem. A* **106** (2002), 982–987.
[104] A. Voet, *Chem. Rev.* **20** (1937), 169–179.
[105] P. Lo Nostro, L. Fratoni, B. W. Ninham and P. Baglioni, *Biomacromolecules* **3** (2002), 1217–1224.
[106] O. Ya. Samoilov, *Disc. Faraday Soc.* **24** (1957), 141–146.
[107] O. Ya. Samoilov, *Structure of Aqueous Electrolyte Solutions and the Hydration of Ions*. New York: Consultants Bureau (1965).
[108] K. D. Collins, G. W. Neilson and J. E. Enderby, *Biophys. Chem.* **128** (2007), 95–104.
[109] Y. Marcus, *Ion Solvation*. Chichester: John Wiley (1985).
[110] W. Melander and C. Horváth, *Arch. Biochem. Biophys.* **183** (1977), 200–215.
[111] Y. Marcus, *J. Chem. Soc. Faraday Trans.* **87** (1991), 2995–2999.
[112] W. Kunz, P. Lo Nostro and B. W. Ninham, *Curr. Op. Coll. Interface Sci.* **9** (2004), 1–18.
[113] Y. Zhang, S. Furyk, D. E. Bergbreiter and P. S. Cremer, *J. Am. Chem. Soc.* **127** (2005), 14505–14510.
[114] Y. Cho, Y. Zhang, T. Christensen, L. B. Sagle, A. Chilkoti and P. S. Cremer, *J. Phys. Chem. B* **112** (2008), 13765–13771.
[115] M. Boström, F. W. Tavares, S. Finet, F. Skouri-Panet, A. Tardieu and B. W. Ninham, *Biophys. Chem.* **117** (2005), 115–122.
[116] M. Boström, D. R. M. Williams and B. W. Ninham, *Phys. Rev. Lett.* **87** (2001), 168103.
[117] M. Boström, D. R. M. Williams and B. W. Ninham, *J. Phys. Chem. B* **106** (2002), 7908–7912.
[118] M. Boström, D. R. M. Williams and B. W. Ninham, *Biophys. J.* **85** (2003), 686–694.
[119] M. Boström, D. R. M. Williams and B. W. Ninham, *Eur. Phys. J. E* **13** (2004), 239–245.
[120] M. Boström, D. R. M. Williams, P. R. Stewart and B. W. Ninham, *Phys. Rev. E* **68** (2003), 041902.
[121] M. Boström and B. W. Ninham, *Langmuir* **20** (2004), 7569–7574.
[122] M. Boström, F. W. Tavares, B. W. Ninham and J. M. Prausnitz. *J. Phys. Chem. B* **110** (2006), 24757–24760.
[123] L. A. Moreira, M. Boström, B. W. Ninham, E. C. Biscaia and F. W. Tavares, *J. Braz. Chem. Soc.* **18** (2007), 221–230.
[124] W. Kunz, P. Lo Nostro and B. W. Ninham, eds. *Curr. Op. Coll. Interface Sci.* **9** (2004), 1–197.
[125] R. J. Gillespie, *Coord. Chem. Rev.* **252** (2008), 1315–1327.
[126] K. H. Gustavson, Specific ion effects in the behaviour of tanning agents toward collagen treated with neutral salts. In *Colloid Symposium Monograph*, ed H. Boyer

Weiser. New York: The Chemical Catalog Company (1926), 79–101, and references therein.
[127] J. Loeb, *Science* **52** (1920), 449–456.
[128] P. Lo Nostro, B. W. Ninham, S. Milani, L. Fratoni and P. Baglioni, *Biopolymers* **81** (2006), 136–148.
[129] J. G. Kirkwood, *J. Chem. Phys.* **5** (1937), 479–491.
[130] J. F. Rabek, *Experimental Methods in Polymer Chemistry*. Chichester: Wiley (1980).
[131] W. Zhang, L. Piculell and S. Nilsson, *Macromolecules* **25** (1992), 6165–6172.
[132] L. Piculell and S. Nilsson, *J. Phys. Chem.* **93** (1989), 5596–5601, 5602–5611.
[133] L. Xu, X. Li, M. Zhai, L. Huang, J. Peng, J. Li and G. Wei, *J. Phys. Chem. B* **111** (2007), 3391–3397.
[134] S. Finet, F. Skouri-Panet, M. Casselyn, F. Bonneté and A. Tardieu, *Curr. Op. Coll. Interface Sci.* **9** (2004), 112–116.
[135] R. Vogel, *Curr. Op. Coll. Interface Sci.* **9** (2004), 133–138.
[136] M. Boström, D. R. M. Williams and B. W. Ninham, *Europhys. Lett.* **63** (2003), 610–615 and references therein.
[137] P. Baglioni, E. Fratini, B. Lonetti and S. H. Chen, *Curr. Op. Coll. Interface Sci.* **9** (2004), 38–42.
[138] M. Boström, D. R, Williams and B. W. Ninham, *Langmuir* **18** (2002), 8609–8615.
[139] Y. Zhang and P. S. Cremer, *Proc. Natl Acad. Sci. USA* **106** (2009), 15249–15253.
[140] P. Lo Nostro, B. W. Ninham, S. Milani, A. Lo Nostro, G. Pesavento and P. Baglioni, *Biophys. Chem.* **124** (2006), 208–213.
[141] G. Emilien, J. M. Maloteaux, A. Seghers and G. Charles, *Arch. Int. Pharmacodyn. Ther.* **330** (1995), 251–278.
[142] M. T. Gladwin, N. J. H. Raat, S. Shiva, C. Dezfulian, N. Hogg, D. B. Kim-Shapiro and R. P. Patel, *Am. J. Physiol. Heart Circ. Physiol.* **291** (2006), H2026–H2035.
[143] R. A. Curtis and L. Lue, *Chem. Eng. Sci.* **61** (2006), 907–923.
[144] V. G. Taratuta, A. Holschbach, G. M. Thurston, D. Blankschtein and G. B. Benedek, *J. Phys. Chem.* **94** (1990), 2140–2144.
[145] R. Aquaron, F. Delange, P. Marchal, V. Lognone and L. Ninane, *Cell. Mol. Biol.* **48** (2002), 563–569.
[146] R. Giorgi, C. Bozzi, L. Dei, C. Gabbiani, B. W. Ninham and P. Baglioni, *Langmuir* **21** (2005), 8495–8501.
[147] B. E. Conway, *Electrochim. Acta* **40** (1995), 1501–1512.
[148] P. Jungwirth and D. J. Tobias, *J. Phys. Chem. B* **106** (2002), 6361–6373.
[149] J. Cheng, C. D. Vecitis, M. R. Hoffmann and A. J. Colussi, *J. Phys. Chem. B* **110** (2006), 25598–25602.
[150] F. J. Millero, F. Huang and A. L. Laferiere, *Geochim. Cosmochim. Acta* **66** (2002), 2349–2359.
[151] E. Demou and D. J. Donaldson, *J. Phys. Chem. A* **106** (2002), 982–987.
[152] D. Rosenfeld, *Science* **287** (2000), 1793–1796.
[153] E. Wernersson and R. Kjellander, *J. Chem. Phys.* **129** (2008), 144701.
[154] E. Wernersson and R. Kjellander, *J. Chem. Phys.* **125** (2006), 154702.
[155] R. Kjellander and B. Forsberg, *J. Phys. A* **38** (2005), 5405–5424.
[156] R. Kjellander, *J. Phys. A* **39** (2006), 4631–4641.
[157] E. Wernersson and R. Kjellander, *J. Phys. Chem. B* **111** (2007), 14279–14284.

[158] D. Horinek and R. R. Netz, *Phys. Rev. Lett.* **99** (2007), 226104.
[159] R. R. Netz, *Phys. Rev. E* **51** (1995), 2286–2294.
[160] R. R. Netz and H. Orland, *Europhys. Lett.* **45** (1999), 726–732.
[161] D. Horinek, A. Serr, D. J. Bonthuis, M. Boström, W. Kunz and R. R. Netz, *Langmuir* **24** (2008), 1271–1283.
[162] M. Lund, R. Vacha and P. Jungwirth, *Langmuir* **24** (2008), 3387–3391.
[163] C. D. Wick, L. X. Dang and P. Jungwirth, *J. Chem. Phys.* **125** (2006), 024706.
[164] P. Jungwirth and D. J. Tobias, *J. Phys. Chem. A* **106** (2002), 379–383.
[165] J. Pittner and P. Jungwirth, *Chem. Phys. Lett.* **321** (2000), 281–286.
[166] M. Boström, D. R. M. Williams and B. W. Ninham, *Langmuir* **217** (2001), 4475–4478.
[167] M. Boström, D. R. M. Williams and B. W. Ninham, *Progr. Coll. Polymer Sci.* **123** (2004), 110–113.
[168] V. Deniz, M. Boström, D. Bratko, F. W. Tavares and B. W. Ninham, *Coll. Surfaces A* **319** (2008), 98–102.
[169] M. Boström, D. R. M. Williams and B. W. Ninham, *Curr. Op. Coll. Interface Sci.* **9** (2004), 48–52.
[170] C. Adamo and V. Barone, *J. Chem. Phys.* **110** (1999), 6158–6170.
[171] C. Adamo, G. E. Scuseria and V. Barone, *J. Chem. Phys.* **111** (1999), 2889–2899.
[172] C. Adamo, M. Cossi, G. Scalmani and V. Barone, *Chem. Phys. Lett.* **307** (1999), 265–271.
[173] R. R. Dagastine, D. C. Prieve, L. R. White and Lee R., *J. Coll. Interface Sci.* **231** (2000), 351–358.
[174] V. A. Parsegian and G. H. Weiss, *J. Coll. Interface Sci.* **81** (1981), 285–289.
[175] L. D. Landau, L. P. Pitaevskii and E. M. Lifshitz. Electrodynamics of continuous media. In *Course of Theoretical Physics*, vol. 8, ed. E. M. Lifshitz and L. P. Pitaevskii. 2nd edn. Oxford: Elsevier-Butterworth Heinemann (1993).
[176] H. Boroudjerdi, Y. W. Kim, A. Naji, R. R. Netz, X. Schlagberger and A. Serr, *Phys. Rep.* **416** (2005), 129–199.
[177] C. Kalidas, G. Hefter and Y. Marcus, *Chem. Rev.* **100** (2000), 819–852.
[178] J. Mahanty and B. W. Ninham, *J. Chem. Phys.* **59** (1973), 6158–6162.
[179] J. Mahanty and B. W. Ninham, *J. Chem. Soc. Faraday Trans. II* **70** (1974), 637–650.
[180] J. Mahanty and B. W. Ninham, *J. Chem. Soc. Faraday Trans. II* **71** (1975), 119–137.
[181] P. Koelsch, P. Viswanath, H. Motschmann, V. L. Shapovalov, G. Brezesinski, H. Möhwald, D. Horinek, R. R. Netz, K. Giewekemeyer, T. Salditt, H. Schollmeyer, R. Von Klitzing, J. Daillant and P. Guenoun, *Coll. Surfaces A* **303** (2007), 110–136.
[182] R. R. Netz and D. Andelman, *Phys. Rev. E* **55** (1997), 687–700.
[183] D. F. Parsons, V. Deniz and B. W. Ninham, *Coll. Surfaces A* **343** (2009), 57–63.
[184] K. D. Collins, *Biophys. J.* **72** (1997), 65–76.
[185] M. V. Basilevsky and D. F. Parsons, *J. Chem. Phys.* **108** (1998), 9107–9113.
[186] M. V. Basilevsky and D. F. Parsons, *J. Chem. Phys.* **108** (1998), 9114–9123.
[187] M. Alfridsson, B. W. Ninham and S. Wall, *Langmuir*, **16** (2000), 10087–10091.
[188] F. W. Tavares, D. Bratko, H. W. Blanch and J. M. Prausnitz, *J. Phys. Chem. B* **108** (2004), 9228–9235.
[189] P. Kékicheff, S. Marčelja, T. J. Senden, and V. E. Shubin, *J. Chem. Phys.* **99** (1993), 6098–6113.
[190] V. E. Shubin and P. Kékicheff, *J. Coll. Interface Sci.* **155** (1993), 108–123.
[191] Y. Marcus, *Pure Appl. Chem.*, **59** (1987), 1093–1101.
[192] M. Boström, V. Deniz, G. V. Franks and B. W. Ninham, *Adv. Coll. Interface Sci.* **123**–126 (2006), 5–15.

[193] D. C. Grahame, *Chem. Rev.* **41** (1947), 441–501.
[194] A. Martín-Molina, J. G. Ibarra-Armenta and M. Quesada-Pérez, *J. Phys. Chem. B* **113** (2009), 2414–2421.
[195] P. Richmond, B. W. Ninham and R. H. Ottewill, *J. Coll. Interface Sci.* **45** (1973), 69–80.
[196] J. N. Munday, F. Capasso and V. A. Parsegian, *Nature* **457** (2009), 170–173.
[197] V. A. Parsegian and G. H. Weiss, *J. Adhesion* **3** (1972), 259–267.
[198] V. A. Parsegian and G. H. Weiss, *J. Coll. Interface Sci.* **40** (1972), 35–41.
[199] E. R. Smith and B. W. Ninham, *Physica* **66** (1973), 111–130.
[200] V. A. Parsegian, *Van der Waals Forces: a handbook for biologists, chemists, engineers, and physicists*. Cambridge: Cambridge University Press (2006).
[201] P. Attard, D. J. J. Mitchell and B. W. Ninham, *Biophys. J.* **53** (1988), 457–460.
[202] P. Richmond and B. W. Ninham, *J. Phys. C* **4** (1971), 1988–1993.
[203] B. Davies and B. W. Ninham, *J. Chem. Phys.* **56** (1972), 5797–5801.
[204] V. A. Parsegian, *J. Chem. Phys.* **56** (1972), 4393–4398.
[205] D. J. Mitchell, B. W. Ninham and P. Richmond, *J. Theor. Biol.* **37** (1972), 251–259.
[206] P. Richmond, B. Davies and B. W. Ninham, *Phys. Lett.* **39A** (1972), 301–302.
[207] B. W. Ninham and S. Lidin, *Acta Cryst. A* **48** (1992), 640–649.
[208] D. W. Thompson, *On Growth and Form.* Cambridge: Cambridge University Press (1917).
[209] H. K. Christenson, J. Fang, B. W. Ninham and J. L. Parker, *J. Phys. Chem.* **94** (1990), 8004–8006.
[210] J. L. Parker, H. K. Christenson and B. W. Ninham, *J. Phys. Chem.* **92** (1988), 4155–4159.

8

Effects of dissolved gas and other solutes on hydrophobic interactions

8.1 Bubble–bubble coalescence

8.1.1 Effect of electrolytes

The Hofmeister effects we have just discussed required a complete revision of theoretical ideas, and a revision of the way we interpret measurements that date back a century or more. We come now to some neglected phenomena that throw the entire theoretical applecart upside down again.

It is an extraordinary fact that an explanation of what appears the simplest conceivable experiment remains elusive. Breaking waves in the ocean are foamy, those in fresh water are not. Bubbles in salt water do not fuse, those in fresh water do. The experiment – illustrated in Figs. 8.1 and 8.2 – quantifies this effect, which is quite dramatic [1–4].

Bubbles of nitrogen pass through a glass frit, as in a fish tank, and ascend a column. As they do they collide and fuse, and the column stays clear. (There is no problem with impurities that might affect matters. The column is self-cleaning.)

If salt is added, nothing much occurs until around 0.1 M. Suddenly the bubbles do not fuse, they remain small and the column becomes a mass of small bubbles. The change can be monitored by a laser diode that measures transmission of light through the column, and the scale gives per cent coalescence vs. salt concentration. (The same experiments can be done with single bubble–bubble interactions.)

A range of salt 1:1, 1:2, 2:1, 3:1 electrolytes give precisely the same curve if coalescence is plotted against Debye length rather than concentration (see Fig. 8.2). For all these salts there is complete bubble–bubble fusion inhibition at an equivalent salt concentration of 0.175 M. But for other salts there is no effect whatever on the bubble interaction! The effect does not change if the water is saturated with different gases, e.g., N_2, or SF_6.

Astonishingly, there is a universal rule that categorizes this behaviour: a characteristic or 'flavour', α or β, can be assigned to each anion or cation, depending on

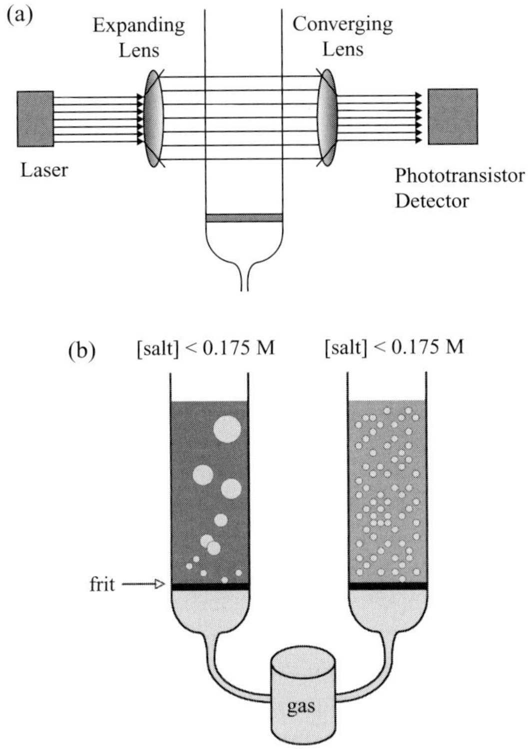

Fig. 8.1. Simple schematic of apparatus illustrating bubble–bubble interactions. Adapted with permission from Ref. 1. Copyright 1993 American Chemical Society.

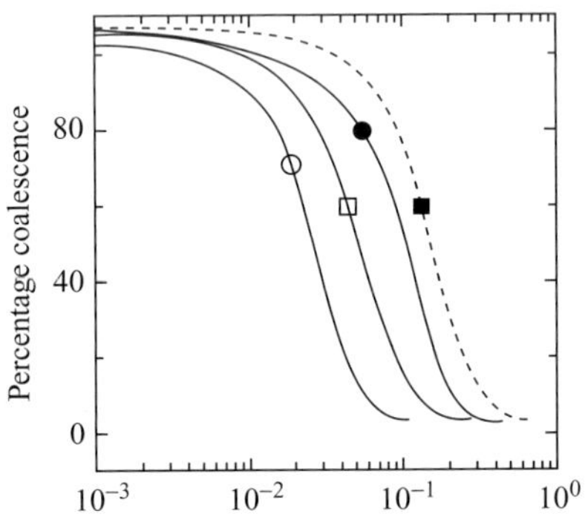

Fig. 8.2. Percentage coalescence as a function of salt concentration for $MgSO_4$ (○), $CaCl_2$ (□), $NaNO_3$ (■) and NaCl (•) [2]. Adapted with permission from Ref. 1. Copyright 1993 American Chemical Society.

its capacity to induce coalescence inhibition or not. Then a universal rule emerges. The rule is that for $\alpha\alpha$ or $\beta\beta$ pairs fusion of bubbles is inhibited following Fig. 8.1 and Table 8.1. On the other hand for $\alpha\beta$ or $\beta\alpha$ pairs there is no effect whatever. Bubbles continue to fuse up to 6 M!

With mixed electrolytes the rules become a little more complicated but remain universal (see Table 8.2). An extensive study of mixtures of different salts [4] which separately give either the $\alpha\alpha$, $\beta\beta$ behaviour and $\alpha\beta$, $\beta\alpha$ phenomena yields a similar and wider systematic set of combining rules. A proposal due to Marcelja attributes the effect to the differences in adsorption of ion pairs at the air–water interface. This might be expected to affect rates of drainage of the film of water between two approaching bubbles. But the range of the effects at which coalescence occurs is hundreds of nm, which appears to rule out such an idea.

8.1.2 Effects of sugars

If correct, such ideas would not seem to explain the same kinds of effects that occur also with different isomeric sugar solutions and their mixtures rather than salts. The bubble–bubble inhibition phenomena here occur over the concentration range 0.5–1 M rather than 0.1–0.175 M as for salts. For glucose the fusion reduction has one form. For fructose it is different in magnitude. The sum of the two effects is identical to that produced by sucrose! But the effect of an equimolar concentration of glucose and fructose equivalent to that of sucrose gives quite different magnitudes for the effects [1–6].

Bulk water 'structure' induced by the sugars is then certainly implicated in this case. (The distance apart of the 'surfaces' of sugar molecules at 1 M is about 3 nm, at most six water molecules.)

Except for the universal scaling of the effect with the Debye length when the effect occurs, there is no correlation whatever with any standard theory or parameter, such as surface tension decrease with salt that reflects ion adsorption. (Direct and indirect effects such as viscosity and surface elasticity due to electrostatic forces would be expected to be screened and reduced with added salt so that fusion should be reduced by these effects.)

The fusion phenomenon involves macroscopic bubbles and takes place over very large distances, at least hundreds of nm.

8.2 Colloid stability and dissolved gas

8.2.1 The role of dissolved gas and other solutes in colloidal interactions

The air bubble fusion effects depend on salt type and concentration, that is they are specific ion effects par excellence. Likewise they depend on dissolved atmospheric gas in the water between the bubbles.

Table 8.1. *Combining rules for bubble–bubble coalescence inhibition due to single electrolytes in aqueous solution at 20°C [2].* $\alpha\alpha$ *or* $\beta\beta$ *inhibit coalescence (✓);* $\alpha\beta$ *or* $\beta\alpha$ *have no effect on coalescence (×). Adapted with permission from Ref. 2. Copyright 2007 American Chemical Society.*

Ions		H^+	Li^+	Na^+	K^+	Cs^+	Mg^{2+}	Ca^{2+}	NH_4^+	$MeNH_3^+$	$Me_2NH_2^+$	Me_3NH^+	Me_4N^+
		β	α	α	α	α	α	α	α	β	β	β	β
OH^-	α	×		✓	✓								
Cl^-	α	×	✓	✓	✓		✓	✓		×	×	×	×
Br^-	α	×		✓	✓	✓							×
NO_3^-	α	×	✓	✓	✓			✓					
SO_4^{2-}	α	×	✓	✓	✓		✓						
$C_2O_4^{2-}$	α	×			✓								
IO_3^-	α			✓									
ClO_3^-	β			×									
ClO_4^-	β	✓		×			×		×				
AcO^-	β	✓		×	×	×	×		×				✓
SCN^-	β			×									

Table 8.2. *Combining rules for bubble–bubble coalescence inhibition due to mixtures of electrolytes in aqueous solution at 20 °C [2].* $\alpha\alpha$ *or* $\beta\beta$ *inhibit coalescence (✓);* $\alpha\beta$ *or* $\beta\alpha$ *have no effect on coalescence (×). Values of the surface tension increment* $(\partial\Delta\gamma/\partial c)$ *at 20 °C for the salt mixtures. Adapted with permission from Ref. 2. Copyright 2007 American Chemical Society.*

Combinations	Electrolyte mixtures	$\partial\Delta\gamma/\partial c$	Coalescence inhibition
$\alpha\alpha,\alpha\alpha$	$KOH+NaCl$	1.6	✓
	$NaCl+Ca(NO_3)_2$		✓
$\alpha\alpha,\alpha\beta$	$KCl+AcONa$	1.01	
	$KCl+NaClO_3$	1.2	
	$NaCl+NaClO_4$		✓
	$NaClO_4+Ca(NO_3)_2$		✓
	$NaCl+NaClO_3$		✓
	$NaNO_3+Ca(ClO_4)_2$		✓
$\alpha\alpha,\beta\alpha$	$NaCl+HCl$	0.45	✓
	$NaNO_3+HCl$	0.28	✓
$\alpha\alpha,\beta\beta$ **or** $\alpha\beta,\beta\alpha$	$NaCl+HClO_4$ or $NaClO_4+HCl$	−0.23	✓
	$KClO_3+HNO_3$	−0.23	✓
	$NaClO_3+HCl$	−0.18	✓
	$KCl+HClO_4$	−0.17	✓
	$Ca(ClO_4)_2+HCl$	−0.32	✓
	$Ca(ClO_4)_2+HNO_3$		✓
	$Ca(ClO_4)_2+HClO_4$		✓
	$NaOH+HClO_4$	0.33	
	$NaOH+AcOH$	0.72	
	$AcONa+HCl$	6.9	
	$KCl+AcOH$	8.0	
$\alpha\beta,\alpha\beta$	$AcOK+NaClO_3$	0.66	×
	$NaClO_3+NaClO_4$		×
	$AcONa+AcONH_4$		×
	$AcONa+Ca(ClO_4)_2$	0.49	
$\alpha\beta,\beta\beta$	$NaClO_3+HClO_4$	−0.69	✓
	$AcONa+HClO_4$	6.3	
	$NaClO_3+AcOH$	9.0	
$\beta\alpha,\beta\alpha$	HNO_3+Me_4NCl		×
	$HCl+HNO_3$		×
$\beta\alpha,\beta\beta$	$HCl+HClO_4$	−9.4	✓
	$Me_4NCl+AcONMe_4$		✓

In searching for a guide to whatever mechanism is involved it seems sensible then to explore effects of dissolved gas on colloidal particle interactions. Again, astonishingly, this had never been considered until recently. The effects of removal of gas are dramatic [7].

The experiment measured flocculation rates of a simple hydrophobic colloid, paraffin particles functionalized by adsorbing and anchoring steric acid surfactants in their surfaces. To all intents and purposes these are innocuous perfectly normal colloids with a reasonable surface (zeta) potential of 40 mV. If one varies the counterion through the Hofmeister series one finds, first, that these flocculation rates in no way correspond to the predictions of DLVO theory. That we have now learnt to anticipate. But more interesting is this: the dissolved gas is evacuated and flocculation rates are studied as a function of added salt. The salt reduces the electrostatic interactions, a matter that in the DLVO theory should *enhance* the flocculation rates. In fact the flocculation rates are reduced by factors of one or two orders of magnitude! Dissolved gas apparently enhances attractive interactions between the 'hydrophobic' particles. Its removal turns the interactions off.

8.2.2 Emulsion stability and dissolved gas

The same effects occur with emulsions made from oil and water only with exclusion of atmospheric gas. The emulsions are normally metastable and droplets of oil rapidly coalesce. With removal of atmospheric gas they do not fuse and the emulsions are stable at least for months. Surfactants adsorbed at the oil drop–water interface are considered essential to stabilize an emulsion. But here there are no surfactants. These emulsions are just oil in water. For an extensive review see Ref. 8.

With emulsion polymerization involving surfactants the same effects occur. It is clear here that the accepted theory of emulsion polymerization that underlies the huge field of latex synthesis is quite wrong. The process depends intimately on dissolved gas [9].

8.2.3 Exploiting gas dependence

Clearly these effects can be exploited in water purification and many other separation technologies. For example with hydrophobic proteins produced by bacteria with genetic engineering, the desired macromolecules are held together by 'hydrophobic forces' and are very difficult to separate. They can be separated into individual macromolecules using charged surfactants. Then the surfactants have to be stripped off in an expensive ion-exchange column. But if such a system is degassed with a simple water pump and gently centrifuged the protein clumps separate into individual macromolecules.

A major problem for workers in protein structure determination via X-ray crystallography is that of forming an ordered crystal. That can be achieved by degassing a suspension by adding a salt in the Hofmeister series such as sodium acetate rather than chloride. The removal of strongly attractive hydrophobic forces from the system and imposition of stronger repulsive forces by the salt makes the molecular suspension stable. Then a slow precipitation rate and ordering of the crystal can be achieved by ion exchange with a more strongly bound ion such as chloride.

8.3 Other phenomena affected by dissolved gas

A number of other experiments all confirm this. A significant effect of dissolved gas was also found in the phase-separation temperature of aqueous dispersion of diC$_8$PC [10], in the formation of supramolecular host–guest complexes (pseudopolyrotaxanes) [11], particle coagulation [12] and solubility of salts [13].

The mystery has evidently deepened.

8.4 Water structure as revealed by laser cavitation: bubble–bubble experiments in electrolytes

A long-standing problem is that of cavitation in liquids. The tensile strength of water is hundreds of atmospheres less than what we would expect from apparently reasonable theoretical estimates. For solids the same is true. There the reason is well known. In solid mechanics the strength of materials is explained by a theory of Griffiths. It is due to the existence of impurities, defects, that enhance crack propagation. The strength of a material is much less than what one would have expected for the pure solid. In the same way it seems that dissolved gas molecules act as defects in the liquid. Further light on the problem of the role of gas in mediating interactions comes from studies of laser cavitation of bulk water. The experiments are simple. Nanosecond neon laser pulses are fired through water. Every 20 or so pulses result in a brilliant flash of light. This reflects a massive dielectric breakdown. But when dissolved gas is pumped out no such breakdown occurs. When the laser pulses are passed near hydrophobic or hydrophilic surfaces the rate of flashes is in the first case enhanced by a factor of 50. In the second case they are reduced dramatically. It is known that gas adsorbs preferentially near hydrophobic surfaces and is reduced near hydrophilic surfaces. Evidently gas induces long-range 'water structure'.

When the water contains NaCl the structure revealed by experiments involving the Rayleigh wings for low-angle light scattering is quite different above and below, the critical concentration (0.1–0.175 M) for bubble–bubble fusion inhibition. With NaOAc, for which no bubble–bubble fusion inhibition occurs, no change in water structure is revealed. There is a cooperativity between water structure that depends on a competition between the ions of the electrolyte salt and dissolved gas molecules [14–18].

8.5 Mechanisms of bubble–bubble and long-range 'hydrophobic' interactions

In the light of the above experiments it seems reasonable to consider dissolved gas molecules of diameter about 0.4 nm as hydrophobic impurities, defects in the dynamic structure of pure water.

These gas molecule impurities will certainly induce changes in the arrangements of water molecules around them. So too at the surface of the bubble, cooperative subcritical fluctuations that impose order on the arrangements of dipole moments of water molecules (if one likes, cracks in the dynamic structure of water) can extend from the surface to a gas molecule. These fluctuating 'cracks' correspond to density lowering that would occur in a cylindrical 'submolecular tube' of effective diameter about 0.01 nm [19].

As far as bulk water molecules involved in such a fluctuation are concerned, a gas molecule presents as a quite large surface. Consequently subcritical fluctuations that normally would extend at most six water molecules between two surfaces can percolate from gas molecule to gas molecule across the macroscopic gap between two surfaces. These subcritical fluctuations in density can then lead to a density lowering in the form of cylindrical microtubes connecting the bubbles. Then fusion will occur in the same manner as it does for fusion of emulsion drops [20]. The effects are larger near a hydrophobic surface than near hydrophilic ones due to the differences in gas density. These notions are consistent with all the experimental observations above, consistent with the known reduction in tensile strength of water due to dissolved gas, and with the corresponding problem in solid-state mechanics.

When salts are added two effects occur. One is that adsorption of some ions – such as bromide – to the bubble interface can enhance or oppose crack propagation. The second is that the local water structure around and between ion pairs can either enhance or diminish the percolation mechanism. The hydration characteristics oppose or enhance long-range propagation of subcritical density fluctuations due to dissolved atmospheric gas. These 'defect-mediated cracks' propagated via long-range ordering of dipoles via gas and salt molecules extend throughout the medium from one surface to another. Once formed, they can open up to form genuine channels between the bubble surfaces (Fig. 8.3); or, with salts that inhibit the bubble fusion process, close the formation of a channel down.

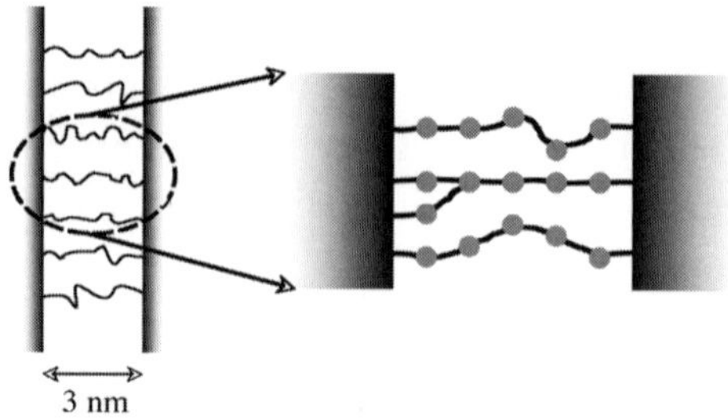

Fig. 8.3. Illustration of possible mechanism of long-range hydrophobic interactions. Dissolved gas molecules or other weakly soluble hydrophobic molecules act as defects that propagate subcritical density fluctuations across a gap between two hydrophobic particles. The resulting lowering of density inside the gap produces the attractive forces. Salt molecules (or other solutes) can either enhance or oppose such fluctuations.

The dynamic fluctuations due to water molecules at a hydrophobic surface are slight rearrangements that can be envisaged as 'cracks' due to dipole and hydrogen bond reorientations of width only about 0.1 Å. These subcritical fluctuations give rise to a lowering of the density in the gap between the surfaces, and the outside pressure drives the surface together, to give a hydrophobic force. The surface-induced fluctuations extend at most about six water diameters, 3 nm, in agreement with simulation studies.

With dissolved atmospheric gas (or other hydrophobic solutes such as methane), the gas molecules are on average 3 nm apart. The molecules act to propagate the cracks and density lowering over much large distances.

When salt is added the local ordering of water around the ions can either oppose or enhance the crack propagation effect, and therefore the range of the attraction.

This mechanism might dictate the ion pair specificity of bubble–bubble interactions.

A coalescence of cracks can lead to a macroscopic cylindrical connection between the bubbles that could drive bubble fusion.

8.6 Bubble–bubble experiments in non-aqueous solvents

Recently the same experiments have been carried out in a variety of non-aqueous solvents: methanol, formamide, propylene carbonate and dimethylsulfoxide [3].

Coalescence inhibition by electrolytes was observed in all the solvents, at a lower concentration range (0.01 M to 0.1 M) than that observed in water. Formamide shows ion-specific salt effects dependent upon ion combinations in a way analogous to the combining rules observed in water. Bubble coalescence in propylene carbonate is also consistent with ion-combining rules, but the ion assignments differ from those for water. In both methanol and DMSO all salts used were found to inhibit bubble coalescence. The results show that electrolytes influence bubble coalescence in a rich and complex way, but with notable similarities across all solvents tested. Coalescence is influenced by the drainage of fluid between two bubbles to form a film and then the rupture of the film and one might expect that these processes will vary dramatically between solvents. The conclusion of the authors is 'that similarities in behavior show that coalescence inhibition is unlikely to be related to the surface forces present. It is perhaps related to the dynamic thinning and rupture of the liquid film through the hydrodynamic boundary conditions'.

That is the mechanism, related to the tensile strength of the liquid–electrolyte–dissolved gas mixture, might be as explained above.

Whatever the explanation the phenomenon is real and dramatic. So long as it remains unexplained, there must remain major questions surrounding present theories that purport to describe the real world [1,5,6].

8.7 Hydrophobic interactions and the hydrophobic effect

There are many kinds of 'hydrophobic' interactions. The term 'hydrophobic' first appeared early in the twentieth century. In the field of protein chemistry, it was coined to describe an intermolecular force associated with the negative entropy of dissolution of nonpolar gases in water.

The free energy of transfer of a hydrophobic molecule from water to oil is qualitatively given in terms of the dispersion self energy of a molecule. The uniform solvent model captures only some features of the free energy transfer mechanism, and probably not the most essential feature.

The words hydrophobic, hydrophilic, amphiphilic describe molecules that have water-hating or water-loving features or embrace elements of both, respectively. Oil is sparingly soluble in water and is hydrophobic; salt molecules dissolve in water and are (usually) hydrophilic, and surfactants, with a hydrophobic (oil-like) tail and a hydrophilic headgroup, are amphiphilic. But the words have never been defined, and probably never can be. Most biological molecules such as amino acids, proteins or biopolymers such as DNA and RNA are made up of molecular groups that have elements of both hydrophilic and hydrophobic properties. Even simple electrolytes such as tetramethyl ammonium chloride have a hydrophobic cation and and hydrophilic anion (chloride). Some 'hydrophobic' ions such as asymmetric quaternary ammonium ions have such an affinity for water that they actually remove water from phosphorus pentoxide!

While the terms cannot be defined, the distinction is useful, provided it is not carried too far.

The energetic–entropic demands of the hydrophobic moiety of surfactant or lipid molecules in water are responsible for their association properties. Such molecules associate and self-assemble only in water. At a sufficiently high concentration they associate to form aggregates like micelles and bilayer membranes. Their hydrophobic hydrocarbon tails form an oil-like interior protected and separated from water by the hydrophilic headgroups. The free energy of transfer of a small oil molecule from bulk oil to water is unfavourable, and comparable in magnitude with the (favourable) Born free energy of transfer of an ion pair from its crystal form to water. The interactions between two small hydrophobic molecules are very short-range and correspond to the squeezing out, or minimization of the, usually, unhappy hydration layers that surround them. They might be better termed attractive hydration forces. These short-range forces are central to any understanding of molecular self-assembly. That is our concern in Part II of this book. There is as yet no satisfactory theory of water, and so no detailed molecular theory of these forces exists.

8.8 Long-range hydrophobic forces and capillary forces: polywater

We are concerned here with interactions between macroscopic 'hydrophobic' surfaces, where something different happens to the kinds of electrostatic and quantum mechanical forces we have been dealing with.

Techniques that made possible direct measurement of surface forces proliferated in the 1980s. Most work was focussed on the DLVO theory and its extensions. But the most pervasive kind of surface interaction, due to capillary bridging, was ignored. The results, and the theory, existed already in the 1800s. These capillary forces are those that, for example, consolidate a sand cake, for the transport of water in trees and in soils. They are responsible for the sintering of powders, for cement, fabric shrinkage on drying, the drying of paints and many hydrophobic effects in biology. The classical theory of capillarity does explain these particular long-range hydrophobic, 'non-DLVO' forces that we now consider further.

Many experiments over the past two decades demonstrated the existence of long-range forces of attraction between 'hydrophobic' surfaces in water. The word 'hydrophobic' here again defied definition. In the present context it means (a) a surface with a quite large contact angle with water, or (b) two surfaces that attract each other from quite large distances, a tautology. These long-range forces between such surfaces could not be accommodated by the usual theories of surface forces. Their range varies from several to hundreds of nanometres. Their magnitude exceeds predictions of the Lifshitz theory by orders of magnitudes. The 'decay length' of these hydrophobic attractions showed no clear correlation with contact

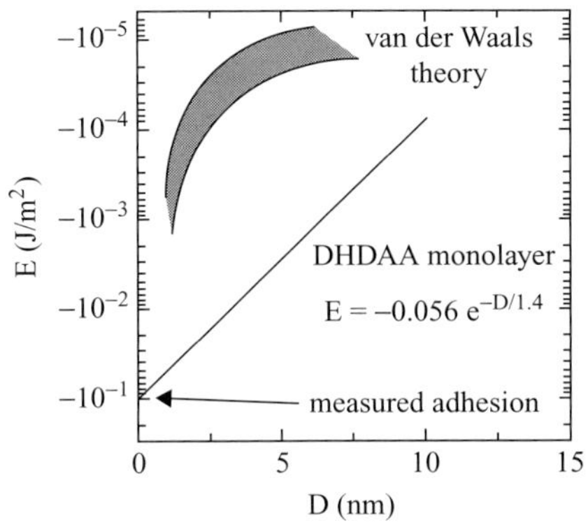

Fig. 8.4. Forces between hydrophobic monolayers of surfactants. Note that the measured force is orders of magnitude greater than van der Waals forces. The adhesion energy is exactly what one expects, precisely twice the interfacial energy of an oil–water interface. From Ref. 23. Reprinted with permission from AAAS.

angle or any other physical parameter. Sometimes the attraction changes to an equally long-range repulsion.

This direct measurement of long-range attractive forces between hydrophobic surfaces has posed a seemingly startling problem for several decades since 1980 [21,22].

It all began with the observation of an exponentially decaying attraction between mica surfaces on which were adsorbed cationic surfactant monolayers from solution (cetyl trimethyl ammonium bromide, CTAB) close to the point of zero charge. The force appeared to be exponential with a decay length of a nanometre (Fig. 8.4), and orders of magnitude larger than the expected DLVO forces. (In fact by any criterion the surfaces could not be considered 'hydrophobic' – the exposed surfaces were predominantly hydrophilic mica with a sparse coverage of hydrophobic surfactant tails.) Subsequently an even longer-range attraction was measured between mica surfaces made hydrophobic by adsorption from a dispersion of an insoluble double-chain cationic surfactant.

The cationic surfactants adsorb head down on the mica surfacc, negatively charged in water, leaving an exposed molecularly smooth oil-like monolayer on each surface [23].

They occur between mica surfaces coated with Langmuir–Blodgett deposited monolayers of lipids and between chemically bound alkylsilane-treated silica surfaces. Later, similar observations occurred with a wider variety of systems. These included polymer films produced by spin-coating, plasma polymerized films, self-assembled monolayers of alkylthiols on gold supports and chemically cross-linked LB films. In most cases mica and silica were used as the base substrates. Some AFM force measurement work was done with particles of bulk polymers. The range of such apparently exponential forces increased monotonically with time and with the large number of following papers. (Most of a large list of measurements, later, via AFM littered with artefacts, can be found in the review by Yaminski cited below.) The 'range' of these forces expanded from 1 nm to a record 300 nm. An empirical 'double exponential' fit became popular and was adopted as standard procedure to analyse experimental results. No systematic trends emerged.

Sometimes, with high-contact-angle surfaces, spontaneous cavitation occurred at short distances of separation of the surfaces, around 1 nm before contact. Such capillary cavitation was first observed by Lord Rayleigh in a brief, forgotten, note [24]. Sometimes, usually, no such cavitation occurred. The forces were smooth, both on approach and on receding.

The forces were observed with one surface hydrophobic and another hydrophilic. There were many more such observations. In the presence of electrolyte the forces could be switched off. Confusion reigned. It still does.

In fact the various 'hydrophobic' interactions form the basis of an entire vast field in its own right, that of capillary phase separation. There is no limit to the range of such effects. It turns out that many if not all the experimental observations on hydrophobic forces can be subsumed under the classical theory of capillarity as described by the Young, Laplace and Kelvin equations. Thermodynamically, no 'new' forces are required. The subject and a host of experimental observations are extensively and exhaustively reviewed by Yaminski et al. with Gibbsian thoroughness in a monumental review which, like Gibbs's work, no one has ever read [25,26].

Much of the work on hydrophobic effects and between surfaces had all been done before in Russia but was ignored [27]; similarly the experiments of Laskowski and Kitchener [28], on the boundary between hydrophobic and hydrophilic surfaces. These might be argued away as structural effects (hydrogen bond rearrangements) limited to the first few layers of water and present theories would be unperturbed. On the other hand the experiments of Deryaguin and Churaev [29] and of Pashley and Kitchener [30] on thick water films on quartz point to some anomalous and large effects at the quartz–water interface.

But while thermodynamics encompasses the phenomena, the molecular mechanisms by which the effects occur depend on the system. Addition of alcohols or other 'structure breakers' or 'structure makers' changes matters qualitatively [25].

Polywater

We remark in passing that further very large forces of variable range, very similar to this hydrophobic attraction, both with and without cavitation steps, were also observed in quite different situations. These apparently have little to do with surface hydrophobicity. For example, interaction forms very typical of 'hydrophobic force laws' occur between freshly molten hydrophilic glass surfaces in water vapour. Here there is a bridging effect of water condensates. However, in this system the result is complicated by the specific surface chemistry of silica glasses. Formation of thick silica gel layers by exposure of soda-rich glasses to water and even moist air has long been known from studies of glass corrosion (this is the effect that caused a great deal of confusion in studies of 'polywater').

As an aside, the polywater saga in the 1960s at the height of the cold war is a disgraceful episode in the history and politics and sociology of science, a most readable account of which is given in the book by Felix Franks, written in 1973 [31]. Unfortunately, the bias remains; polywater was dismissed, quickly dead and buried even by Deryaguin himself, its discoverer, who withdrew his claim. But he was discredited in the West. The story outlined by Franks is also incorrect. Polywater was roundly dismissed as an artefact due to impurities. Science politics aside, and many reputable scientists who jumped on the polywater bandwagon were later embarrassed, the very long-range structured water phenomenon envisaged as 'polywater' does exist. Real water does contain impurities, such as dissolved gas! Science politics has held up progress in our understanding of real water for more than 50 years.

The whole question of very long-range effects in dilute gels periodically emerges and is periodically dismissed. But such phenomena keep recurring and pose a huge challenge. We simply do not know the answers. See, for example, a beautiful historical and scientific discourse on the state of water in living systems from the liquid state to the jellyfish by Henry [32], and the book by G. Pollack [33]. The claims for the existence of ordered water over enormous distances seem absurd. But the experiments from which the conclusion is reached are not so easily dismissed, as for bubble–bubble interactions.

Some of the phenomena do fall under the umbrella of the classical theory of capillary phase segregation [34,35].

8.9 Molecular basis of long-range 'hydrophobic' interactions

The fact that all results on hydrophobic interactions are encompassed by the surface thermodynamic description of the Young–Laplace–Kelvin equations is satisfactory. But it in no way explains the molecular mechanisms operating. For example, the original long-range 'hydrophobic' force between mica surfaces on which CTAB surfactants are adsorbed from water is due to a two-dimensional gas–liquid-like phase transition of each of the adsorbed CTAB monolayers. This is driven by the double-layer electrostatic force between the layers. The long-range cementing force between freshly molten glass is due to bridging of the surfaces by a sodium silica gel leached from the glass surfaces, and so on. A dimensional argument seems to

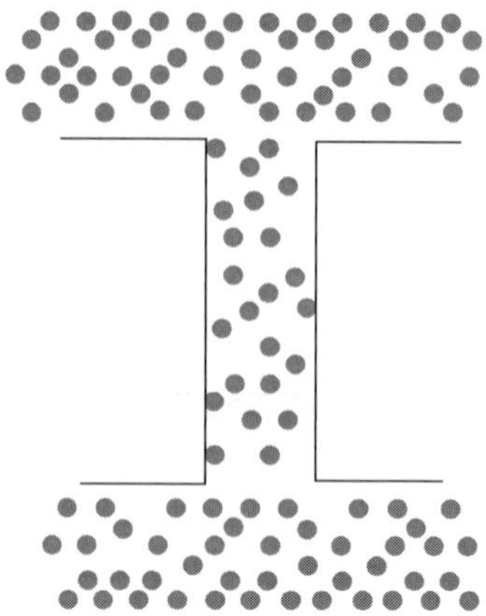

Fig. 8.5. Cavitation at high contact angles – very hydrophobic surfaces also seem to be covered by such a mechanism [19]. Adapted with permission from Ref. 19. Copyright 1993 American Chemical Society.

capture how subcritical fluctuations in water between two hydrophobic surfaces gives rise to a lowering of the density in the gap between the interacting surfaces. The slightly higher density of water outside pushes them together (see Fig. 8.5).

How the range of such forces can be extended by dissolved gases or other sparingly soluble 'hydrophobic' molecules such as methane has already been discussed above [36–40].

What we would like to understand are the origins of several key observations.

In summary:

Whether for molecularly smooth surfaces or hydrophobic proteins, and other situations:

1 'hydrophobic' forces can be an order of magnitude or more larger than any conceivable van der Waals force;
2 cavitation can occur at a distance of about 1 nm [41];
(This is a phase transition that somehow harnesses cooperatively all the weak van der Waals and other physical forces operating to replace a water film between surfaces by a vapour cavity. It is a cooperative effect by which these weak forces are inveigled to produce a chemical energy. We might anticipate, as in Section 6.4, that this cooperative effect might be used by enzymes to drive their catalysis.)
3 divalent salts affect the interactions by adsorbing at the interfaces, so changing them to hydrophilic surfaces [42].

All such effects are dominated by dissolved gas and the interplay between gas and other hydrophobic solutes on the one hand, and electrolytes and electrolyte type on the other. There are two consequences. Whether we understand them at a molecular level or not, the phenomena, like the specific bubble–bubble interactions and their specificity, do exist. This means that these large forces can be exploited in many applications.

Any molecular modelling or simulation of hydrophobic interactions that ignores dissolved gas is irrelevant.

8.10 Speculations on possible implications for Burgess Shale pre-Cambrian and other geological extinctions

Whatever the ultimate explanation of the bubble–bubble phenomenon, it exists and has application at the very least to understanding the phenomenon of the 'bends', decompression sickness [6].

Massive extinctions that mark geological eras are not understood except perhaps one, in terms of meteoric collisions like that held to be responsible for the dinosaurs' demise.

But another amusing speculation might be allowed: complete bubble–bubble fusion inhibition occurs at an equivalent salt concentration of 0.175 M. It is perhaps not coincidental that this is precisely the concentration of salt in the blood in ourselves and in all other land animals. It is also the known concentration of the Permian ocean 200 million years ago at and around the time when land animals emerged. The salt concentration in the present ocean is much higher than then at 0.4 molar. It varies depending on the particular geological era. This is generally held to be due to ice ages. Withdrawal of ice in massive amounts to the poles of the earth leads, paradoxically, to a concentration of salt, especially in inland enclosed oceans from which land forms emerged. This then led to precipitation of salt in massive thick beds, which is well known.

The result then would be that after an extensive ice age, when the polar caps melt there follows a scene like that in the Baltic today. That is, the Baltic sea is brackish, with a salt concentration less than our magic 0.15 M. As for the phenomenon of the bends, microbubble fusion in organelles of phytoplankton is no longer inhibited. The phytoplankton would all become extinct, and indeed this is known to have happened.

As they are the first stage in the food chain, the consequence would be the extinctions that define and demark different geological eras.

That may or may not be so, but in the absence of any other proposition it might be worth a thought. Stephen J. Gould, defender of Darwinism and writing on the Burgess Shales extinctions, famously wrote on contingency in evolution. What occurred was the massive extinction of about 24 different phyla, leaving only about 4 remaining, which include ourselves (whether its 4 or 5 or 6 is a matter for the paleontologists to dispute). The same happened in the preceding Precambrian Ediacara extinctions, and subsequent Permian extinctions 200 million years ago, when 90% of species died. Gould's proposition is that this so-called contingency was pure chance. If one ran the clock backwards and repeated the whole experiment of evolution it would have all been different.

But the fact that all extinctions are coupled to ice ages and reduction in salt levels might suggest that the real story may be more complicated than just contingency.

References

[1] V. S. J. Craig, B. W. Ninham and R. M. Pashley, *J. Phys. Chem.* **97** (1993), 10192–10197.

[2] C. L. Henry, C. N. Dalton, L. Scruton and V. S. J. Craig, *J. Phys. Chem. C* **111** (2007), 1015–1023.

[3] C. L. Henry and V. S. J. Craig, *Langmuir* **24** (2008), 7979–7985.

[4] C. L. Henry, L. Parkinson, J. R. Ralston and V. S. J. Craig, *J. Phys. Chem. C* **112** (2008), 15094–15097.

[5] B. W. Ninham, *Adv. Coll. Interface Sci.* **16** (1982), 3–15.

[6] V. S. J. Craig, B. W. Ninham and R. M. Pashley, *Nature* **364** (1993), 317–319.

[7] M. Alfridsson, B. W. Ninham and S. Wall, *Langmuir* **16** (2000), 10087–10091.

[8] R. M. Pashley, M. J. Francis and M. Rzechowicz, *Curr. Op. Coll. Interface Sci.* **13** (2008), 236–244.

[9] M. E. Karaman, B. W. Ninham and R. M. Pashley, *J. Phys. Chem.* **100** (1996), 15503–15507.

[10] M. Lagi, P. Lo Nostro, E. Fratini, B. W. Ninham and P. Baglioni, *J. Phys. Chem. B* **111** (2007), 589–597.

[11] P. Lo Nostro, L. Giustini, E. Fratini, B. W. Ninham, F. Ridi and P. Baglioni, *J. Phys. Chem. B* **112** (2008), 1071–1081.

[12] Z. Dai, D. Fornasiero and J. Ralston, *J. Chem. Soc. Faraday Trans.* **94** (1998), 1983–1987.

[13] H. R. Corti, M. E. Krenzer, J. J. De Pablo and J. M. Prausnitz, *Ind. Eng. Chem. Res.* **29** (1990), 1043–1050.

[14] O. I. Vinogradova, N. F. Bunkin, N. V. Churaev, O. A. Kiseleva, A. V. Lobeyev and B. W. Ninham, *J. Coll. Interface Sci.* **173** (1995), 443–447.

[15] N. F. Bunkin, A. V. Kochergin, A. V. Lobeyev, B. W. Ninham and O. I. Vinogradova, *Coll. Surfaces A* **10** (1996), 207–212.

[16] N. F. Bunkin, O. A. Kiseleva, A. V. Lobeyev, T. G. Movchan, B. W. Ninham and O. I. Vinogradova, *Langmuir* **13** (1997), 3024–3028.

[17] B. W. Ninham, K. Kurihara and O. I. Vinogradova, *Coll. Surfaces A* **123** (1997), 7–12.

[18] N. F. Bunkin, A. V. Lobeyev, G. A. Lyakhov and B. W. Ninham, *Phys. Rev. E* **60** (1999), 1681–1690.

[19] V. V. Yaminski and B. W. Ninham, *Langmuir* **9** (1993), 3618–3624.

[20] A. S. Kabalnov and H. Wennerström, *Langmuir* **12** (1996), 276–292.

[21] J. N. Israelachvili and R. M. Pashley, *Nature* **300** (1982), 341–342.

[22] J. N. Israelachvili and R. M. Pashley, *J. Coll. Interface Sci.* **98** (1984), 500–514.

[23] R. M. Pashley, P. M. McGuiggan, B. W. Ninham and D. F. Evans, *Science* **229** (1985), 1088–1089.

[24] Lord Rayleigh, *Scientific Papers by Lord Rayleigh*. New York: Dover (1964), vol. IV, p. 430.

[25] V. V. Yaminsky, S. Ohnishi and B. W. Ninham, Long-range hydrophobic forces due to capillary bridging. In *Handbook of Surfaces and Interfaces of Materials*, ed. H. S. Nalwa. New York: Academic Press (2001), vol. 4, Ch. 3.
[26] V. V. Yaminsky and B. W. Ninham, *Adv. Coll. Interface Sci.* **83** (1999), 227–311.
[27] V. A. Pchelin, *Vestn. Mosk. Univ.* **13** (1972), 131–142; **14** (1973), 131–141 (in Russian).
[28] J. Laskowski and J. A. Kitchener, *J. Coll. Interface Sci.* **29** (1969), 670–679.
[29] B. V. Deryaguin and N. V. Churaev, *J. Coll. Interface Sci.* **49** (1974), 249–255.
[30] R. M. Pashley and J. A. Kitchener, *J. Coll. Interface Sci.* **71** (1979), 491–500.
[31] F. Franks, *Polywater*. Cambridge, MA: MIT Press (1983).
[32] M. Henry, *Cell. Mol. Biol.* **51** (2005), 677–702.
[33] G. H. Pollack, *Cells, Gels and the Engines of Life: a new, unifying approach to cell function*. Seattle: Ebner and Sons (2001).
[34] V. V. Yaminski, B. W. Ninham and R. M. Pashley, *Langmuir* **14** (1998), 3223–3235.
[35] D. T. Atkins and B. W. Ninham, *Coll. Surfaces A* **129** (1998), 23–30.
[36] V. V. Yaminsky, B. W. Ninham, H. K. Christenson and R. M. Pashley, *Langmuir* **12** (1996), 1936–1943.
[37] V. V. Yaminsky, C. Jones, F. Yaminsky and B. W. Ninham, *Langmuir* **12** (1996), 3531–3535.
[38] R. M. Pashley, P. M. McGuiggan, B. W. Ninham, D. F. Evans and J. Brady, *J. Phys. Chem.* **90** (1986), 1637–1642.
[39] P. M. Claesson, P. C. Herder, C. E. Blom and B. W. Ninham, *J. Coll. Interface Sci.* **114** (1986), 234–244.
[40] V. S. J. Craig, B. W. Ninham and R. M. Pashley, *Langmuir* **15** (1999), 1562–1569.
[41] H. K. Christenson and P. M. Claesson, *Science* **239** (1988), 390–392.
[42] H. K. Christenson, J. Fang, B. W. Ninham and J. L. Parker, *J. Phys. Chem.* **94** (1990), 8004–8006.

Part II

Self assembly

9

Self assembly: overview

This and succeeding chapters deal with how it is that molecular forces conspire with the size and shape of molecules called surfactants to associate spontaneously in solution into a myriad multimolecular aggregates. They can be ephemeral entities called micelles. Typical micelles formed by short-chained surfactants exist as entities for times of around 10^{-5} seconds. On the other hand they can be as long-lived as three or more months. That is so for membrane mimetic long-chained phospholipids that form complex single-walled vesicles and multi-bilayered structures. These self-assembled aggregates provide the organized microstructural scaffolding that forms the basis of biological cell membranes. Self-assembled entities direct biochemical cell traffic.

We have tried to identify conceptual developments in self assembly as they emerged over the past three decades. The result of a great deal of theorizing and experimentation is that some simple rules emerge. These allow the prediction of microstructure, as a function of components, component ratios and physicochemical solution conditions. This, combined with an understanding of how to change molecular forces via specific ion effects, gives some insights into the astonishingly complex background self-organization that occurs in biology. (The genius of DNA, RNA and proteins in biology is not in dispute. What is not generally recognized is that their work takes place within a hidden framework built from and involving the lipids, which are not just passive bystanders.)

It is important for industrial formulation too – surfactants are at the core of a vast number of industries, a few of which are drug delivery, soaps, cosmetics, paper, polymers, coatings and mineral flotation. Progress occurred in theories, experimental technique and measurement of forces in parallel with a better understanding of both ion specificity and hydration.

9.1 Surfactants and lipids

We have first to explain what we mean by the word surfactant. It is an abbreviation for surface-active agent. The word lipid embraces biological surfactants and fats. It is usually reserved for insoluble double-chained surfactants that are essential components of biological membranes. For definiteness, and necessary for an understanding of what follows, a detailed account of common surfactant molecules and lipids and some of their characteristics is given in the Appendix to this chapter.

These molecules have one part, a headgroup, that is hydrophilic and prefers to be in water, like a salt molecule. Another part is hydrophobic, an oil-like tail. Typically the tail is a hydrocarbon chain, which prefers to associate with its fellows or to be in oil. But other surfactants with industrial and biological applications, e.g., as gas carriers or blood substitutes, have fluorinated tails [1]. This circumstance confers a duality in properties that is embraced by the alternative word amphiphile. Many proteins, polymers and other complex molecules are also amphiphilic. They also have separated hydrophobic and hydrophilic moieties that confer surface-active properties.

Surfactants can be single-chained, like soap molecules, double-chained like most of the lipids that form cell membranes, or triple-chained. A detailed molecular description of specific molecules and an account of characterization of surfactant behaviour is given in the Appendix.

A brief schematic view of events that take place at surfaces or in bulk solution with increasing surfactant concentration, explained in more detail in the Appendix, is as follows: at an air–water or at an oil–water or solid hydrophobic surface, surfactant molecule hydrocarbon chains that are unhappy in water escape to form monolayers. These lower interfacial tension. With increasing concentration, the monolayer fills up. With nowhere else to go, the surfactants then accommodate their opposing amphiphilic requirements by self-assembling into aggregates in a way that minimizes the exposure of oil-like regions to water. This energetic requirement is opposed by the entropy available to the monomers if they remain in the bulk dilute solution. The chain of initial events is reflected in the interfacial tension as depicted in Fig. 9.1.

With increasing surfactant, γ decreases to a minimum. Thereafter it remains essentially constant. The monolayer has filled up. At this *critical micelle concentration* (cmc), aggregates form in the solution.[1] Beyond the cmc, the new 'micellar phase' is in equilibrium with monomers at the concentration of the cmc.

[1] This is too limited a characterization. It applies mainly to single-chained surfactants that have been most studied. For non-ionic surfactants, the interfacial tension increases somewhat after the cmc rather than decreases. For some double-chained surfactants the interfacial tension gradually decreases and there is no abrupt cmc. These different behaviours reflect more complicated microstructures at the interface that are in equilibrium with more

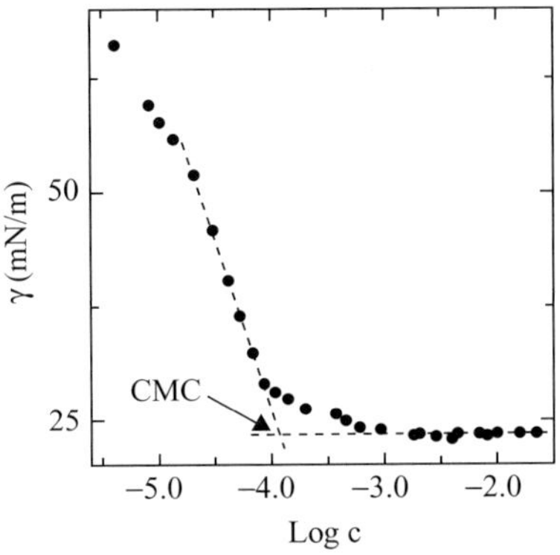

Fig. 9.1. Typical variation of the surface tension of water (γ) as a function of the surfactant concentration (plotted on a logarithmic scale). The figure shows how the cmc is obtained from the intersection of the straight lines for high and low amphiphile concentrations.

The simplest kind of structure that forms with single-chained surfactants, at the critical micelle concentration, is called a micelle. It is usually visualized as in Fig. 9.2a as a sphere of typically 50 surfactant molecules. Their headgroups face water and the hydrocarbon tails form a liquid oil-like interior. This standard picture is somewhat misleading, conveying as it does the notion of a static entity when it is in fact a highly dynamic and temporary association, as represented in Fig. 9.2b.

So, for example, for a common surfactant like SDS (sodium dodecyl sulfate), with 12 carbons in its chain, the lifetime of a molecule in an aggregate is estimated to be 10^{-8} seconds, that of the whole micelle about 10^{-5} seconds.

By contrast, another structure that forms spontaneously from double-chained lipids is a bilayer (cf. Fig. 9.3). The residence time of a lipid in a bilayer or cell membrane can be of the order of months, about that of the lifetime of a red cell.

With further increase of surfactant concentration, or with addition of salt for ionic micelles, or with change of salt type, or with increasing hydrocarbon chain length or numbers of chains or changes in temperature or with mixed surfactants, the micelles transform to other phases. These can consist of cylinder-like structures

complicated, polydisperse bulk solution aggregates like vesicles [2,3]. In general the kinds of (two-dimensional) aggregates, and their phase transitions, that form at interfaces mimic in complexity those that form in bulk solution. The cmc for a given surfactant depends on temperature, a matter not much explored. For example, for the anionic sodium dodecyl sulfate the cmc has a minimum at 24 °C. If solid surfaces are involved at which surfactants adsorb, hemi-micelles and complex mono- and bilayers can and do form. This vast area of research, and that of wetting, we do not consider. We refer for an entrée to the subject to the papers of V. V. Yaminski, H. K. Christenson, P. Kekicheff and their colleagues. All studies of surface-active phenomena are acutely dependent on surface and surfactant impurities, a matter convincingly shown by Lunkenheimer et al. [4]. Very probably they also depend on dissolved atmospheric gas, a matter that is universally ignored. (The concentration of dissolved atmospheric gas around 10^{-2} M is the same as that of typical cmcs of single-chained surfactants, and molecular gas may well associate with hydrocarbon tails.) For some further illustrative examples, surveys and detailed theories of micelle formation see Refs. 2–7.

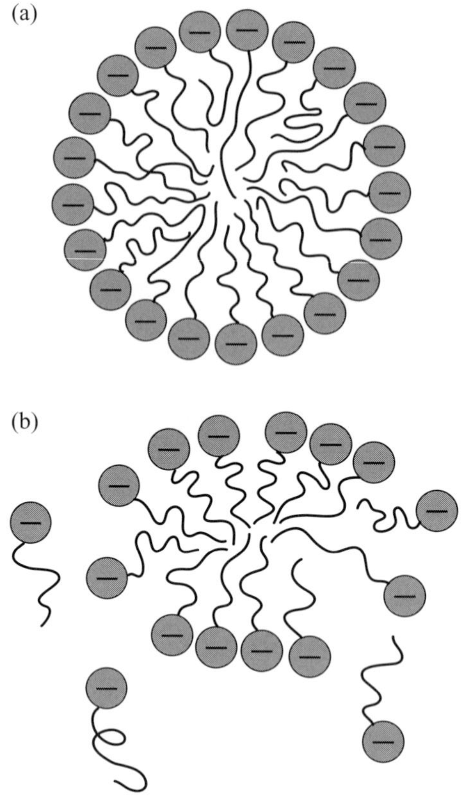

Fig. 9.2. Idealized spherical micelle (a). Less idealized micelle (b).

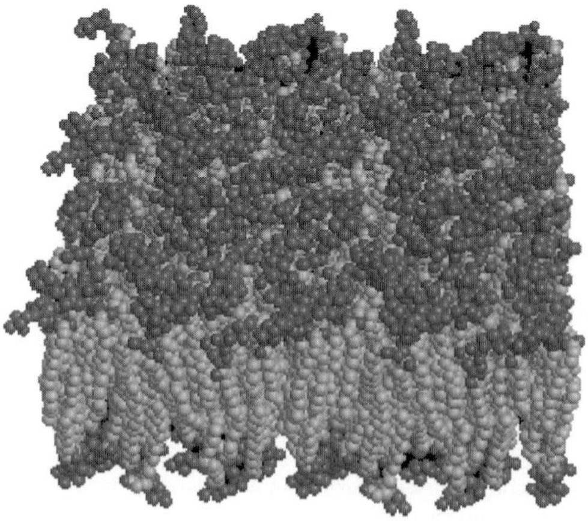

Fig. 9.3. Lipid molecules organized in a bilayer. This arrangement is typically found in biomembranes.

(hexagonal phase), bicontinuous structures (cubic phases), vesicles, giant vesicles or multi-bilayers (lamellar phases). Eventually inverse structures form (reverse phases), with surfactant tails surrounding water.

The complexity of these aggregates is normally summarized by phase diagrams described in the Appendix at the end of this chapter. It will be our task to elucidate and predict the circumstances in which different aggregates form.

9.2 Emulsions and microemulsions

With a third component of oil added to the mixture of surfactants and water the system can form emulsions. Here the surfactant adsorbs at the water–oil interface to segregate oil from water into macroscopic pools (Chapter 12).

The surfactant can adsorb at the oil–water interface as a monolayer, or it can consist of multiple layers. Because of their size, emulsions usually scatter light and appear opaque. The pools of water in oil, or of oil in water, often contain within them smaller complicated oil–water–surfactant microstructures. Emulsions are usually thermodynamically unstable. That is, the droplets will coalesce with time and phase-separate into macroscopic oil and water regions separated by a complex interfacial region. It used to be thought that emulsions always required work (stirring, sonication) for their formation. This is not so. Some do form spontaneously and are thermodynamically stable; cf. Chapter 12. Some unstable emulsions, with no surfactant at all, can be stabilized by removing dissolved gas or other sparingly soluble hydrophobic molecules. Dissolved gas has been shown to enhance coalescence of droplets through long-range hydrophobic interactions. (cf. Chapter 8). This extraordinary development is controversial, of enormous potential use, and turns conventional ideas upside down.

Microemulsions are clear solutions of oil, water or other aqueous solutions, and surfactant, with a rich variety of microstructures. They are thermodynamically stable; that is, for example, reversible to temperature cycles. They are transparent, as the liquid microstructures have dimensions so small that they do not scatter light. Sometimes they comprise droplets of oil in water (called normal phase) or of water in oil (reverse phase). The interface of a droplet is usually separated by a single surfactant layer. The microstructure is commonly more reminiscent of bicontinuous connected cylinders of oil in water or of water in oil. There are a huge diversity of other microstructures available, as outlined in Chapter 12. They can be one, two, three and more phase systems. The formation of microemulsions with single-chained surfactants usually requires a *cosurfactant*. This is a small surfactant-like molecule, typically an alcohol, with a small hydrophilic end group. The association of cosurfactant and surfactant in the interfacial region effectively gives it the characteristics of a double-chained surfactant.

The distinction between emulsions and microemulsions and other self-assembled aggregates is somewhat artificial. It is important to remark again that these are all dynamic microstructures, with a vast range of lifetimes depending on the surfactant solubility.

9.3 Order from complexity: theoretical challenges and bicontinuity

The brief survey above and the greater detail in the Appendix only hints at a bewildering diversity of microstructures that can and do form.

In trying to make sense of it all, a number of fixed ideas hampered progress. We have already explored several reasons for the difficulties that confronted the development of systematic theory.

They are worth rehearsing again:

It seems that surfactant self-assembly occurs only in water or some other hydrogen-bonded liquids. Micelles form in hydrazine, N_2H_4. This rocket fuel is a hydrogen-bonded liquid with thermodynamic properties very similar to those of water. Micellization is supposed to depend on the unique 'hydrogen bonding' character of water. Micellization in hydrazine at 35 °C is very similar to that in water at 135° and 160 °C, where water loses its entropy enthalpy compensation and a high degree of three-dimensional order [6,8].

Theories of intramolecular forces acting between surfactant molecules in interfacial layers or between headgroups in micellar systems involve hydration. But detailed molecular theories were too crude and still are.

Further, theories of intermolecular forces between aggregates taken over from colloid science were deficient, as we have seen. The classical theories for interactions between surfactant molecules or between aggregates did not take account of specific ion (Hofmeister) effects. Nor did they deal with specific ion effects on pH and buffers properly.

They took no cognizance of the complexity of the Debye length in multicomponent systems; i.e. the range of stabilizing electrostatic forces between aggregates in concentrated dispersions or mixed electrolytes. There was the bugbear of mysterious long-range hydrophobic interactions (Chapter 8). Then there was and remains the problem of the role of dissolved gas, universally ignored by theories but not by nature.

However, what was equally inhibiting to progress was another fixed idea. It was that simple Euclidean geometries, spheres, ellipsoids, cylinders, planes, were sufficient to describe self-assembled aggregates. We now know that bicontinuous structures are more the rule than the exception. These can be random connected cylinders of water coated with a surfactant layer of constant average curvature in oil that occur in microemulsions. These are bicontinuous, like Swiss cheese or honeycombs. They can be ordered structures of zero average curvature (called cubic phases because of their cubic symmetry) (Chapters 11 and 12). They can transform from one form to another with extravagant ease [9]. There exist a variety of other complex structures. Some of these can be seen in the websites of Stephen T. Hyde [10], in the biological cells of Yuru Deng [11] and the much more complicated structures elucidated by Sten Andersson [12].

9.4 Evolution of theoretical ideas

The bewildering diversity of microstructures that form became clearer over the last few decades, as new experimental probes became available. As recently as 1986 a leading German researcher into microemulsions claimed that microstructure did not exist! Not surprisingly then, and given the paucity of knowledge of molecular forces, predictive theories that made sense of surfactant–water–oil systems took some time. Developments took place over the last three or four decades at a time when colloid chemists and condensed-matter physicists took up the challenges in interdisciplinary areas at the boundaries of physical and biological sciences. A large literature came to the field from developments in the physics of phase transitions in simple liquids via statistical mechanics, and in molecular simulation. These theories tended to focus on thermodynamic generalities, like critical phenomena in phase transitions, as opposed to the particular. By and large such theories, taken over from physics, ignored the size and shape and flexibility of molecules and the particular forces between them. So they lacked the capacity to design and predict the microstructures that form with surfactant molecules.

Some sought to capture the essence of the problem of self assembly via thermodynamics. By so doing one skirts over the problems of water, such as cooperative hydrogen bonding and hydration, that are still imperfectly understood. But thermodynamics is a tautology and can only go so far. A predictive theory at some point has to invoke statistical mechanics, molecules and their interactions. This itself cannot be carried too far, or else all insight is lost in a mass of detail.

Since formal analytical statistical mechanics rapidly becomes too hard, the availability of bigger and better computers led to a focus on insights that could be gained from molecular simulation. Here again we were hampered by lack of knowledge of the molecular forces operating.

On the experimental side a proliferation of bigger and more sophisticated and more expensive scattering devices emerged to probe the microstructure of self-assembled solutions. The suite of tools is large. They include X-rays, small-angle X-ray scattering, dynamic and static light scattering, neutron scattering, nuclear magnetic resonance, transmission, scanning and cryo electron microscopy, and so on. This was all well and good. But nearly all involve artefacts, both experimental and theoretical. There are artefacts that occur with microscopy. 'Fixing' structures by staining, or by freezing, changes the microstructure of a sample. Apparently non-invasive, neutron-scattering measurements often were, and still are, bedevilled by the fact that the interpretation of scattering data is not unique.

To illustrate: an experimental scattering result can be interpreted in different ways. For example, a scattering experiment can be fitted to a microstructure that might be, say, polydisperse spherical micelles. The fitting then requires that the micelles interact with an adjustable (but often physically unreasonable) force between the aggregates. But the real microstructure of the solution is often something completely different, that of a bicontinuous medium! Such ambiguities can sometimes be resolved by the use of simpler complementary techniques – a conductivity or NMR diffusion measurement can determine whether a microstructured fluid is bicontinuous or not. The ambiguities can also be resolved by resort to the neglected, irrefutable fact that spheres cannot pack above volume fractions more than about 0.74, the maximum limit for close-packed spheres. Many published data on microstructure have to be viewed with a jaundiced eye, credible inversely to the cost of the experiment.

Indeed it turns out that in most applications the microstructure of these complex fluids can often be pinned down by the first principle of the chemical engineers: mass balance, volume fractions that impose global packing constraints on allowed structures, alone. It turns out that geometric global packing constraints together with the knowledge we now have of how to manipulate intra- and interaggregate molecular forces such as specific ion effects are sufficient to make a good deal of progress. Together, with luck, they are enough to determine and prescribe microstructures of a desired shape and size, or porosity and physical properties such as viscosity.

9.5 Supraself assembly

Theoretical characterization of microstructure adapted ideas from Gibbs's thermodynamics. So one thought of a 'phase' made up of a suspension of homogeneous aggregates. These could be, for example, spherical micelles. Or they could be vesicles, 'balloons' of uniform size, with a water interior bounded by a lipid bilayer. This is too simple a picture. Within the boundaries of a single 'phase', the component 'subunits' like micelles themselves vary continuously in microstructure as concentration changes.

At a higher level, we now understand and are able to predict the spontaneous formation of what we can call 'supraself-assembled' aggregates. Their microstructure is not homogeneous at all. They form even in simple two, three or more component systems.

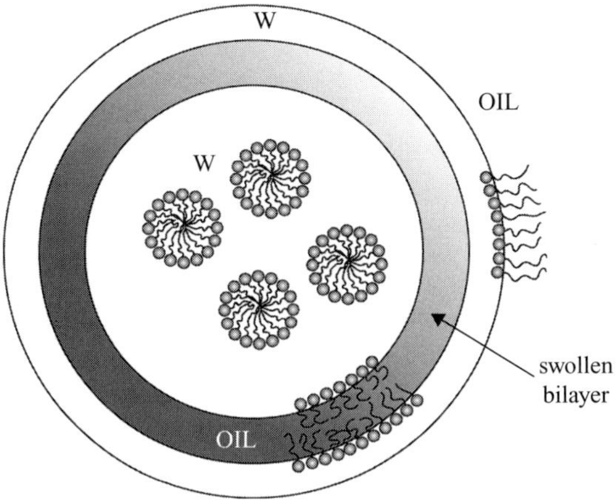

Fig. 9.4. Schematic of one of many supraself-assembled aggregates that can occur in an emulsion. Multiple bilayers of surfactant swollen by oil surround macroscopic water droplets. The interior water region contains micelles.

To visualize such structures take the following example in Fig. 9.4. In a system of vesicles formed by one or more bilayer membranes, physicochemical conditions can be different inside and outside. In a homogeneous system of vesicles the bounding bilayers separate interior and exterior aqueous phases. In a supraself-assembled system the interior and exterior of the aggregate are quite different [2]. Supra aggregates can have an interior of bicontinuous, cylindrical or micellar phases surrounded by protective layers of bilayers.

The interior might be micelles, or it can be a (bicontinuous) cubic 'phase'. Both are protected by bilayer lamellae, with an exterior that can have a different microstructure structure again. These supraself-assembled systems form spontaneously. They can be stable for months, like red blood cells. Or else they can transform from one state of self assembly to another with extravagant ease as physicochemical conditions – temperature, salt, concentration – are changed. The vesicles involved in transmission of calcium across synaptic junctions are such supraself-assembled aggregates.

New concepts on structure evolved once it was recognized that bicontinuous, hyperbolic, geometries play an important role. Fig. 9.5 shows the interior of an Australian termite nest. It is a perfect bicontinuous structure, everywhere saddle-shaped, of zero average curvature, made from its excreta.

This macroscopic structure, channel diameter about 1 cm, is identical to that of cubic phases formed with surfactants in water and in microemulsions where the pore sizes are typically 1–2 nm [9].

The idea of supraself-assembly takes older ideas on what constitutes a 'phase' in small system thermodynamics to a different level.

Fig. 9.5. Computer tomogram of interior of a termite nest. See text. Reproduced with permission, courtesy of T. Senden and R. Corkery.

The design of microstructures for nanotechnological applications then becomes possible, when coupled to progress in understanding better the nature of molecular forces in solution.

9.6 Microstructures of self-assembled aggregates

Microstructures of typical aggregates that form with the simplest system, a surfactant of just one kind only in water, are illustrated in the figures that follow (Figs 9.6–9.10). Which structure forms and under what conditions for a particular surfactant can be predicted reasonably well in terms of a calculable surfactant parameter.

9.7 Local interfacial curvature a determinant of microstructure

The first attempts to build a general theory of self assembly based on statistical mechanics for dilute systems were built on the ideas of Tanford [14] and Israelachvili, Mitchell and Ninham [15,16]. The second of these [16] is the more honest in the sense that it explores what we can say about when and what aggregates form given the most general, minimal assumptions. Both, however, do not consider bicontinuous structures and interactions between aggregates.

It was possible to give a reasonable estimate of cmcs as a function of chain length and temperature by invoking measured hydrophobic free energies of transfer of oil to water. There emerged a semi-predictive useful parameter to characterize

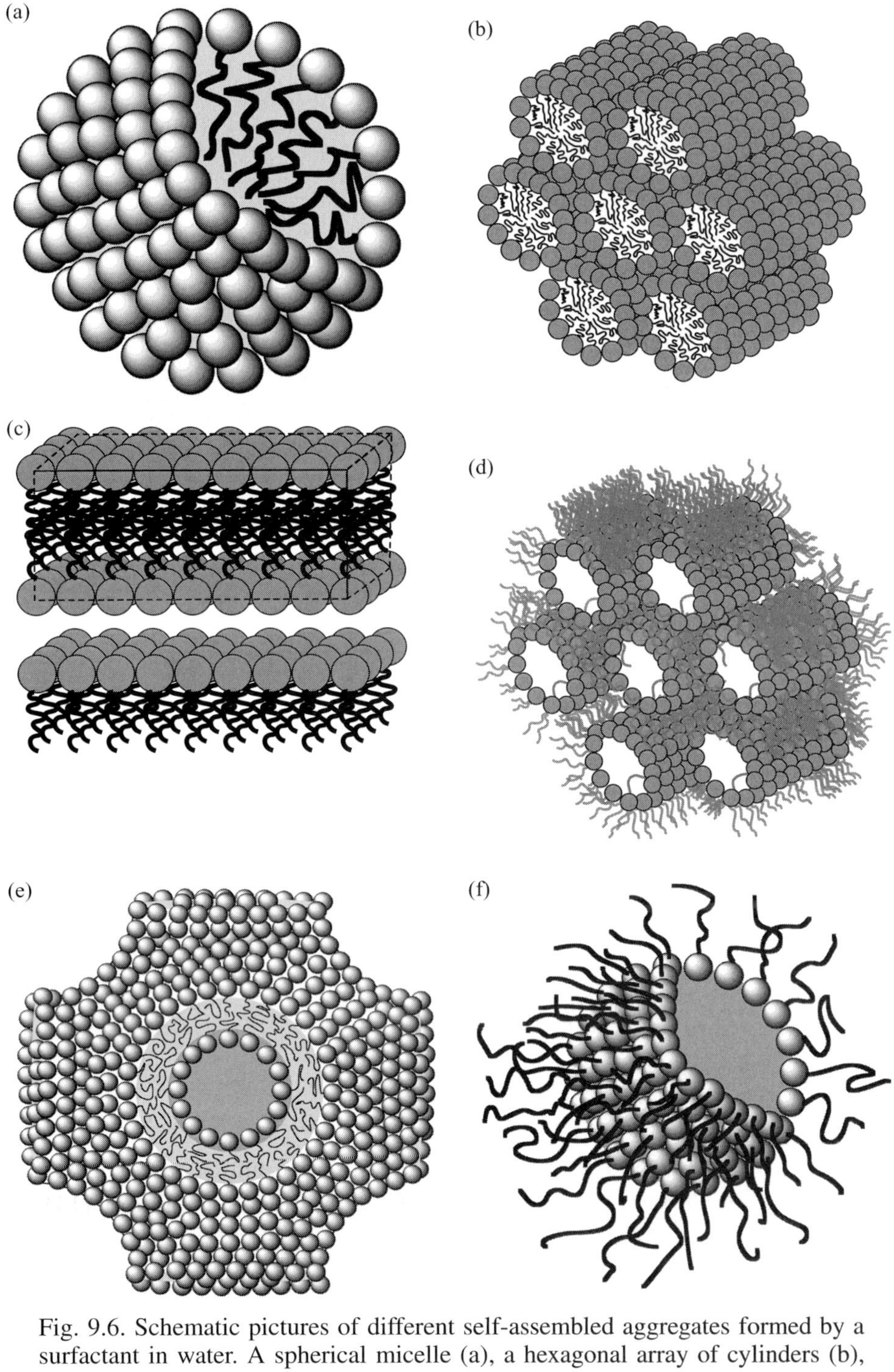

Fig. 9.6. Schematic pictures of different self-assembled aggregates formed by a surfactant in water. A spherical micelle (a), a hexagonal array of cylinders (b), a lamellar bilayer (c), a reversed hexagonal packing of rod-like aggregate (d), a cubic phase (e) and a reverse micelle (f). In the last a cylinder or sphere of water is surrounded by surfactant headgroups the tails of which interpenetrate, or which in microemulsions are in oil. For cubic (bicontinuous) phases see later chapters. Which kind forms when and why can usually be predicted from knowledge of the surfactant parameters.

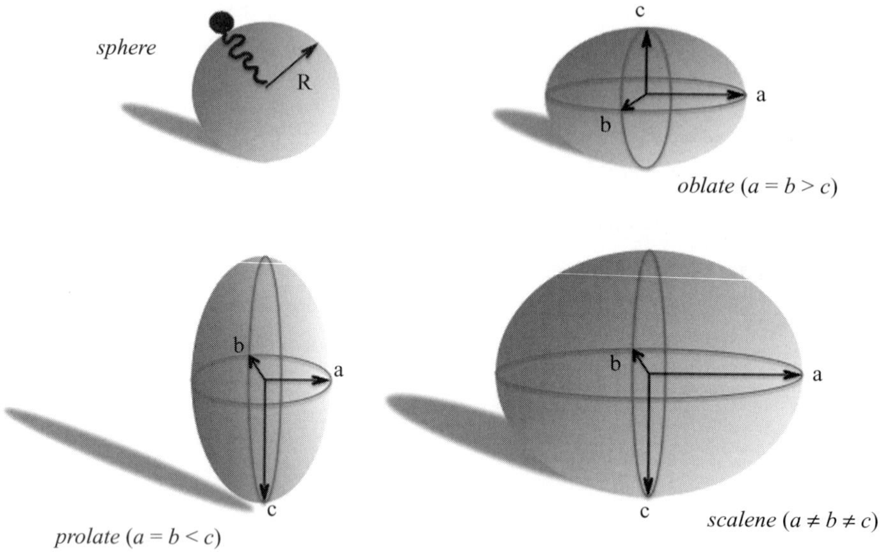

Fig. 9.7. Spherical and ellipsoidal micelles. Prolate and oblate micelles can usually pack with mixtures of surfactants of differing headgrou areas.

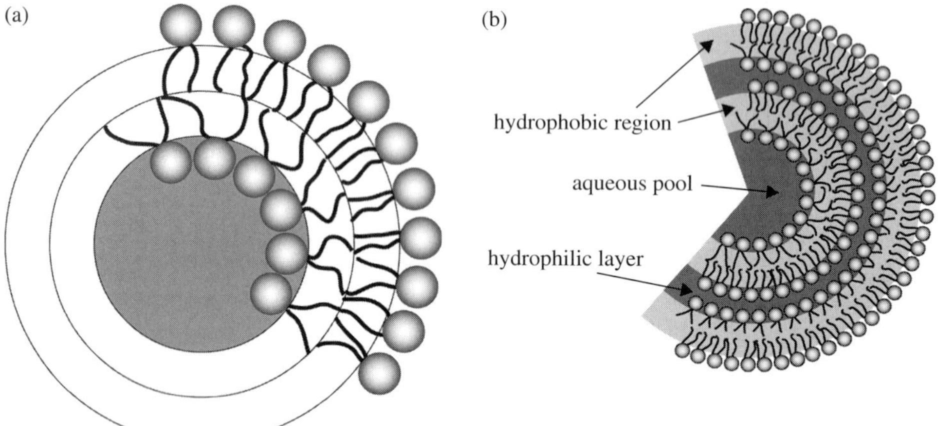

Fig. 9.8. (a) Single-walled vesicle formed from double-chained surfactants that pack happily into a planar bilayer. In a vesicle the chains are stretched in the outer layer of the bilayer, compressed on the inner layer and show strong asymmetry. This tension means that for typical double-chained surfactants or lipids there is a limit to the radius of the inner layer. Typically such vesicular balloons have an aqueous core radius of around 100 nm as a result. (b) Multiple-walled vesicle or lamellar phase. Dark grey indicates the hydrophilic layer, light grey represents the hydrophobic compartment.

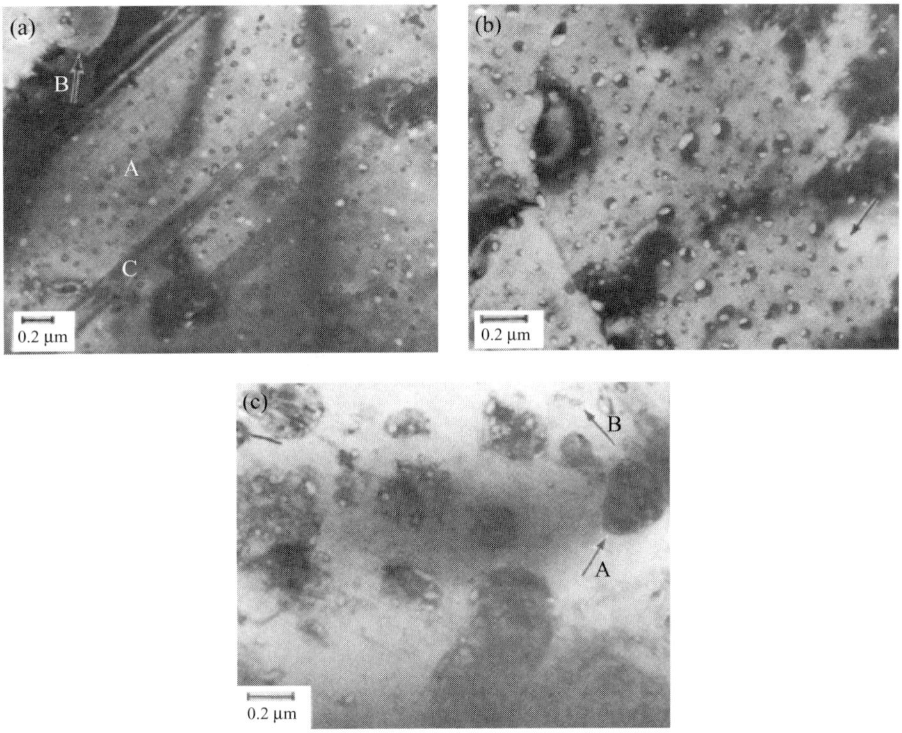

Fig. 9.9. Supra-aggregate: vesicle with micelles inside (see Fig. 9.3). Seen by cryo transmission electron microscopy. The double-chained surfactant, didodecyl dimethyl ammonium acetate in water (see the Appendix), forms spontaneous vesicles. Typical diameters are around 1–200 nm. At high concentrations (10^{-2} M), the interior contains micelles. At lower concentrations (10^{-3} M), the interior contains small vesicles [13]. (a) Frozen aqueous dispersion of 0.014 M $(C_{12}H_{25})_2(CH_3)_2NOH$. Note vesicles (A), grain boundary (B) and stacking faults in the ice matrix (C). (b) Frozen samples of an aqueous dispersion of 0.014 M $(C_{12}H_{25})_2(CH_3)_2NOH$ that was partially neutralized (25%) with HBr. Note the large vesicles (arrow). (c) Frozen sample of an aqueous dispersion of 0.014 M $(C_{12}H_{25})_2(CH_3)_2NOH$ that was partially neutralized (50% with HBr). Note liquid crystalline particles (A) and vesicles (B). Reprinted from Ref. 13 with permission from AAAS.

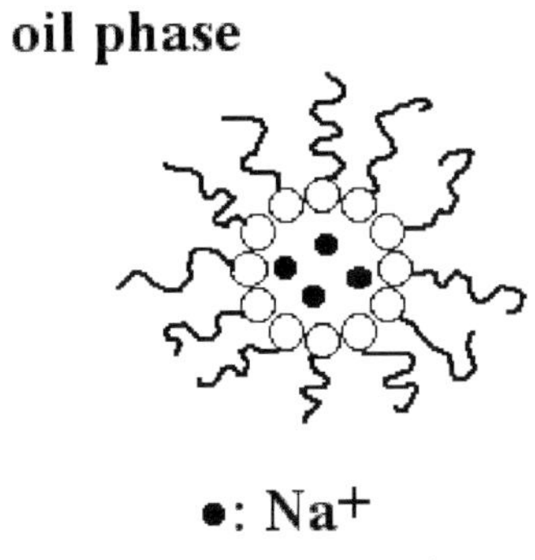

Fig. 9.10. Reverse-phase ionic micelles with Na^+ counterion. The hydrophobic tails can be in oil or interpenetrate.

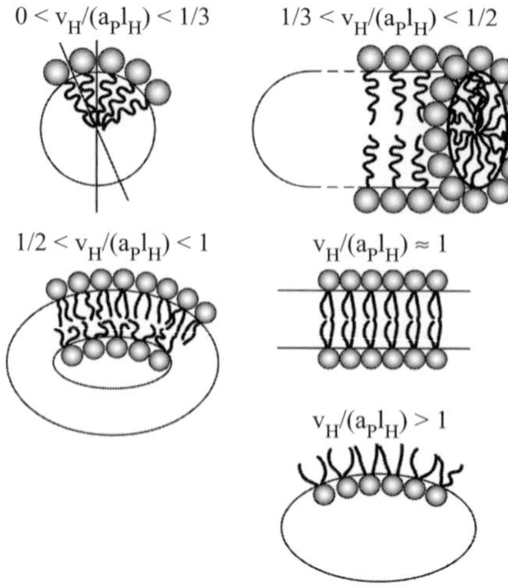

Fig. 9.11. The geometry of the self-assembled microstructures depends on the surfactant parameter. From top left: a spherical micelle, a cylinder, a curved bilayer, a planar bilayer and a reverse micelle.

aggregate formation. This replaced an older characterization termed hydrophobic-lipophilic balance (HLB) number.

The surfactant parameter is $v_H/(a_P l_H)$, where v_H and l_H are the hydrophobic tail's volume and length, and a_P is the cross-section area of the polar headgroup [15]. In practice, l_H can be taken as roughly 80% of the fully extended chain length. For estimates see Refs. 14 and 15. The sequence of microstructures that form as the surfactant parameter varies is shown in Fig. 9.11.

As already remarked the surfactant parameter depends on surfactant concentration, or addition of salt, or change of salt type,or increasing hydrocarbon chain length or numbers of chains, or changes in temperature, or mixed surfactants. Our job will be to be able to change and quantify these changes.

Some rules that can be distilled out for dilute systems emerge after an extensive statistical mechanical analysis based on minimal assumptions [15,16]. They are: if the surfactant parameter of a molecule satisfies

$$v_H/(a_P l_H) < 1/3, \text{ then micelles form}$$
$$1/3 < v_H/(a_P l_H) < 1/2, \text{ cylinders (hexagonal phases) form}$$

When the surfactant parameter increases further there are more complicated situations. These emerge from theory as a result of further, local and global packing constraints which will be discussed later.

The simplest situation as $v_H/(a_P l_H)$ increases from the value 1/2 is this [16]:

for stiff hydrocarbon chains:

$1/2 < v_H/(a_P l_H) \rightarrow$ multi walled bilayers (liposomes) $\rightarrow$ single walled vesicles $\rightarrow$ liposomes < 1

for flexible chains:

$1/2 < v_H/(a_P l_H) \rightarrow$ vesicles $\rightarrow$ several walled vesicles $\rightarrow$ liposomes $v_H/(a_P l_H) > 1$, reverse phases are formed

The question of what we mean by *stiff* and *flexible* and the emergence of cubic phases is matter of some moment to which we return.

These rules ignore bicontinuous phases that occur alternatively and competitively with single and multiwalled vesicles, which we explore later.

The schematic pictures shows the surfactant tails as packing together rigidly. This seems inconsistent with the notion of an oil-like interior on which the theoretical ideas are based. In fact in a first-order mean field theory there is no conflict (cf. Chapter 10).

For vesicles or any curved bilayer outer surfactants have chains that are stretched. On the inner side the chains are compressed. The bilayer has to be asymmetric (Fig. 9.8). The tension between the two necessary requirements gives rise to bicontinuous microstructures.

9.8 Mixed surfactants and illustration of local packing constraints

Essentially, single-chained surfactants usually satisfy $v_H/(a_P l_H) < 1/3$ and form micelles – because they can. (From statistical mechanics [15,16], the smallest aggregate is most favoured, if it can pack.) Double-chained surfactants characteristically have $1/2 < v_H/(a_P l_H) < 1$, and so form bilayers or vesicles or more complicated structures.

The simplest illustration of these rules involves mixed surfactants. A mixture of single- and double-chained surfactants of the same hydrocarbon chain length and headgroup area will have a combined surfactant parameter that depends on the ratio chosen.

The sequence of aggregates that form can be demonstrated as follows. Begin with a test tube containing a dilute solution of a single-chained micelle-forming surfactant $v_H/(a_P l_H) < 1/3$ solution above its cmc. Then titrate in successively an increasing concentration of a corresponding double-chained surfactant, $v_H/(a_P l_H) \sim 1$.

For example, we take the surfactants dodecyl-trimethyl ammonium bromide (DTAB) (1) and didodecyldimethyl ammonium bromide (DDAB) (2).

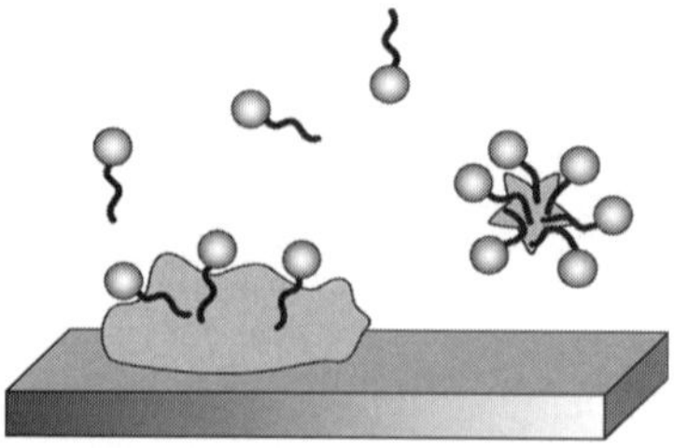

Fig. 9.12. The soil (light grey) is removed by the surfactant by solubilization in the hydrophobic micellar core.

The composite *effective* surfactant parameter is:

$$\left(\frac{v_H}{a_P l_H}\right)_{\text{eff}} = \frac{(x_1/3) + x_2}{x_1 + x_2}$$

where x are mole fractions of each. As $[v_H/(a_P l_H)]_{\text{eff}}$ increases on addition of the double-chained surfactant, the mixture can easily be seen to transform successively from freely flowing micelles to viscous cylinders. When $[v_H/(a_P l_H)]_{\text{eff}} > 1/2$, it suddenly becomes a very stiff bicontinuous cubic phase. With further increase it transits to a bluish tinged suspension of freely flowing vesicles, and then to a white suspension of colloidal sized liposomes. These transformations in microstructure can be seen by visual inspection: growth to cylinders and interactions between them make the still clear solution viscous. The bicontinuous cubic phases are extremely stiff with the texture of vaseline. The vesicles, around 200 nm diameter, are freely flowing and take on a bluish tinge due to light scattering. The much larger suspension of multilamellar liposomes is white. (These changes can be quantified further with crossed polar glasses, and a magnetic stirrer, to show up viscosity changes. These quantify the different phases more explicitly.)

Similarly if we mix a single-chained micelle forming anionic with a corresponding cationic surfactant, the combined mixed surfactant has $v_H/(a_P l_H) \sim 1$ and forms vesicles. The vesicles have to bear some excess charge to impose repulsive electrostatic forces to stop them from associating.

9.9 Detergency

Much the same principle underlies detergent action. Solid and oil-like hydrophobic dirt attached to surfaces is removed by surfactants. These adsorb at heterogeneous solid surfaces to be cleaned and on to the attached colloidal particles. The desorption of particles, now coated with a physisorbed monolayer of surfactant, is assisted by agitation. For non-ionic surfactants it is the hydration forces imposed by the adsorbed monolayers of surfactant that assist the removal process (see Fig. 9.12) [17].

The higher temperatures often used in detergency are necessary to liquefy the oil-like soil contaminants. That assists surfactant tail adsorption. More effective detergency is often achieved by using less, rather than excess, surfactant. The reason is that multiple layers will form at the interface at high concentrations and inhibit removal of soil particles.

An antidetergent principle underlies the use of double-chained cationics as softeners for clothes and hair conditioners. Adsorption of long double-chained (diC_{18}) cationic surfactants above the Krafft (chain melting) temperature (see the Appendix) results in a rigid monolayer at room temperature with a smooth hydrophobic surface. This yields the desired soft texture. Cationics are used because of their positive charge. That assists adsorption onto negatively charged substrates.

9.10 Bactericidal action

The same principle behind the changing of curvature through mixed surfactants underlies the mechanism of bactericidal action. The most commonly used surfactants for sterilization in hospitals are cationics such as cetyl-trimethyl-ammonium chloride, or cetyl-pyridinium chloride (cf. Appendix). They have a multitude of everyday applications: in toothpastes, contact lenses, cold relief.

The mechanism of sterilization can be demonstrated by adding the single-chained surfactants to a suspension of biological cells. From a physicochemical point of view the membranes of the cells are a very dilute suspension of large vesicles with membranes made up of proteins, cholesterol and a mixture of double-chained phospholipids for which $v_H/(a_P l_H) \sim 1$.

The cell membranes form a system of dilute bilayers. Consequently, at the cmc, the mixed system of single- and double-chained surfactants has an effective parameter of $[v_H/(a_P l_H)]_{eff} \ll 1$ as illustrated in the mixed system above. The bilayer membranes are immediately solubilized and the cells destroyed. It is noteworthy that the cmc at which the catastrophic disruption of the bacterial membrane occurs is that appropriate to the cationic surfactant at physiological concentrations of salt (0.15 M). This is different from the cmc in water. The optimal choice of surfactant is achieved by choosing the maximum chain length compatible with Krafft (chain melting) temperature, which must be less than body temperature, 37 °C. This (maximum possible chain length) ensures maximum hydrophobic free energy of transfer and mixing of the lipids and detergent molecule, and therefore maximum uptake by the bacterial membrane. Further, maximizing the chain length lowers the cmc and therefore the required drug dosage. The surfactant is more effective if it is impure. This is because the impurities act as hydrophobes to induce micellization at lower concentrations than for the ordinary cmc.

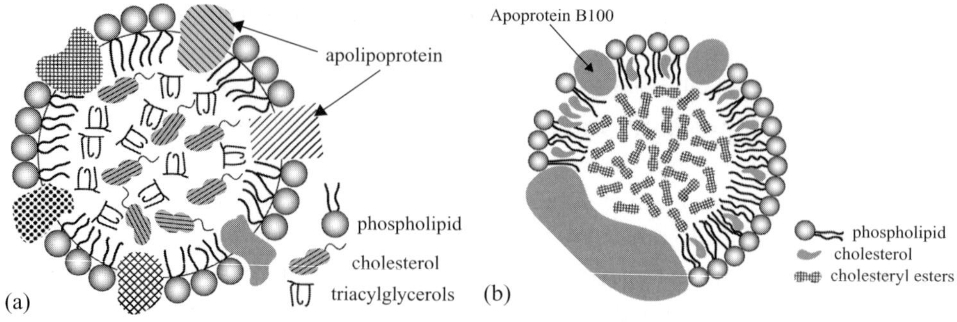

Fig. 9.13. Schematic picture of HDL (a) and LDL (b) assemblies.

For contact-lens sterilization, lower chain lengths are used because of the need to more easily access bacteria that grow in the less accessible pore of the contact lenses. With spermicides, now banned, the same principles are involved in destroying sperm cell membranes.

Cationics are most effective surfactants because the trimethyl ammonium or similar headgroups are compatible with the choline group of PC and other phospholipids. The reason that anionics such as SDS or non-ionics are not so effective seems to be that their headgroups are not so compatible with the water structure of the membrane phospholipids and do not easily mix. But this is not yet properly understood.

9.11 Biocides

Standard biocides such as hexachloridine work in a different way. They are inert, and being hydrophobic are physisorbed into the membrane and disrupt it. Surprisingly this seems not to be known in the microbiology literature. Claims of magic effectiveness for 'natural' oils appear dubious.

9.12 Detergency in other biosystems

Plaque formation on teeth, due to the building by bacteria of biocompatible protein substrates, is reduced somewhat by cationics in toothpastes for much the same reasons (detergency) as above.

9.13 High-density vs. low-density lipoproteins

It is tempting to speculate further: diagnostics of arterial health involve measurements of cholesterol contained in blood plasma protein particles, with the ratio of HDL (good) vs. LDL (bad) a measure of plaque formation (Fig. 9.13). LDLs deposit cholesterol through and on arterial walls and eventually block arteries. HDLs do not. The difference is a matter of curvature that determines adsorption and consequent curvature. The mixed cholesterol, protein and lipid aggregates differ, from a physicochemical viewpoint, principally in the ratio of double-chained C_{14} to longer-chain C_{16} lipids. This ratio is genetically determined. But the ratio is crucial.

With a preponderance of longer-chained lipids, lipoproteins adsorb at and inside arterial walls and build up thick plaque comprising multiple layers of (solid) lipid plus cholesterol. This leads eventually to blockage of blood flow. With a major component of shorter-chained lipids, which cannot so easily form multiple bilayers, and the chains of which are more fluid, the lipoproteins behave rather as detergents and inhibit plaque formation. LDLs contain more diC_{16} lipids, and tend to form and deposit bilayers. HDLs contain lower-chain-length lipid $<$ diC_{14}. They tend to solubilize and remove cholesterol from arterial walls. The mix of lipids, HDL/LDL, is genetically determined.

Excess alcohol, a cosurfactant, partitions and adsorbs into the phospholipid headgroups of the lipoprotein particles. This certainly occurs and would assist detergent action. Presumably alcohol increases headgroup curvature of phospholipids, and decreases adhesion, so inhibiting plaque formation. The dramatic effects of alcohol adsorption on cationic bilayer interactions have been explored and measured [18].

It is a curious fact that autopsies performed on genuine alcoholics show that they generally have cleaner arteries – but of course, compensatingly destroyed livers!

9.14 Local anaesthesia

A different kind of effect due to mixtures of surfactants is that associated with local anaesthesia.

The most popular local anaesthetic molecule, for half a century, was and is lidocaine. This is a cationic amphiphilic molecule with a bulky positively charged hydrophilic headgroup and a small hydrocarbon tail. In typical dental applications it is injected in excess into the region required to be anaesthetized. The law of mass action demands that the hydrocarbon tail anchors the hydrocarbon into the hydrophobic core of the membrane. The cationic charge of the adsorbed lidocaine molecule headgroup alters the electrostatic (action) potential of the nerve membrane and switches off the conduction of the nervous impulse.

As an aside the same (electrostatic) effects occur with some snake venoms, stinging jellyfish and lethal funnel-web spiders and octopuses that use polycationic polymeric poisons. Synthetic curare used as muscle relaxant in anaesthesia operates in the same way. The effects can be neutralized by exploiting specific ion effects through acetic acid or sodium acetate that appear to inhibit polymer attachment to nerves.

The change in electrostatic potential induced by adsorption of the polycation triggers what can be lethal nervous activity.

One sad illustration of adhesion of cationic polymers to anionic surfaces is that of extensive work over a decade done by Australian Government laboratories to isolate and then develop antibodies to spider polycationic venom. Attempts to purify the venom, collected at great pains, failed. The venom always, mysteriously, disappeared. The reason was finally identified. Cationic polymers adsorb strongly to the surface of the (anionic) glass vessels used. The use of hydrophobic vessels solved the problem.

Again, in all such cases there are no specific chemical reactions involved.

9.15 Global packing restrictions and interactions

These rough rules for determining aggregate structure are useful. They appear to be obvious and are apparently tautological. Local curvature set predominately by the balance of intramolecular interactions at the (oil-like) surfactant–water interface sets the allowed microstructure.

But local curvature also depends on interactions *between* the aggregates. Thus if intermicellar interactions cause a change in headgroup area, a, this will affect chain length, l, and the surfactant parameter $v_H/(a_P l_H)$. So if salt is added to a suspension of ionic micelles, one expects intuitively that electrostatic headgroup repulsion will be screened so that the headgroup area, a, will reduce. Then the surfactant parameter $v_H/(a_P l_H)$ will increase until spheres are no longer allowed to pack by geometry, and the micelles should grow, to ellipsoids or cylindrical shapes. That picture of events often does occur. In reality it is again far too simple a characterization for several reasons.

The meaning of a micellar interface can be defined thermodynamically, but can be nebulous in reality (cf. Fig 9.2B). 'Water structure' around non-ionic headgroups depends on temperature and the nature of the headgroup. So does that between aggregates. With ionic surfactants, the nature of headgroup repulsion depends on ion specificity and 'binding' of counterions to the micelle. This affects not just local water structure but interactions, in a way that seemed impossible to include in an overall predictive characterization.

However, the problem can be at least partially resolved once global packing constraints are recognized. Then a more systematic picture starts to emerge.

9.16 Global packing constraints and 'dressed' micelles

To see this we need some preliminary remarks on packing.

Consider a system of hard spheres, of volume faction V_s. These can pack in a simple cubic array up to a volume fraction of $V_s = 0.5$. At that point the spheres are

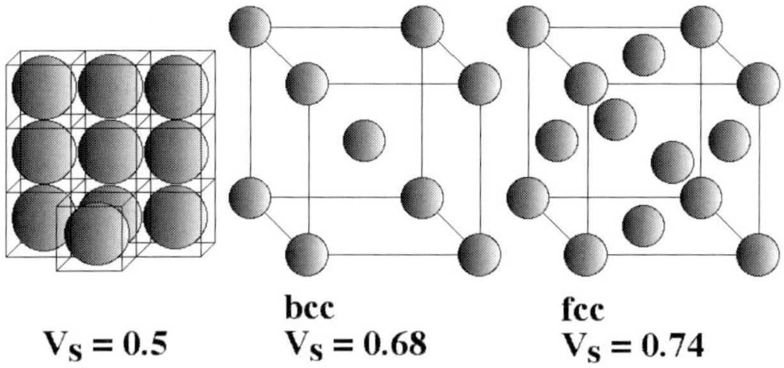

Fig. 9.14. Cubic arrays for spheres. Simple cubic, body-centred cubic (bcc); face-centred cubic (fcc).

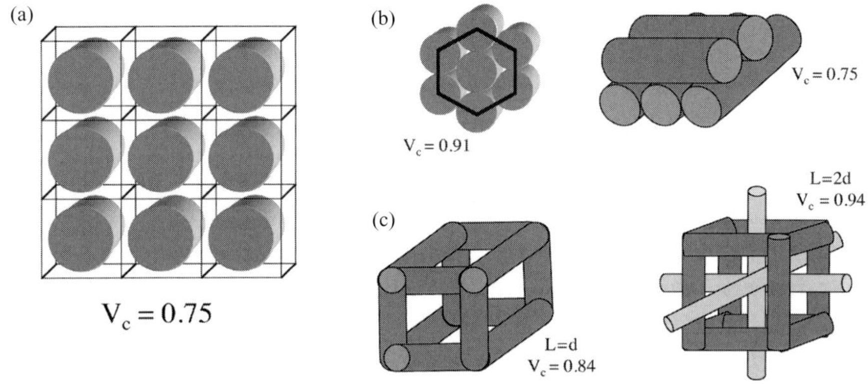

Fig. 9.15. Illustrative packings: square array of cylinders (a), square hexagonal and orthogonal array of cylinders (b), and orthogonal packings (c) [19–21]. Reprinted from Ref. 21. Reproduced by permission of the Royal Society of Chemistry on behalf of the Centre National de la Recherche Scientifique.

touching. Closer packings are allowed at higher volume fractions. The maximum packing fraction for a body–centred cubic array is $V_S = 0.68$. For random close packing or face–centred cubic symmetries it is 0.74 (Fig. 9.14).

Packings of parallel cylinders are allowed at volume fractions V_c up to 0.75 for a square lattice array and 0.91 for hexagonal close packing. For a square array the layers can be orthogonal at the same packing fraction.

If the cylinders are arranged orthogonally in intersecting overlapping alternate layers, the maximum packing volume fraction is about 0.84. A more complicated, interpenetrating packing illustrated can pack up to a volume fraction of 0.94 (Fig. 9.15).

For planar layers there is no limit on the volume fraction allowed.

For ionics packing limits refer to the 'dressed' micelles discussed below. When packing limits are reached the surfactant aggregates are forced to rearrange. Thus spherical micelles grow to form cylinders. Cylinders rearrange to form bilayers, or develop interconnections that grow into bicontinuous cubic phases. When multi-bilayer lamellar (liposomes) form due to attractive forces, the onion-like interior layers at some point must collapse – the curvature of the inner monolayer of internal surfactant layers simply becomes too great. The interior 'onions' then collapse to form unhappy micellar or other microphases inside the onions. When multiwalled liposomes reach a close packing limit, they have to precipitate to form infinite lamellar bilayers.

Such rearrangements depend on physicochemical conditions such as ionic strength inside and outside surfactant containers. We shall explore global packing in more detail in succeeding chapters.

With reverse curvature surfactants the same considerations apply. This is illustrated explicitly for reverse micellar surfactants in Refs. 16, 19 and 20.

9.17 Packing of spherical ionic micelles

With these constraints in mind, consider a system of spherical micelles. Suppose first that they are ionic. At the cmc practically all the surfactants are in micellar aggregates.

Typically, e.g. for SDS, with cmc 8 mM, the volume occupied by surfactant molecules is insignificant compared with the volume of water, being only a few per cent. However, what determines what happens is the *range* of interactions between the micelles. Here, for ionic micelles, it is the Debye screening length Λ. The *effective* sphere radius of such a 'dressed' micelle is that of the bare micelle itself with its associated 'bound' counterions, *plus* the size of its ionic cloud of free counterions. Here it turns out that about 90% of the counterions are 'bound' and the aggregation number is about 50 monomers per micelle. So the 'free' counterions have a concentration of around one tenth the cmc, i.e. 0.8 mM. This corresponds to a Debye length of about 9 nm. For measurements of the Debye length in ionic micellar systems see Ref. 22. This is much larger than the radius of a micelle. With chain length l for the surfactant molecule of about 1 nm, the radius R of the effective sphere is $(\Lambda + l) \sim 10$ nm! Then the *effective* volume V_{eff} occupied by micelles at the cmc can be estimated. The volume fraction of effective micelles is then $V_{\text{eff}}/(V_{\text{water}} - V_{\text{eff}}) \sim 0.6$. So the effective micelles are pretty much close packed already at the cmc (the volume fraction for close packing for a simple cubic array is 0.5, for fcc packing 0.74).

At the boundary of the effective cell occupied by each micelle, the repulsive electrostatic potential$\left[ze^2/(\varepsilon_r r)\right] \exp\left[-r/(\Lambda + l)\right] \approx k_B T$, where z is the effective charge of the micelle, e the unit charge, ε_r the dielectric constant of water, k_B the Boltzmann constant and T the absolute temperature.

If the concentration of surfactant is increased by a factor of 10, the consequent change in Debye length compensates. Roughly the same conditions apply. If background salt is added, say 10^{-1} M, the cmc is lowered and again not much change occurs in conditions. The same applies to cationic micelles such as CTAB [23].

An explicit quantitative analysis which predicts practically the entire phase diagram of water, sodium decyl sulfate and octyl alcohol illustrates the point elegantly [24]. We can infer from this analysis that both local curvature and interaggregate interactions are involved in the self-assembly process.

They depend on each other. Even at the apparently low concentration of the cmc, these systems are quite delicately poised, close to the packing limits allowed by geometry. Local curvature in ionic surfactants (and interactions between aggregates) depends very delicately on ion 'binding'. Thus if bromide replaces chloride as counterion in CTAB, the system forms cylindrical, not spherical, micelles. This cannot be due to changes in electrostatic interactions, and must result from headgroup area change between bromide and chloride. Again a 'dressed cylinder' description captures the essentials (interactions are here much more complicated; see Part I, Section 3.6, on effects of divalent ions and ion fluctuation forces).

9.18 Non-ionics and cloud points: water structure

Rather different forces guide the aggregation of non-ionic surfactants and zwitterionic lipids. Typical non-ionic surfactants are the polyoxyethylene systems discussed in the Appendix. Extensive phase diagrams for these are given in Ref. 25.

Consider the surfactant C_8E_5. This forms micelles in water at low concentrations (cmc ≈ 7 mM at 25 °C) [26]. But if we consider the phase diagram in the Appendix, with increasing concentration they exhibit a cloud point, a phase separation at 35 °C. In the single-phase region the micelles are spherical, like those for single-chained ionics. The aggregation number is about 70 at 25 °C [26]. In the two-phase region, the upper phase contains cylindrical aggregates, the lower phase is water [27].

For a phase-separation transition to occur, there must be an attractive force operating above the cloud point temperature. Here the surfactants are non-ionic. There are no electrostatic forces, and van der Waals forces between the micelles are insignificant. The headgroups are hydrated, and surface-induced hydration can extend only about three molecular layers. If we consider 'dressed micelles', micelles with any hydration water, we see that the global packing considerations relevant to ionics just discussed are irrelevant. Something else must be going on to cause the phase transition. Indeed it is. NMR measurements show that above the cloud point the headgroups lose two water molecules of hydration per PEO group [28]. The headgroup area of a surfactant molecule in an aggregate then reduces, and

$v_H/(a_P l_H)$ increases to above 0.5. Cylindrical micelles form. Simultaneously, an attractive force between such objects, which are necessary for the phase separation, turns on above 35 °C. This can be seen by direct measurements of forces between monolayers of PEO surfactants across water, adsorbed onto mica [29]. Below 35 °C the forces are repulsive. Above the phase transition cloud point the forces are attractive. Nobody knows why. But the existence of any weak attractive force between (effectively) infinitely long cylindrical aggregates will cause their phase separation.

Again the effective dressed cylindrical aggregates are close to the packing limit for cylinders. When, with increasing concentration, the packing limit is exceeded, the cylindrical phase can no longer pack and the system transits to lamellar phases.

Similar forces due to water structure operate in many other systems. Their nature depends on the characteristic hydration of the system. For example, in the interconnected networks produced by diC_8PC in water dispersions, dehydration of the polar headgroups plays a major role in the phase-separation process [30].

Another familiar example of this specificity of hydration forces occurs in sugar solutions. In sucrose solutions at 1 M the activity coefficient is zero, the system behaving as a perfect gas. The reason is that the large contribution to the virial coefficient from hard-core interactions is cancelled by an attractive exponentially decaying hydration force due to water structure that again extends only over about three molecular layers from each sugar molecule. For other isomers of sugars the virial coefficients at the same concentrations vary enormously, reflecting the fact that these hydration forces can be either repulsive or attractive. They depend on the particular anisometry and alignment of OH groups.

The same global geometric constraints, taken with the forces between aggregates as the determinant of self assembly, applies to micelles formed with less common surfactants with different counterions; e.g. if we ion-exchange Br^- with OAc^- or OH^-. With the latter, cmcs are twice as high, aggregation numbers half as large, but there is no 'binding'. The changes in measured parameters can be shown to compensate to give essentially the same energetic and effective packing conditions [3].

9.19 Renormalized variables for phase behaviour

A further remark is that if a phase diagram is made in which $[v_H/(a_P l_H)]_{eff}$ replaces temperature on the ordinate, and volume fractions replace concentration on the abscissa, the phase diagram has the same form as that for non-ionic surfactants with temperature vs. concentration as the variables (see the following chapter for the phase diagrams). For the non-ionic surfactants, $v_H/(a_P l_H)$ *decreases* with temperature. For single-chained ionic surfactants, $v_H/(a_P l_H)$ *increases* with temperature.

So they appear to show different thermodynamic behaviour. But the replacement of temperature by a renormalized variable representing local curvature, and concentrations by a variable representing interactions (global packing), results in universality of phase behaviour. This can be demonstrated by using the mixed cationic system discussed above to change curvature. The system now exhibits a cloud point.

This suggests that more appropriate proper thermodynamic variables to characterize phase behaviour in microstructured fluids are local curvature versus interactions (global packing), not the Gibbsian variables *P*, *T*, *V* and density.

Appendix

Specific surfactants and lipids, molecular characteristics

The term 'surfactant' (surface-active agent) is usually restricted to molecules that have separated hydrophobic and hydrophilic regions, since water is by far the most common dispersing phase. Actually this concept holds for hydrogenated and fluorinated chains as well: due to the mutual insolubility of fluocarbons and hydrocarbons, these chains are incompatible, and produce self-assembled aggregates in selective solvents [31]. And diblock semifluorinated copolymers, e.g. $F(CF_2)_8(CH_2)_{16}H$ (in brief F_8H_{16}), are referred to as 'primitive surfactants'.

This molecular duality is sometimes embraced by the alternative descriptor, amphiphile. Lipids are usually double-chained, usually highly insoluble, surfactants that are integral to biomembranes.

The conjoined antipathy between water and hydrophobic substances generates radically different behaviour to that displayed by the individual surfactant component moieties. In order to avoid contact with water molecules, and depending on its structure, a surfactant molecule can adsorb at interfaces (air–water, oil–water or solid–water), or it can self-assemble exposing the polar segment to the aqueous environments and confining the apolar block in the aggregate core, or it can fold into a coiled structure (especially for polymers). In any case, this process can be envisaged as a pseudo-micro-phase separation.

The hydrophobic moiety, typically one or more hydrocarbon chains, prefers to escape from water and adsorb at a hydrophobic solid or oil, or air–water interface. The hydrophilic headgroup prefers to be in water. Adsorption decreases the interfacial energy between the two phases. Surface activity is a property of many proteins, polymers and other molecules, even some 'hydrophobic' ions like acetate or quaternary ammonium ions. Our focus is on the behaviour and self assembly of surfactants in aqueous solutions, and surfactant–water–oil systems. To illustrate that behaviour we need some familiarity with the molecular structure of typical surfactants.

(1)
(2)
(3)
(4)
(5)
(6)

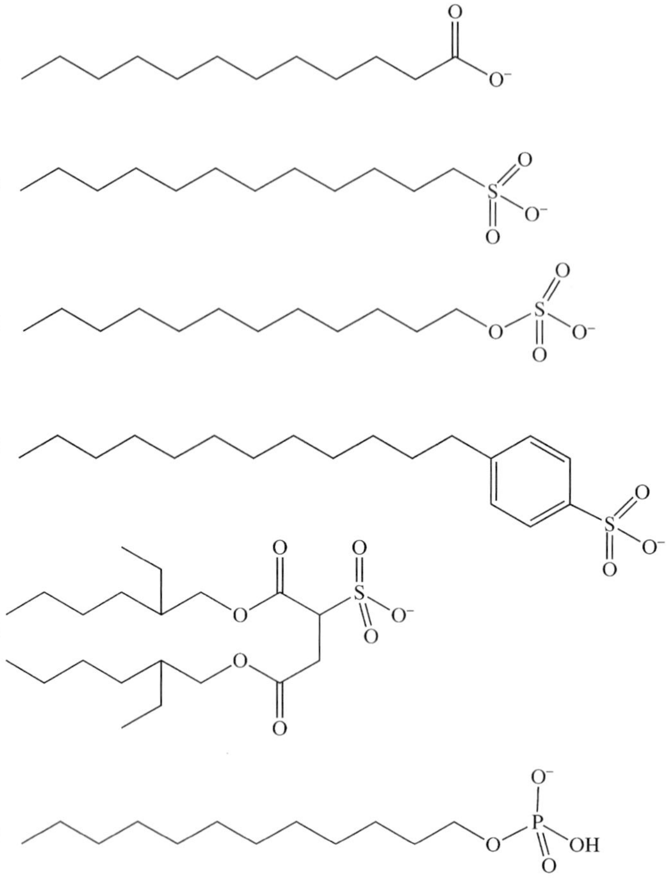

Fig. 9.16. Chemical structure of a dodecanoate (1), dodecyl sulfonate (2), dodecyl sulfate (3), dodecyl benzene sulfonate (4), sulfosuccinate (AOT) (5) and a dodecyl phosphate (6).

For definiteness we focus on surfactant molecules that have a hydrophilic polar headgroup and one or more oil-like hydrophobic tails. Depending on the nature of the polar groups they can be subdivided into different categories: anionic, cationic, non-ionic, zwitterionic and amphoteric (this last means that the headgroup can be positive or negatively charged depending on pH).

Anionic surfactants

These are by far the most common. The headgroup is negatively charged. Figure 9.16 shows some common anionic species.

Other anionic surfactants contain phosphorus-based ionic species (phosphonates and phosphates). The counterion is usually an alkali metal such as sodium or potassium, or an alkaline earth such as calcium.

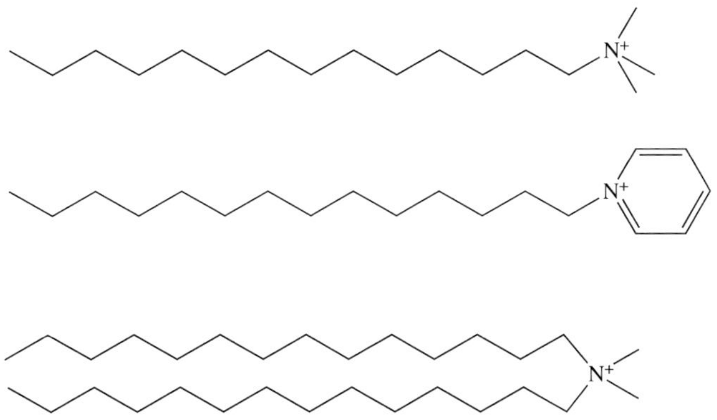

Fig. 9.17. Chemical structure of cationic amphiphiles. From top to bottom: tetradecyl trimethyl ammonium, tetradecyl pyridinium and ditetradecyl dimethyl ammonium.

Cationic surfactants

The positively charged polar headgroup carries a quaternary ammonium ion, or a pyridinium moiety, as in the figure (Fig. 9.17).

The counterion X is usually a halide (chloride or bromide) that can be ion-exchanged for a nitrate or acetate ion, for example.

Due to their positive charge, cationic surfactants adsorb strongly at all natural interfaces, which are overwhelmingly negatively charged.

Non-ionic surfactants

Representative members of this class are derived from ethylene oxide. They are therefore poly(oxyethylene) ethers, namely polyethoxylated fatty alcohols (in brief C_iE_j) and polyethoxylated alkyl-phenols (cf. Fig. 9.18).

Interestingly, if the poly(oxyethylene) residue forms a macrocycle, e.g. in octyl-18-crown-6 (in brief $C_8\Phi_6$), then the headgroup can selectively bind some cations (typically K^+), so turning it into a cationic surfactant [32]. Other non-ionic surfactants are the alkyl glucosides and sorbitan alkanoates. These bear more complex headgroups, usually derived from saccharides.

Non-ionic surfactants are less sensitive to the effect of electrolytes than their ionic counterparts, while the effect of temperature is by far more significant (see cloud point later).

Zwitterionic surfactants

The headgroup contains two opposite charges and is therefore dipolar, but it is electroneutral as for non-ionic species. Typical examples are phospholipids where

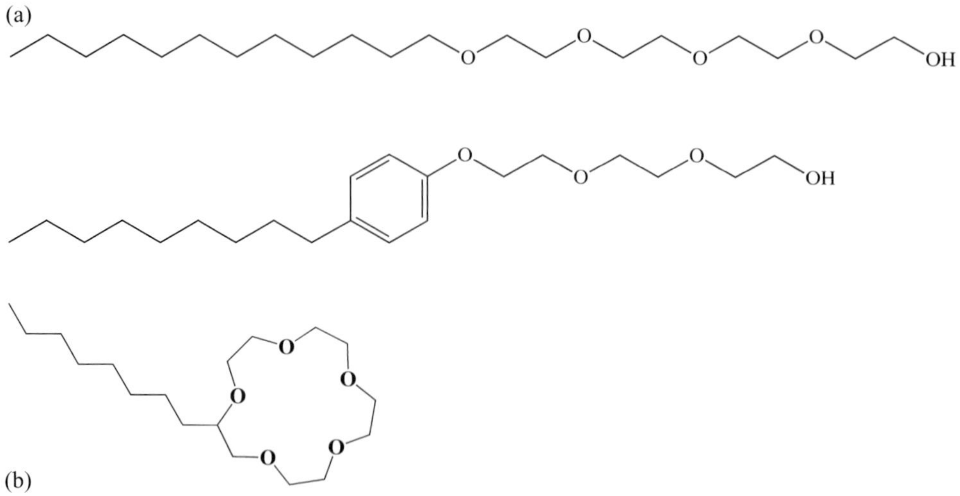

Fig. 9.18. Molecular structure of $C_{12}E_4$, C_9-Ph-E_3 (a) and $C_8\Phi_6$ (b).

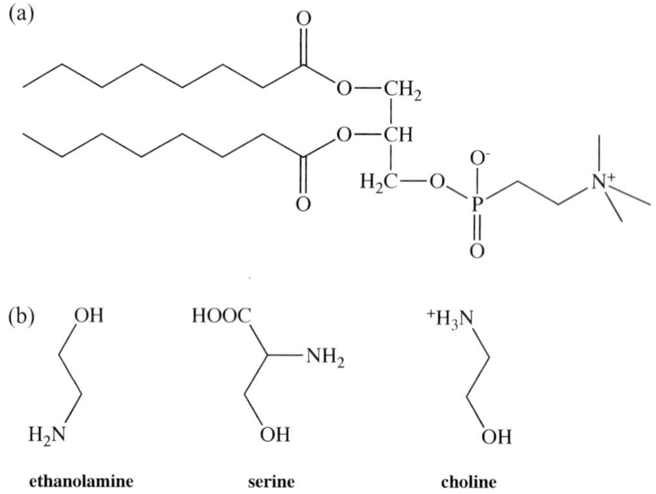

Fig. 9.19. Chemical structure of dioctanoyl phosphatidylcholine (a), ethanolamine, serine and choline (b).

the polar headgroup derives from a glycerol unit bound to a phosphate group and to another segment that can be a choline or ethanolamine (Fig. 9.19).

Other examples of zwitterionic surfactants are the sulfobetaines (Fig. 9.20).

As opposed to non-ionic surfactants, zwitterions are more soluble at higher temperatures. While the structural properties of zwitterionic surfactants do not change significantly with pH, amphoteric amphiphiles can change their charge depending on the pH of the bulk phase in which they are dispersed.

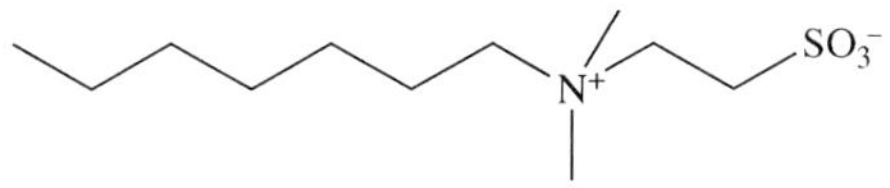

Fig. 9.20. Chemical structure of a sulfobetaine.

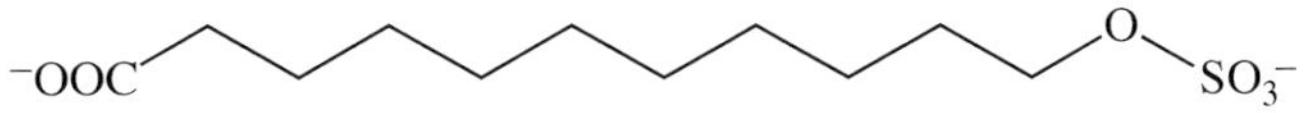

Fig. 9.21. Chemical structure of an asymmetric bolaform amphiphile.

The hydrophobic chains of surfactants can be saturated or can carry one or more double or triple bonds. They may contain aromatic spacing groups, or branched alkyl chains. The most common surfactants are either single-chained or double-chained molecules. But, for example, the very important biologically active triglycerides possess three aliphatic chains bound to a glycerol moiety that acts as a polar headgroup. And there are even fancier amphiphiles, the so-called 'bola-amphiphiles', that bear two polar headgroups held together by one or more hydrophobic chains (Fig. 9.21).

In the figure we show a bola-amphiphile constituted by a COO- (left) and a sulfate (right) group, connected by a C_{10} aliphatic chain. In nature, more complex bola-amphiphile molecules are found in the membranes of several Archaebacteria organisms that are adapted to live in severe conditions of temperature and pH, such as in volcanic areas.

Phospholipids are ubiquitous membrane-forming lipids, common to all mammals and other multicelled organisms. They are formed most commonly from choline, serine or ethanolamine (Fig. 9.19). The free –OH residues can then react with a phosphate group and generate phosphatidylcholine (PC), phosphatidylserine (PS) and phosphatidylethanolamine (PE), respectively. The chain lengths vary from 8 carbons to the more usual 14, 16 and 18.

Common lipids that form fats are the mono-, di- and triglycerides, pictured in Fig. 9.22, where the hydrophobic residues R are saturated or unsaturated aliphatic chains. The $\omega-3$, $\omega-6$ and $\omega-9$ lipids have unsaturated hydrocarbon tails and are major components of the brain (Fig. 9.23). For comparison we picture oleic and linoleic acid in Fig. 9.24. Cholesterol is an essential rigid molecular spacer in biomembranes (Fig. 9.25).

Model surfactants that we shall have much use of are the double-chained cationics, didodecyl dimethyl ammonium bromides and other counterions (DDAB), single-chained cationics, cetyl (C_{16}) trimethyl ammonium bromide (CTAB) or pyridinium derivatives, single-chained anionics such as sodium dodecyl sulfate (SDS) or Aerosol OT (AOT), depicted above in Fig. 9.16.

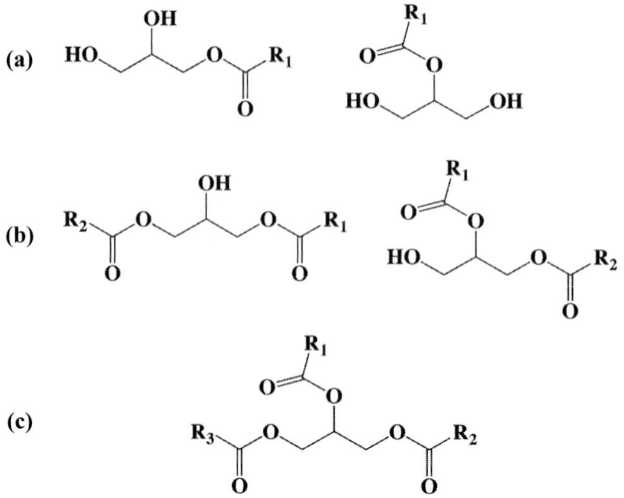

Fig. 9.22. Structure of a monoglyceride (a: a single acyl chain in position 1 or 2), diglyceride (b: two acyl chains in positions 1,2 or 1,3) and a triglyceride (c: three acyl chains).

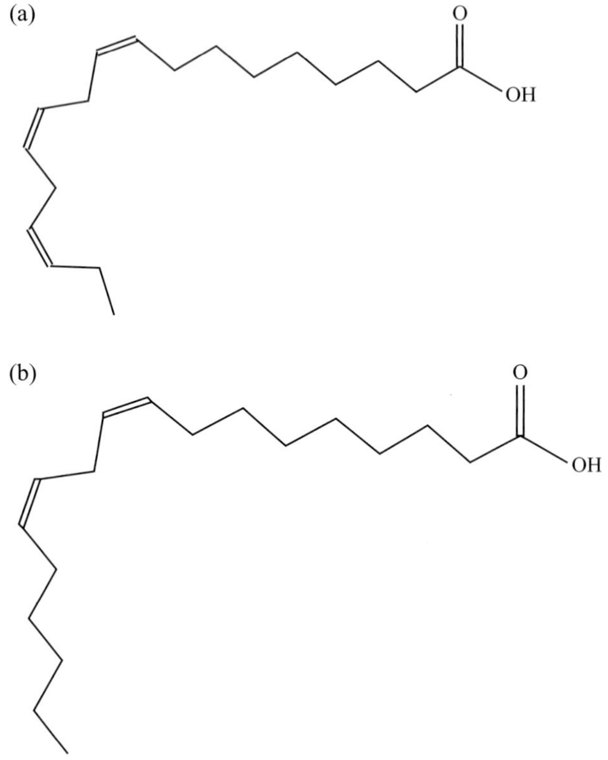

Fig. 9.23. Chemical structure of $\omega-3$ (a) and $\omega-6$ (b) lipids.

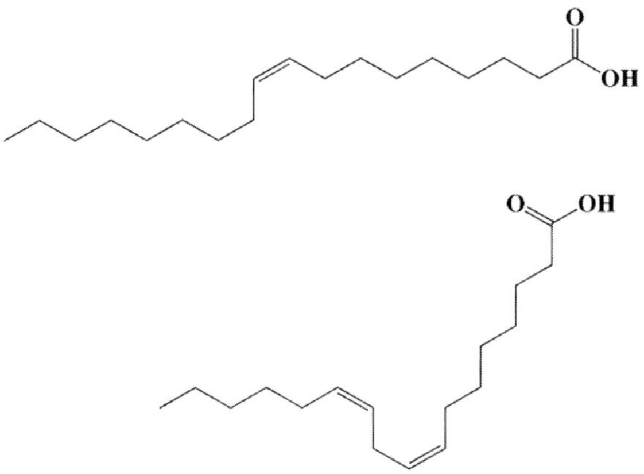

oleic acid

linoleic acid

Fig. 9.24. Chemical structure of oleic and linoleic acids.

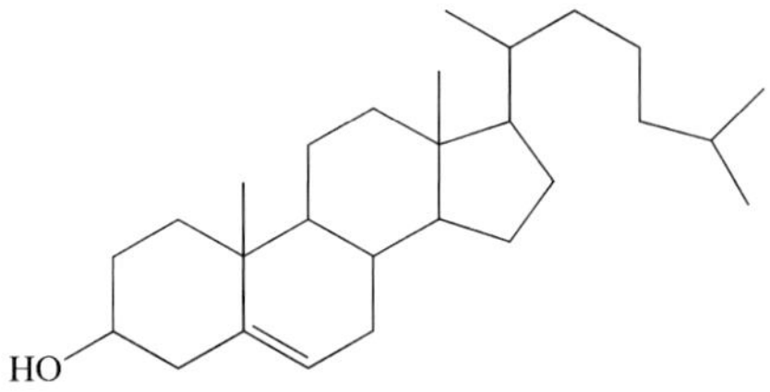

Fig. 9.25. Chemical structure of cholesterol.

Bio-inspired amphiphiles with nucleotide or nucleoside moieties embedded in the polar head group (see Fig. 9.26) constitute an excellent tool for the study of molecular recognition in self-assemblies [33] and DNA–membrane interactions [34].

For detailed technical expositions on surfactant chemistry, usages, synthesis and biochemistry we refer the reader to some excellent books [35–37].

Macroscopic manifestations of self assembly

As surfactants are added successively to water the interfacial tension drops and at a certain point shows an abrupt break, thereafter remaining constant, as depicted in Fig. 9.1. Beyond the break point, the critical micelle concentration (cmc), a monolayer of adsorbed surfactant at the interface is close-packed. With further addition of surfactant, the molecules assemble into small aggregate micelles. The same behaviour is shown with conductivity of ionic surfactants.

The abrupt behaviour is not always sharp – some double-chained surfactants show only a gradual increase of conductivity, reflecting gradual assembly of polydisperse aggregates. Non-ionic surfactants show an increase of interfacial tension above the cmc.

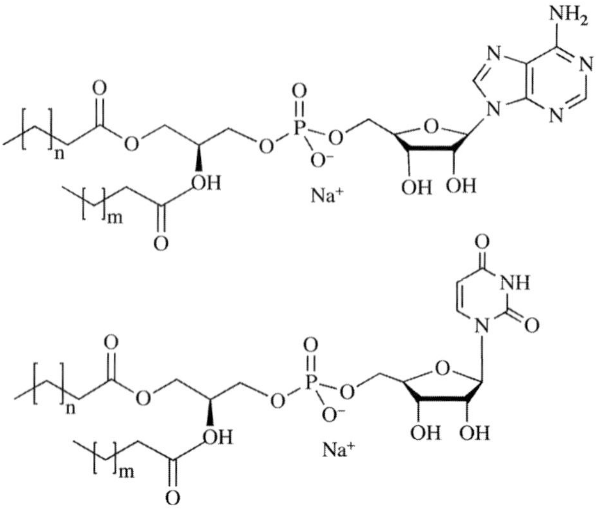

Fig. 9.26. Chemical structures of phosphatidyladenosine (top) and phosphatidyluridine (bottom) amphiphiles, examples of bio-inspired functional surfactants.

What really happens at the interface is in fact extremely complex, and usually also fraught with problems of contamination.

In the bulk solution a mapping of 'phase' behaviour as a function of temperature vs. concentration is more revealing. The aqueous mixtures of surfactants form a variety of mesomorphic phases depending on the chemical structure, composition and temperature. Phase boundaries between different kinds of aggregates are generally sharp. The microstructure can be and is studied by many techniques, simple inspection between crossed polar plates, viscosity or conductivity being often the most revealing. X-ray scattering seems to have been most reliable. Neutron and dynamic and ordinary light scattering, differential scanning calorimetry, infrared spectroscopy, electron spin resonance, cryoTEM (cryo-transmission electron microscopy), ordinary electron microscopy and nuclear magnetic resonance are powerful sledgehammer techniques. These are often beset by theoretical software limitations and experimental artefacts, so that complementarity in techniques becomes mandatory.

We are not concerned with these specialized techniques and refer the reader to the literature. Our concern is to draw out some design principles that enable us to prescribe the microstructure of self-assembled aggregates.

In Figs. 9.27 and 9.28 we give typical phase diagrams of ionic and non-ionic surfactants. The diagrams reflect the change of the surfactant packing parameter that characterizes the shape of the surfactant in an aggregate.

In the case of the cationic surfactant CTAB (Fig. 9.27), the most extended region comprises the hexagonal phase, formed by long rods packed in a 2-D

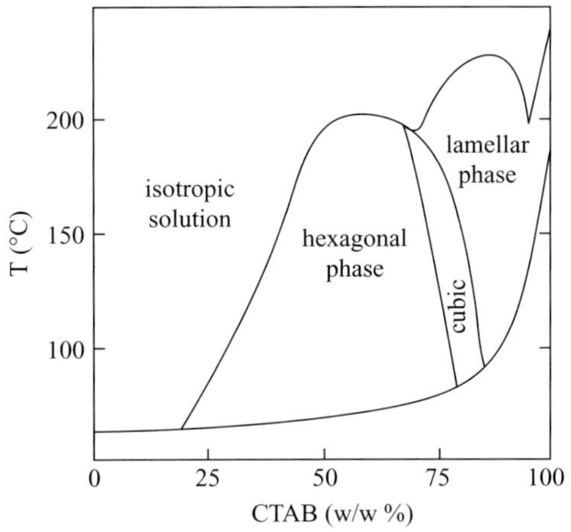

Fig. 9.27. Phase diagram of CTAB/water dispersions. Adapted from Ref. 38. Copyright 1988 with permission from Elsevier.

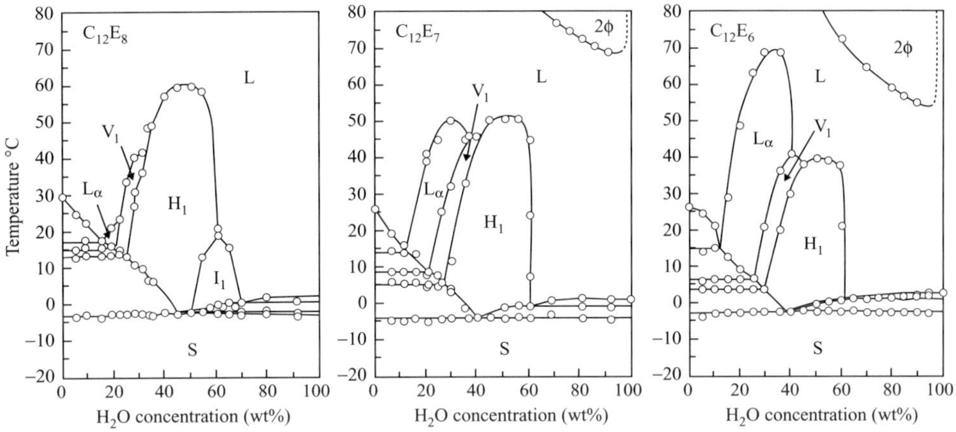

Fig. 9.28. Phase diagram of $C_{12}E_8$/water, $C_{12}E_7$/water, and $C_{12}E_6$/water dispersions. L, micellar phase; I_1, normal discontinuous-type cubic; H_1, normal hexagonal; V_1, normal bicontinuous-type cubic; $L\alpha$, lamellar; S, solid; 2ϕ, biphasic system. Reprinted with permission from Ref. 39. Copyright 2002 American Chemical Society.

hexagonal array. At higher surfactant concentrations the lamellar phase appears, and in between the bicontinuous cubic phase. A typical feature of ionic surfactants is that the phase boundaries do not vary with temperature.

Instead, non-ionic amphiphiles are much more sensitive to temperature changes. Fig. 9.28 illustrates the phase diagrams of three C_iE_j compounds in water. The sequence of phases looks similar to the cationic/water system at low temperatures. But at higher temperature the area per polar headgroup is significantly reduced and

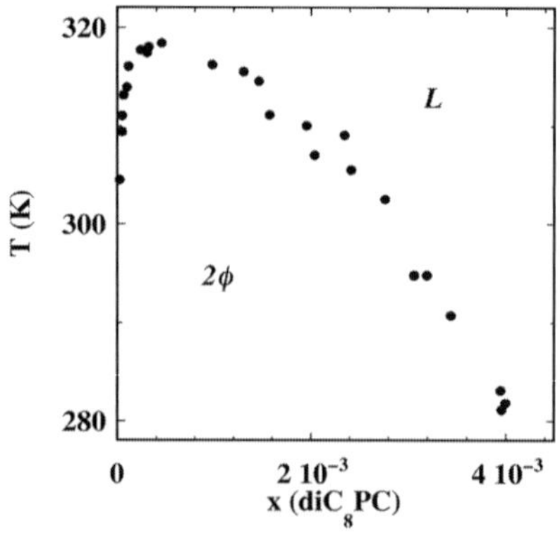

Fig. 9.29. Phase diagram of diC_8PC in water. The upper region L comprises a liquid micellar dispersion, while in the bottom 2ϕ section of the diagram a two-phase system is formed.

the surfactant packing parameter increases, due to a change in water–surfactant interactions.

A big difference in the phase diagrams of non-ionic surfactants with respect to cationic amphiphiles is the presence (see the upper right corner in Fig. 9.28) of a biphasic region (2ϕ) that is separated from the liquid micellar dispersion (L) by a coexistence curve. Upon heating, aqueous dispersions of C_iE_j surfactants phase-separate and form two coexisting clear phases. In the case of zwitterionic lipids, e.g. dioctanoyl phosphatidylcholine (diC_8PC), the coexistence curve is reversed (see Fig. 9.29) and the system phase-separates upon cooling [40].

The Krafft point line is the temperature below which the surfactant tails crystallize to form a solid liquid crystal subphase that precipitates out of solution. At the Krafft point the solubility of a particular surfactant equals its cmc. At temperatures lower than the Krafft point an increase in the amphiphile concentration leads to the precipitation from the solution. At the Krafft point the surfactant crystals melt and form micelles. The micelles are very soluble, and at the Krafft point the solubility curve increases steeply. Experimentally, the Krafft point is determined by plotting the logarithm of the molar solubility versus the reciprocal of the absolute temperature (Fig. 9.30).

With three-component systems, phase behaviour is best plotted at a fixed temperature in a ternary diagram. One such that we shall have much recourse to is the system water, alkane, DAAB.

The phase diagram below is with dodecane as alkane.

The phase behaviour is usually plotted as a function of weight fractions of components. (Samples are generally prepared by titration and weighing into subsequently sealed containers.) It is trivial in principle to make a phase diagram, but in practice a difficult matter that takes much skill and patience.

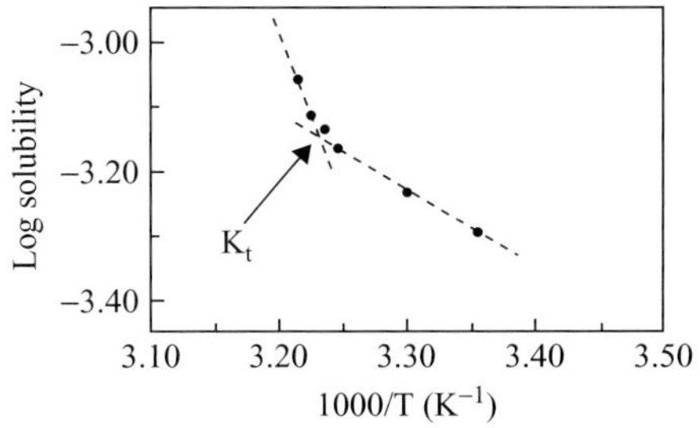

Fig. 9.30. Determination of the Krafft point (K_t) for an anionic surfactant from the solubility–temperature plot.

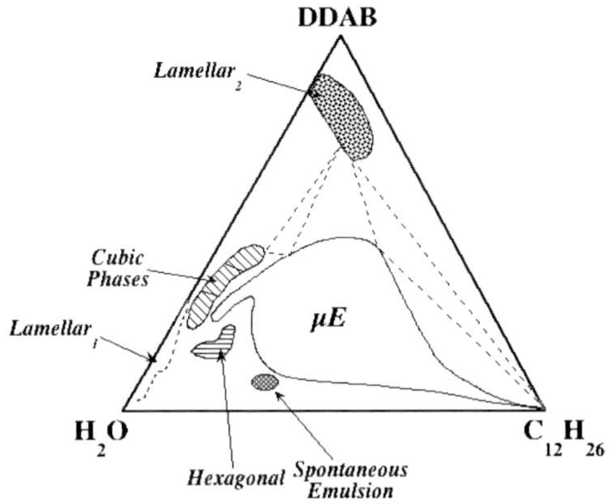

Fig. 9.31. Phase diagram of the DDAB/dodecane/water system [7, 41]. Readapted from Ref. 41.

In the diagram in Fig. 9.31, single phase boundaries are marked. A line parallel to the oil surfactant axis represents a fixed weight fraction of water. A line from a point on that oil surfactant edge to the water corner represents a line of constant oil-to-surfactant weight ratio. Two and three phase regions and their proportions are constructed by a Maxwell construction (lever rule).

Macroscopic manifestations of self assembly are several. In some cases, a simple visual inspection of a surfactant dispersion reveals that, upon adding small quantities of surfactant to a volume of water, suddenly above a certain concentration the dispersion becomes opaque and scatters light (Tyndall effect), provided the size of the particles is large enough. This phenomenon, as the others that will be reviewed below, depends on the fact that the first molecules of surfactant will dispersed in the monomer state by the solvent, as depicted schematically in Fig. 9.32. However, if one continues to add more and more surfactant, its molecules will move at the air–water interface first, and once the interface is saturated they will start forming small aggregates. The process depends on some parameters such as the nature and

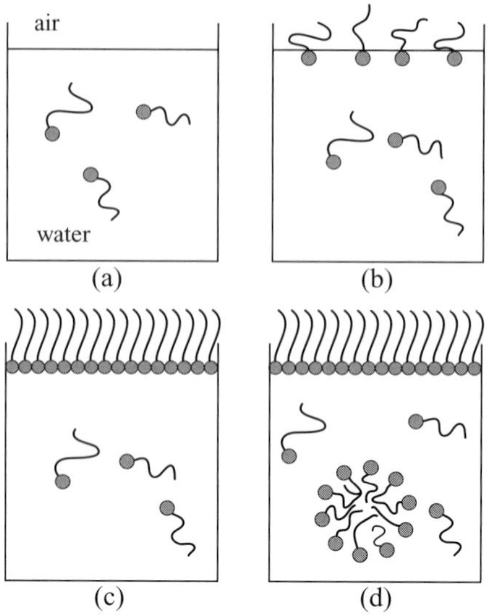

Fig. 9.32. Chain of events occurring upon addition of small quantities of a single-chained surfactant to water.

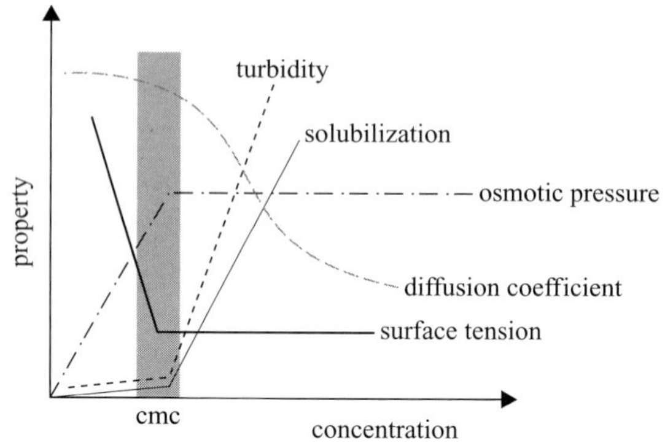

Fig. 9.33. Variation of different properties with increasing surfactant concentration: surface tension, osmotic pressure, solubilization, turbidity and diffusion coefficient. The grey area represents the concentration range for the cmc.

concentration of the surfactant, temperature and presence of other additives in the solution.

One signature for the change in aggregation state with increasing amphiphile surfactants is the change in surface tension. Similar abrupt changes with increasing surfactant concentration can be found when measuring other properties, such as conductivity, viscosity, osmotic pressure, density, light scattered intensity, diffusion coefficient and so on. What is obtained is usually a plot like that shown in Fig. 9.33. The cmc values obtained with different techniques do not coincide. The concept of a concentration range is a more appropriate descriptor.

If the surfactant is soluble in water, its destiny will be different. As depicted in Fig. 9.32, the monomers will try to escape from contact with the solvent and accumulate at the interface, where the so-called adsorption layer will be formed as a result of a phase separation. In this new situation, termed a spreading Langmuir monolayer, the headgroups will keep interacting strongly with the aqueous phase, while the hydrophobic tails are confined in the gas phase and interact among themselves. If one keeps adding more and more surfactant, the molecules will go to the interface, occupying more and more of the available area, and decreasing the surface tension γ. Fig. 9.1 shows the variation of γ with the surfactant concentration (expressed as Log$c$).

For small amounts of surfactant the surface tension decreases quite steeply (left side) from its value for pure water (about 72 mN/m at 25 °C). Then the decrement becomes less pronounced and eventually γ attains a more or less constant value, meaning that self-assembled structures are formed in the bulk liquid. Further addition of monomers produces only an increase in the number of the aggregates, while the concentration of the monomers in solution remains constant. This constant value is the 'critical micellar concentration'. Sometimes the γ vs. Log$c$ plot gets more complicated but for these details the reader is invited to refer to the more specialized literature [42,43].

Another relevant feature here is that the slope of the γ vs. Log$c$ for $c <$ cmc provides very important information, the area per polar headgroup. In fact the Gibbs–Duhem equation applied to an adsorption monolayer finally results in the formula:

$$\Gamma = -\frac{1}{2.303nRT}\left(\frac{\partial\gamma}{\partial Logc}\right) \tag{9.1}$$

$$a_P = \frac{1}{6.023 \cdot \Gamma} \tag{9.2}$$

where Γ, R, T are respectively the surface excess of the surfactant, the gas constant ($R = 8.31$ J/mol·K) and the absolute temperature (K). For non-ionic surfactants $n = 1$, and for monovalent ionic amphiphiles $n = 2$. If γ is measured in mN/m, a_P is given in Å^2/molecule. From Γ it is easy to calculate a_P. This experimental evaluation of the cross-section of the polar heads allows then the calculation of the packing parameter, $p = v_H/(l_H a_P)$ if the volume (v_H) and length (l_H) of the hydrocarbon tail are known, for example through the phenomenological Tanford formulae [44]:

$$v_H = 27.4 + 26.9n_C \tag{9.3}$$

$$l_H = 1.5 + 1.265n_C \tag{9.4}$$

n_C is the number of carbon atoms embedded in the hydrophobic core. The value of p can be used to predict the shape of the micellar aggregate that a particular

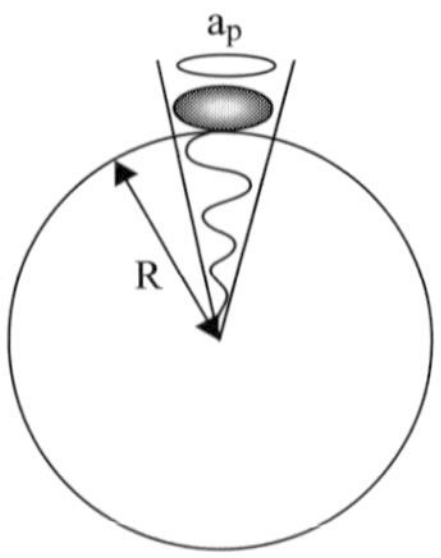

Fig. 9.34. Schematic structure of an ideal spherical micelle. The hydrocarbon chains – in a liquid state – fully occupy the spherical core (radius R), the polar heads cover the surface of the aggregate.

surfactant is going to produce. Let us take a spherical aggregate, such as that depicted in Fig. 9.34. Each surfactant that belongs to this micelle will have the polar headgroup occupying a portion of the outside surface (a_p), while the aliphatic chain fills the hydrocarbon core of the aggregate. If we assume that the hydrocarbon chain in the micelle has the same density as the pure corresponding hydrocarbon (e.g. a dodecyl chain and dodecane), then the length and volume of the surfactant lipophilic block can be obtained from Tanford's rules. Since there is no vacuum inside the micelle, this means that all hydrophobic chains must fully occupy the core of the aggregate, assuming a somehow curled conformation, whose length will be necessarily smaller than that in the fully stretched conformation (Fig. 9.34).

Therefore for the volume of the core, if g represents the aggregation number (i.e. the average number of monomers per micelle), we have $4\pi R^3/3 = g v_H$ and for the core surface we have $4\pi R^2 = g a_P$.

From these relationships we obtain $R/3 = v_H/a_P$. But if l_H is the length of the aliphatic tail in its fully stretched conformation, $R < l_H$, and therefore $v_H/(l_H a_P) < 1/3$. Packing parameters lower than $1/3$ produce spherical micelles.

We can repeat a similar calculation for rods, and then for vesicles (curved bilayers), planar layers, inverted structures and cubic phases, and obtain different values (see Fig. 9.11).

Typical surfactants can be single-chained like soap molecules, double-chained like the lipids that form cell membranes, or triple-chained. The resulting opposing tendencies cause the molecules to self-assemble into aggregates of molecules in a way that minimizes the exposure of oil-like regions to water. This energetic requirement is opposed by the entropy available to the monomers in a dilute solution. With increasing addition of surfactant molecules they can minimize the unfavourable interaction with water first by filling up an air–water or oil–water interface. When this is filled by a layer of surfactants, a monolayer, different structures tend to form in the bulk water.

The chain of events is reflected in the interfacial tension, which decreases to a minimum and thereafter remains constant.

The simplest kind of such structure that then forms with single-chained surfactants at a critical micelle concentration is called a micelle. It can be visualized as a sphere of say 50 molecules with their headgroups facing water and the hydrocarbon tails forming an oil-like interior.

This classical, textbook, picture of micellization is vastly oversimplified. It is increasingly clear that corresponding to a myriad microstructures that exist in bulk solution, which we explore subsequently, there are just as many or more two-phase systems that form at the air–water interface and at other surfaces.

References

[1] P. Lo Nostro, S.-M. Choi, C.-Y. Ku and S.-H. Chen, *J. Phys. Chem. B* **103** (1999), 5347–5352.
[2] B. W. Ninham and D. F. Evans, *Faraday Discuss. Chem. Soc.* **81** (1986), 1–17.
[3] J. E. Brady, D. F. Evans, G. G. Warr, F. Grieser and B. W. Ninham, *J. Phys. Chem.* **90** (1986), 1853–1859.
[4] K. Lunkenheimer, F. Theil and K. H. Lerche, *Langmuir* **8** (1992), 403–408.
[5] D. F. Evans, S. Mukherjee, D. J. Mitchell and B. W. Ninham, *J. Coll. Interface Sci.* 93 (1983), 184–203.
[6] D. F. Evans, M. Allen, B. W. Ninham and A. Fouda, *J. Solution Chem.* **13** (1984), 87–101.
[7] D. F. Evans, D. J. Mitchell and B. W. Ninham, *J. Phys. Chem.* **90** (1986), 2817–2825.
[8] M. S. Ramadan, D. F. Evans and R. Lumry, *J. Phys. Chem.* **87** (1983), 4538–4543.
[9] S. T. Hyde, S. Andersson, K. Larsson, Z. Blum, T. Landh, S. Lidin and B. W. Ninham, *The Language of Shape. The role of curvature in condensed matter physics, chemistry and biology*. Amsterdam: Elsevier (1997).
[10] http://wwwrsphysse.anu.edu.au/~sth110/sth.html
[11] http://medicine.nus.edu.sg/phys/FacultyMembers_Deng.html
[12] http://www.sandforsk.se/
[13] B. W. Ninham, Y. Talmon and D. F. Evans, *Science* **221** (1983), 1047–1048.
[14] C. Tanford, *The Hydrophobic Effect*. New York: John Wiley (1973).
[15] J. N. Israelachvili, D. J. Mitchell and B.W. Ninham, *J. Chem. Soc. Faraday Trans. II* **72** (1976), 1525–1568.
[16] D. J. Mitchell and B. W. Ninham, *J. Chem. Soc. Faraday Trans. II* **77** (1981), 601–629.
[17] K. Bäckström, B. Lindman and S. Engström, *Langmuir* **4** (1988), 872–878.
[18] R. M. Pashley, P. M. McGuiggan and B. W. Ninham, *J. Phys. Chem.* **90** (1986), 5841–5845.
[19] A. Filankembo, P. André, I. Lisiecki, C. Petit, T. Gulik-Krzywicki, B. W. Ninham and M. P. Pileni, *Coll. Surf. A* **174** (2000), 221–232.
[20] P. André, A. Filankembo, I. Lisiecki, C. Petit, T. Gulik-Krzywicki, B. W. Ninham and M. P. Pileni, *Adv. Mat.* **12** (2000), 119–123.
[21] P. André, B. W.Ninham and M. P. Pileni, *New J. Chem.* **25** (2001), 563–571.
[22] R. M. Pashley and B.W. Ninham, *J. Phys. Chem.* **91** (1987), 2902–2904.

[23] D. F. Evans, M. Allen, B. W. Ninham and A. Fouda, *J. Solution Chem.* **13** (1984), 68–84.
[24] B. Jonsson and H. Wennerström, *J. Phys. Chem.* **91** (1987), 338–352.
[25] D. J. Mitchell, G. J. T. Tiddy, L. Waring, T. Bostock and M. P. McDonald, *J. Chem. Soc. Faraday Trans. I* (1983), 975–1000.
[26] G. D'Errico, D. Ciccarelli, O. Ortona, L. Paduano and R. Sartorio, *J. Coll. Interface Sci.* **270** (2004), 490–495.
[27] K. Weckström and A. C. Papageorgiou, *J. Coll. Interface Sci.* **310** (2007), 151–162.
[28] P. G. Nilsson and B. Lindman, *J. Phys. Chem.* **87** (1983), 4756–4761.
[29] P. M. Claesson, R. Kjellander, P. Stenius and H. K. Christenson, *J. Chem. Soc. Faraday Trans. I* **9** (1986), 2735–2746.
[30] P. Lo Nostro, S. Murgia, M. Lagi, E. Fratini, G. Karlsson, M. Almgren, M. Monduzzi, B. W. Ninham and P. Baglioni, *J. Phys. Chem. B* **112** (2008), 12625–12634.
[31] P. Lo Nostro, *Adv. Coll. Interface Sci.* **56** (1995), 245–287.
[32] M. Campagna, L. Dei, C. M. C. Gambi, P. Lo Nostro, S. Zini and P. Baglioni, *J. Phys. Chem. B* **101** (1997), 10373–10377.
[33] D. Berti, *Curr. Op. Coll. Interface Sci.* **11** (2007), 74–78.
[34] S. Milani, F. Baldelli Bombelli, D. Berti and P. Baglioni, *J. Am. Chem. Soc.* **129** (2007), 11664–11665.
[35] O. G. Mouritsen and O. Mouritsen, *Life – As a Matter of Fat*. Berlin: Springer-Verlag (2005).
[36] S. Friberg, K. Larsson and J. Sjoblom, *Food Emulsions*. 4th edn. New York: Marcel Dekker (2004).
[37] B. Jonsson, B. Lindman, B. Kronberg and K. Holmberg, *Surfactants and Polymers in Aqueous Solution*. Chichester: John Wiley (2003).
[38] T. Wärnheim and A. Jönsson, *J. Coll. Interface Sci.* **125** (1988), 627–633.
[39] Z. M. Suzuki, T. Inoue and B. Lindman, *Langmuir* **18** (2002), 9204–9210.
[40] P. Lo Nostro, N. Stubicar and S. H. Chen, *Langmuir* **10** (1994), 1040–1043.
[41] K. Fontell, A. Ceglie, B. Lindman and B. W. Ninham, *Acta Chem. Scand. A* **40** (1986), 247–256.
[42] A. W. Adamson, *Physical Chemistry of Surfaces*. New York: John Wiley (1976).
[43] D. F. Evans and H. Wennerström, *The Colloidal Domain: where physics, chemistry, biology, and technology meet*. New York: VCH (1994).
[44] C. Tanford, *J. Phys. Chem.* **76** (1972), 3020–3024.

10

Self assembly in theory and practice

10.1 Ideas and defects of theories of self assembly

As already remarked, the first attempts to build such a unified molecular theory of self assembly that could embrace all possible aggregates began 30 years ago [1].

It began with Tanford's ideas on opposing intramolecular forces between surfactant molecules at interfaces [2].

A theory via statistical mechanics was developed based on minimal assumptions. This theory is just a characterization of self assembly. It is more complicated than, but at the same level as, nucleation theory for gas–liquid phase transitions. (Nucleation theory considers only the growth of *spherical* aggregates, droplets of liquid in a vapour. With surfactants we have to consider a multitude of possible aggregates of different shapes.) Nonetheless, as we have seen, the characterization that emerged gave out a set of simple design rules to predict microstructure for dilute systems.

In outline the ideas involved are these:

In the processes involved in formation of an aggregate there are several factors that can be identified.

There is a hydrophobic free energy of transfer of a monomer surfactant hydrocarbon tail, from water to the assumed bulk oil-like environment in its associated state. In an aggregate at its interface the forces between the hydrocarbon tails oppose the intra-aggregate headgroup interactions (see Fig. 10.1). These forces could be ionic, hydration, zwitterionic or steric forces. This competition at the interface sets the interfacial curvature. There is then an interfacial free energy contribution augmented by curvature contributions.

There is a free energy associated with the entropy of monomers, and, less important, an entropy of aggregates. These tend to oppose self assembly. Then there are interactions between aggregates.

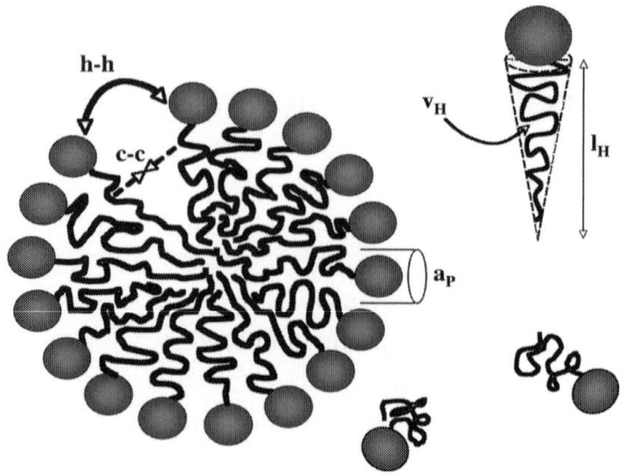

Fig. 10.1. Idealized micelle. Opposing forces at a micellar interface. The real 'interface' is more or less 'fuzzy' and dynamic depending on hydrocarbon chain length (surfactant solubility). At a mean surface of tension, headgroup repulsions are indicated by the full line 'h-h'. They depend on factors such as charge, hydration and molecular geometry. The dotted segment 'c-c' represents opposing attractive van der Waals forces between hydrocarbon tails, with a hard-core interaction set by packing constraints. In the monomers the lipophilic chains are coiled in the attempt to escape from the contact with the solvent molecules; in the aggregate the hydrophobic tails stretch out as they experience a more friendly compatible microenvironment. v_H, l_H, and a_P represent the volume and the length of the hydrophobic segment, and the area per polar headgroup (which accounts for water molecules of hydration and other repulsive interactions). Together the opposing forces at the interface set local curvature.

If we accept these assumptions, then a partition function Z of equilibrium statistical mechanics can be defined. It is a weighted sum over the energies of all possible aggregates of N amphiphiles, and a sum over all possible *shapes* of aggregates of N amphiphiles.

From this, a free energy can be formally derived.

The aggregates in the 'pseudophase' that forms under given conditions (concentration, temperature, electrolyte) are those of minimum free energy.

Such a theory is in principle quite general. But a number of assumptions are necessary before a partition function can even be written down [3].

These concern:

1 The question of units. Whether one should use mole fraction or molar volume units is unresolved. It is of no consequence for dilute systems, but for concentrated systems it is quite a profound issue. It bears on the problem of how we think of a surfactant and solvent molecule as separate identities, and the distinction between a 'chemical' and 'physical' bond, especially with hydration involved.
2 The definition of an aggregate is a problem, insuperable for a physicist, and obvious to a chemist. Some technical assumptions have to be made here.

3 The notion of an interface between an aggregate and water is taken as given. There is no difficulty in principle with this for a bilayer formed from insoluble membrane-forming lipids, or for long double-chained insoluble (e.g. diC_{18}) surfactants. Monomers of these take three months to flip from one side of the bilayer to the other – the lifetime of a red cell. But with short-chained surfactants like SDS the lifetime of a molecule in micelle is 10^{-8} s, and the idea of an interface is blurred (cf. Fig. 9.2b).
4 The idea of an oil-like interior of an aggregate seems inconsistent with packing constraints faced by associated monomers. But this can be shown to be consistent in such a mean field approximation. In comparing and selecting optimal states only aggregates consistent with local packing constraints are allowed.
5 The partition function considered only simple Euclidean shapes, spheres, cylinders, planes. Bicontinuous shapes were not considered in the first theories of self assembly of dilute systems [1–3].
6 Formally, interactions between aggregates were ignored. Micelles were allowed to pass through each other.
7 But interactions do occur *between* aggregates, and have to be taken account of in determining optimal aggregates.

In any event, given these assumptions, an analysis of the partition function enables one to select the optimal aggregates. It produces cmcs for micelles as a function of chain length and temperature of the right order of magnitude. (The definition and calculation of cmcs even in dilute systems is a matter of some complexity: see Ref. 4.)

It produces quantitative results for more refined models of ionic micelles that we shall consider below.

An entity called the surfactant parameter that we have already described emerged as a semi-predictive parameter which replaced the older characterization termed hydrophobic-lipophilic balance (HLB) number. This parameter allowed some prediction of which microstructures form under different conditions.

Technically, the notions on local curvature and interactions that we have outlined are a typical form of a mean field theory of statistical mechanics. It really is a characterization, not a theory proper. The molecular parameters v_H, a_P and l_H (see Chapter 9) required to set microstructure are connected. Such a theory is an extension of, and has the same deficiencies as, nucleation theory for gas–liquid transitions in simple liquids. The complication, for surfactants, comes from the size and shape and amphoteric nature of the surfactant molecules. That is, in selecting for comparison which aggregates form, out of all those possible containing N surfactant molecules, one chooses those of lowest surface energy (in nucleation theory these are spheres); or ellipsoids if spheres cannot pack. If these cannot pack, then cylinders; if cylinders cannot pack then planar bilayers. This procedure for surfactants has the same limitations as standard nucleation theory. That is, critical phenomena are not included quantitatively. The transitions between micelles, cylinders and bilayers are themselves critical phenomena. The reduction of the formal statistical mechanics of these first attempts [1–3] to tease out a set of predictive rules takes no account of a set of microstructures that are not just spheres, cylinders or planes and can be, for example, bicontinuous. These we will come to later.

There is an apparent absurdity in treating an aggregate as if it has an oil-like interior, yet still the hydrocarbon chains repel each other. The characterization apparently did not explain 'ion binding', which we will come to later in Chapter 12.

However, as we have seen, the simple seemingly tautological rules that emerge do work. Local curvature set by a balance of headgroup and tail forces at an aggregate interface together with a knowledge of inter-aggregate forces is enough to capture some essentials of self assembly.

This theory of self assembly was extended and generalized to include oil–water surfactant systems of three and more components [3]. This theory went to great lengths to minimize the use of parameters. It was a far better theory but much more difficult.

The apparent contradiction between treating the hydrocarbon interior of a micelle as an oil-like continuum and the surfactant parameter was resolved [5]. The conditions required to make microemulsions were resolved, as was the role of cosurfactants. Microemulsions we will discuss in detail in Chapter 12. Phase inversion, a switch from oil-in-water to water-in-oil microstructures, was explained predictively [3] in terms of local curvature set by intramolecular forces, controlled by temperature or cosurfactant. Phase changes with volume fractions in reverse-phase microemulsions were shown to be a simple consequence of global packing.

Later work on microemulsions brought a new range of controllable flexibility, and later work still brought in bicontinuous geometries in cubic and random phases in both two and three dimensions that turned out to be much more general and ubiquitous in nature.

The objections to such a unified theory of micelles, vesicles and microemulsions encapsulated in the simple rules were weighty. The main objection is that the ideas that emerged to describe a host of self-assembled structures were too simple.

Nonetheless such a characterization does work for prediction, satisfactorily at most levels. Succeeding examples will make out the case.

10.2 Global packing restrictions and interactions

These rough rules for determining aggregate structure are useful. They appear to be obvious and are apparently tautological. Local curvature set predominantly by the balance of intramolecular interactions at the (oil-like) surfactant–water interface sets the allowed microstructure.

But local curvature also depends on interactions *between* the aggregates. Thus if intermicellar interactions cause a change in headgroup area, a_P, this will affect chain length, l_H, and the surfactant parameter $v_H/(a_P l_H)$. So if salt is added to a suspension of ionic micelles, one expects intuitively that electrostatic headgroup repulsion will be screened so that the headgroup area will reduce. Then the surfactant parameter $v_H/(a_P l_H)$ will increase until spheres are no longer allowed to pack by geometry, and the micelles should grow to ellipsoids or cylindrical shapes. That picture of events often does occur. In reality it is again far too simple a characterization for several reasons.

The meaning of a micellar interface can be defined thermodynamically, but can be nebulous in reality (cf. Fig. 9.2b). 'Water structure' around non-ionic headgroups depends on temperature and the nature of the headgroup. So does that between aggregates. With ionic surfactants, the nature of headgroup repulsion depends on ion specificity and 'binding' of counterions to the micelle. This affects not just local water structure but interactions, in a way that seemed impossible to include in an overall predictive characterization.

However, the problem can be at least partially resolved once global packing constraints are recognized. Then a more systematic picture starts to emerge.

10.3 The question of vesicles: predictions and limitations of theory

10.3.1 The question of vesicles

In the theoretical development so far outlined here and in Chapter 9, the treatment of single-walled vesicles posed a problem. (Vesicles form when micelles and cylinders are forbidden by packing constraints, as for double-chained surfactants and lipids.) The question whether thermodynamically stable self-assembled single-walled vesicles could exist or not has caused much dispute. One view held for many years by one camp was that single-walled vesicles could not exist as equilibrium structures. This was adduced on 'thermodynamic' grounds. A single-walled (bilayer) vesicle could not be in equilibrium with a monolayer at the air–water interface. Therefore, it was argued, lamellar phases, multiwalled vesicles, (with a reduced free energy per monomer due to attractive interbilayer interactions) had to be the stable state of self assembly in equilibrium with the monolayer at the air–water interface. Vesicles, it seemed, had to be metastable. This appeared consistent with the then methods of production of vesicles from phospholipids by, for example, sonication, or other methods that required the input of energy, implying that the vesicles, like emulsions, were metastable.

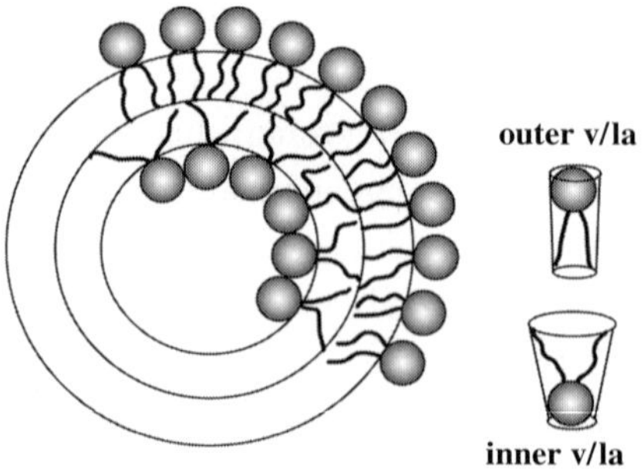

Fig. 10.2. Schematic of a vesicle formed from a double-chained surfactant. The bilayer is necessarily asymmetric.

But this apparently irrefutable argument, we shall see, turns out to be full of holes, literally. The assumption that surfactants form only monolayers at an air–water interface is erroneous. There are as many surface phases possible as there are three-dimensional bulk phases.

Parallel ideas were maintained in membrane biology, where the concept of a simple Danielli–Davson model of a bilayer from the phospholipid components of membranes (see Fig. 9.3) formed the backdrop for thinking. So, it was held, single-walled vesicles could not exist as equilibrium structures. The issue is important in understanding the protective bilayer membranes of cells and cell organelles and their transitions from one form to another.

10.3.2 Conditions for formation of single-walled vesicles: asymmetry of interior and exterior of vesicles: constraints due to chain packing

We now explore the conditions required to form stable vesicles. It had earlier been thought that all, at least biological double-chained lipids, form only multi-bilayer vesicles (lamellar phases). Single-walled vesicles were supposed always to be unstable. (The analysis of this problem in Ref. 2 is incorrect.)

It turns out, and is a prediction of the theory, which emerges after a very complicated analysis, that some double-chained surfactants *will* form stable vesicles [3]. Others will not. The complication comes about as follows. In a curved bilayer, even one formed from a single species of double-chained surfactant or lipid, the outer and inner layers are different (Fig. 10.2).

The outer (concave) and inner (convex) monolayers adjust the numbers and areas and average lengths of their surfactant molecules to minimize the total energy of the aggregate. In this minimization, not just headgroup areas but hydrocarbon chain lengths change. For the outer layer the chains are stretched out. For the inner layer, the chains are compressed. The surfactant parameters are different with positive curvature for the outer layer and negative curvature for the inner monolayer.[1]

Whether one forms vesicles or lamellar phases as the favoured state then depends on chain stiffness, squishiness if one likes, besides the surfactant $v_H/(a_P l_H)$ average shape taken up in the bilayer. Which is the favoured state is not a priori obvious. When vesicles form from the usual double-chained surfactants or membrane lipids (12 to 18 methylene groups per chain), the minimum size is about 100–200 nm diameter, as is observed. So, theoretically, depending on the system, single-walled vesicles can be *either* stable or metastable.

We can summarize matters in terms of the external surfactant curvature parameter $v_H/(a_P l_H)$. Here $v_H/(a_P l_H)$ refers to the surfactant in the exterior monolayer of the bilayer configuration. The sequence of microstructures that form given in Section 9.3 as the external curvature parameter $v_H/(a_P l_H)$ varies, $1/2 < v_H/(a_P l_H) < 1$, can now be understood. Of the range of microstructures possible, with increasing $v_H/(a_P l_H)$ it is:

for 'stiff' chains

$1/2 < v_H/(a_P l_H) \rightarrow$ multiwalled liposomes $\rightarrow$ vesicles $\rightarrow$ liposomes < 1

for 'flexible' chains

$1/2 < v_H/(a_P l_H) \rightarrow$ vesicles $\rightarrow$ several walled vesicles $\rightarrow$ liposomes
$v_H/(a_P l_H) > 1$ reverse phases.

The requirement $v_H/(a_P l_H) > 1/2$ is necessary for formation of all bilayer structures. It is generally satisfied for all double-chained surfactants and lipids. It is a necessary condition for formation of single-walled vesicles, but not a sufficient condition.

It turns out that 'stiff' means that the surfactant tails on the inside layer of the curved vesicle bilayer cannot be compressed to more than about half their fully extended chain length. (This also accounts for the typical size of the water core of vesicles and mutibilayer liposomes.)

The important point is that there are necessary asymmetries for any topologically closed bilayer. The curvature of an outside layer is 'normal', the inside has a

[1] The connection between the local curvature surfactant parameter and the geometry of a surface characterized by its principal curvatures is given in Ref. 2. To a good approximation it is $v/a = l\left[1 - (l/2)\left(R_1^{-1} + R_2^{-1}\right) + (l^2/3R_1R_2)\right]$. Normal curvature surfaces are concave like those of a sphere, and taken to be positive. Reverse curvatures are convex and taken to be negative. The equation is exact for a sphere $R_1 = R_2$, for a cylinder $R_2 = \infty$, and for a plane $R_1 = R_2 = \infty$. It holds to high accuracy for any surface. The average curvature is $H = 1/R_1 + 1/R_2$, and the Gaussian curvature $K = 1/(R_1R_2)$ [6]. The equation can be extended to give a phenomenological equation including bending and torsional energies of membrane. There is extensive literature associated with the application of theories built on such an equation. But these assume that the bilayer itself is homogeneous, an assumption which violates the extreme packing conditions inside and outside a curved membrane.

'reversed' curvature. This asymmetry and chain stiffness account for the size of vesicles normally observed, around 100–200 nm radius for typical double-chained surfactants and phospholipids, whether they are single-walled structures or not. (The size shows up in a bluish hue typical of a suspension containing single-walled vesicles or by viscosity measurements.) Below this size the tension imposed on the inner side of the bilayer by compression of the inner-layer chains imposes a limit on the minimal radius possible. There are significantly different numbers of surfactants in the inner and outer layers of the energetically optimal state [3]. The tension cannot be relieved by adjusting these numbers as the radius shrinks because the energy required to flip from one side to the other is too high. This conclusion is quite general. The characteristic parameters, radii and numbers of surfactant molecules on the inner and outer monolayers of optimal vesicles formed by typical double-chained surfactants and lipids for different assumptions of chain stiffness as $v_{\mathrm{H}}/(a_{\mathrm{P}}l_{\mathrm{H}})$ varies are given in the tables of Ref. 3.

10.3.3 Vesicles and cubic phases

The opposing curvatures in a vesicle formed by a double-chained surfactant set restrictions on size. There is a lower limit to the radius of both a vesicle and multi-bilayer aggregates. The asymmetry within a curved bilayer and the resulting tension due to internal packing constraints within the bilayer can be relieved in another way. Imagine a suspension of spherical single-walled vesicles or cylindrical worm-like micelles formed by a double-chained surfactant or lipid. As the concentration increases, the system eventually runs into a global packing limit. This occurs when the volume fraction of spheres, which includes the internal water contained within the vesicle, is between 0.5 and 0.74.

Faced with such a dilemma, our anthropomorphic vesicles have several choices. Either they can lower the volume fraction by removing some from the solution. They can do this by adding more layers through fusion of vesicles if the van der Waals and other forces between the bilayers are favourable. (As concentration increases still further the multi-bilayers will again run into the same global packing problem.) An alternative energetically favourable solution is for the vesicle membranes to fuse, buckle and form a bicontinuous cubic phase of zero average curvature $v_{\mathrm{H}}/(a_{\mathrm{P}}l_{\mathrm{H}}) = 1$. Such a geometry is depicted in Fig. 10.3. These are called cubic phases because of their cubic symmetry. We come back to them in Chapter 11.

One such structure (the P surface) is shown in Figs. 10.3 and 11.1. In X-ray crystallography, without extremely careful work, it looks like cylinders arranged in alternate orthogonal layers (see Fig. 9.15). These pack at a surfactant volume fraction of about 0.84. Densely packed cylinders, with increasing volume factions

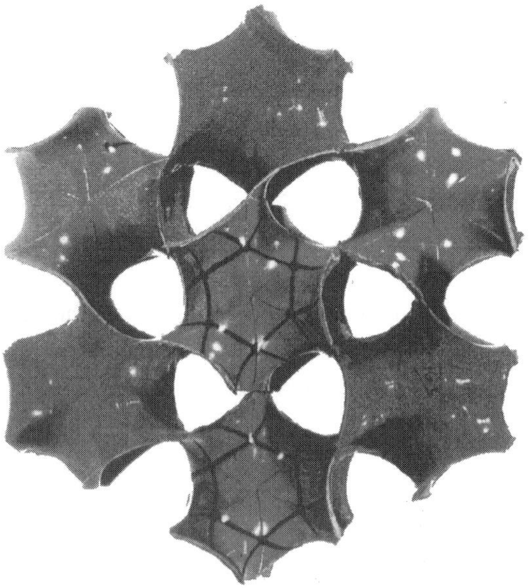

Fig. 10.3. Bicontinuous cubic phase of double-chained surfactants. The surfactants lie on either side of the bilayer and everywhere the mean of the principal radii of curvature is zero. The Gaussian curvature, which is the product of the principal radii of curvature, varies continuously over the curved bilayer. The bilayer divides up connected channels of water. Note similarity to the termite nest in Fig. 9.5. Pore dimensions are around 1–2 nm for lipids [7]. Reproduced with permission, courtesy of S. Andersson and S. T. Hyde.

of surfactants, face the same problems as do close-packed vesicles. It is in precisely this range of a phase diagram that cubic phases often form.

The sequence of transitions in form as $v_H/(a_P l_H)$ varies from $1/2$ to 1 described above now has to be augmented by the additional structures embraced by the totality of bicontinuous phases. That a transition from a vesicular state of self assembly to a bicontinuous state is energetically required and occurs is evident from the phase diagram of a lipid suspension. *How* it occurs is a matter of some considerable interest and difficulty. This is because fusion of membranes, endocytosis and exocytosis require the same kinds of rearrangements. In real biology the processes are facilitated by the availability of mixtures of lipids, with very different surfactant parameters and proteins available. These can facilitate changes in curvature by lateral diffusion and separation within a membrane (so-called raft models). And of course biochemistry comes to the rescue by splitting double chains into less hydrophobic single chains that can engage in flip-flop and other molecular gymnastics to effect the desired rearrangements.

We are interested in physical driving mechanisms. There are at least two molecular arrangements that are required. The topological changes in such transitions are very difficult to account for. Some insights into this that have been obtained and their consequences will be mentioned in Chapters 11 and 12.

Another way that a release of tension can be achieved in a vesicle or any other curved lipid membrane under tension like that in the cartoon of Fig. 10.2 is this: suppose that the inner chains are allowed to stretch so that the outer and inner chains now interdigitate somewhat. The unfortunate vesicle is confronted with another problem. The internal volume per surfactant molecule has increased. The surface area is fixed more or less by hydration requirements. It can satisfy this extra constraint by buckling to give a saddle-shaped surface, with zero average curvature still, but varying Gaussian curvature. So the local conditions for formation of a cubic phase (see Fig. 10.3) are easily satisfied.

The same physical effect, shrinking of surface area at constant internal volume, is at work in the making of potato chips and popadoms, in the bark of trees, or the regular cracking of the surface of dried muds.

10.3.4 Constraints due to charge asymmetry

Apart from the necessary bilayer asymmetry that occurs with vesicles, there is another condition that affects the existence of any topologically closed self-assembled aggregate. This can be illustrated by considering ionic surfactants.

A typical dilute suspension of double-chained surfactants that form vesicles or lamellar phases will have a concentration and a counterion concentration of say 10^{-5} M. The vesicle radii are around 100–200 nm. For ionic surfactants a typical headgroup area is around 0.5 nm^2. The counterion concentration exterior to the vesicles is then 10^{-5} M, or as low as 10^{-6} M if ion 'binding' is taken into account. It is a straightforward matter of geometry and the condition of charge neutrality to show that by contrast the interior concentration of counterions is radically different, as high as 10^{-1} M!

The physicochemical conditions and the binding of counterions in the aqueous regions inside and outside a vesicle will then be quite dramatically different. This affects the optimal conditions for aggregate formation also [8].

10.4 Supraself assembly: formation of spontaneous vesicles illustrated

Variation of the surfactant parameter at a fixed chain length to give the appropriate conditions for vesicle formation can be accomplished by changing headgroup area, and through this the chain-packing parameter $v_H/(a_P l_H)$. To illustrate, first consider the single-chained surfactant cetyl trimethyl ammonium bromide (CTAB) illustrated in the Appendix in Chapter 9. It forms micelles. We ion-exchange the bromide counterion to acetate (or hydroxide or nitrate). For the new surfactant,

CTAAc, the headgroup area of the micelles is twice as large, the aggregation numbers half as large [9].

This specific ion effect is reflected in the behaviour of the corresponding double-chained surfactants, didodecyl dimethyl ammonium bromide (DDAB) and acetate (DDAAc). These have the same headgroups and the same differences in headgroup area as those of the single-chained analogues. DDAB has a measured headgroup area in a bilayer around 25 Å^2. Its volume is around 800 Å^3. Its chain length in a bilayer is about 16 Å, so that $v_H/(a_P l_H) \sim 0.8$–0.9. The corresponding area for DDAAc or OH^- is 50 Å^2, so that here $v_H/(a_P l_H) \sim 0.5$ [10].

The chains are quite flexible for both surfactants (Krafft point > 50 °C).

With bromide as counterion, $v_H/(a_P l_H) \sim 0.8$, DDAB forms (multiwalled vesicles). With acetate as counterion, or hydroxide or nitrate, $v_H/(a_P l_H) \sim 0.5$ and DDAAc forms unilamellar vesicles spontaneously [8,11–13].

10.4.1 Mixed surfactants and catanionic mixtures

The essential condition for single or multilamellar vesicle formation is that the normal curvature as measured by the surfactant parameter satisfies $1/2 < v_H/(a_P l_H) < 1$.

It is possible to make vesicles with mixtures of surfactants that, individually, form micelles $v_H/(a_P l_H) \sim 1/3$. One takes an equimolar mixture of single-chained anionic (e.g. SDS) and cationic (e.g. CTAB) surfactants (The mixed surfactants would be net electrically neutral. The system of vesicles formed is stabilized by using an excess of SDS to charge the vesicles.) The SDS–CTAB pairs hydrogen-bond to give, as far as packing is concerned, what is effectively a double-chained surfactant with parameter of $v_H/(a_P l_H)_{\text{eff}} > 0.6$, the necessary value for vesicle formation. The difference from the insoluble biolipids is that for these fusion is difficult due to hydration forces. For the soluble single-chained SDS and CTAB transfer of surfactant from one vesicle to another is easy. Stabilization of such suspensions is achieved by arranging for the mixture to have an excess of SDS to impose repulsive electrostatic forces between the vesicles [14].

With complicated mixtures in real biomembranes, zwitterionic and charged, asymmetry in interior and exterior physicochemical conditions allows a redisposition of surfactants so that outside and inside lipid species distributions are different on either side of the membrane. An example is that provided by, e.g., T cells of the immune system where negatively charged phosphatidylserine resides on the inner side of the cell plasma membrane.

The discrimination and predisposition of a surfactant system to form single-, double- or multi-walled aggregates is hardly exhaustive. Other states of self assembly can emerge. They are all consistent with the simple rules.

10.4.2 Catanionic surfactants

If the bromide and sodium ions are removed from a mixture of the single-chained SDS–CTAB system a net effectively neutral double-chained surfactant is formed. On cooling below the Kafft temperature of such a system defects separate into edges and some extraordinary virus-sized icosahedral particles form spontaneously [15].

10.4.3 Giant vesicles and the beginnings of supraself assembly

Hofmeister effects, interactions and ion binding

The vesicle story gets more interesting. Electron micrographs of spontaneous vesicles formed from DDA hydroxide, acetate, etc. show single-walled or double-walled vesicles, the interiors of which were empty at low concentrations of the hydroxide surfactant. But at higher concentrations, (10^{-2} M), their interiors appeared to contain micelles (cf. Fig. 9.9) [11].

In such an ionic container the interior has a much different physicochemical environment than the exterior. For example, as already remarked, for vesicles of diameter 200 nm the interior charge of counterions is around 0.1 M, whereas the exterior counterions are at a concentration of 10^{-5} M. How this state of self assembly comes about is of much interest. A later study confirmed these apparently strange states of supraself assembly [15].

10.4.4 Vesicles with different physicochemical conditions inside and outside

That differences in ion binding could be exploited to produce other structures became apparent. The suspensions appear to be quite stable. They are reversible to temperature cycles. If with such a suspension of vesicles one neutralizes the exterior by titration with HBr, the vesicles grow rapidly to become giant vesicles of micron-sized diameter. They look to all intents and purposes like, and have the same size as, red blood cells, slithering past each other and never fusing, for months. Subject to shear, they turn into worm-like structures and then, without shearing, back again to giant vesicles. They can be studied by video-enhanced differential interference contrast microscopy, and cryo transmission electron microscopy. Leakage of the external bromide counterions across the membrane of such giant vesicles can by tracked by forming them with added indicator, in their interior. Leakage takes some weeks at least, so the vesicles are tight containers [16–19]. Spontaneous vesicles form from the same double-chained surfactant if the bromide counterion is exchanged for hydroxide, acetate or nitrate, and other anions of the Hofmeister series. They also form with similar ion exchange made with the anionic double-chained surfactant Aerosol AOT (cf. Appendix in Chapter 9). It forms a multilamellar phase with Na^+ as counterion. On ion exchange to H^+ the system transforms to spontaneous vesicles [16].

The role of molecular forces in the self organization of amphiphiles and some biological implications for matters such as endocytosis have been reviewed in Ref. 19.

10.4.5 Hofmeister effects on self organization and ion binding

These complicated states of self assembly involve 'ion binding' which follows a Hofmeister series. The order of the Hofmeister sequence reflects differences in degree of binding that depends on the headgroup, as well as counterion. But it is also controlled by interaction between aggregates. The interactions also reflect charge of the aggregates determined by 'ion binding'. The interactions have a strong effect on supraself assembly.

We have seen in Chapter 4 that the measured forces between bilayers of these cationic surfactants differ greatly depending on counterion. With Br^- as counterion or with added NaBr, the Br^- ion appears to be strongly 'bound'. With OH^- or OAc^- or with added NaOAc there is no binding. The surface charge is 10-times larger and the forces between surfaces 50-times larger than for bromide.

The net result is that as surfactant concentration increases, the single-walled vesicles, a state preferred at low concentrations, experience large repulsive forces. This (global packing) condition forces the system to rearrange to put surfactant into multiple bilayer vesicles. The internal multiple surfactant bilayers, facing impossible asymmetric packing conditions, then collapse to form micelles. They may be unhappy micelles, in what is probably very differently structured water, but at least the system can satisfy the global packing conditions. The observed state of self assembly with vesicles containing micelles then is a consequence of global packing conditions.

Another similar startling example of changing curvature to change structure with ionic double-chained surfactants is provided by the double-chained surfactant sodium dihexadecyl sulfosuccinate. It is completely insoluble in water. But on ion exchange, the lithium salt is soluble at high concentrations, and forms spontaneous vesicles [20].

Two factors are involved. The local curvature is different due to specific ion binding (Na^+ vs. Li^+). But at the same time the repulsion between bilayers is orders of magnitude larger for Li than for Na counterions.

There are similar differences to be expected in physicochemical conditions such as water structure inside and outside vesicles for any topologically closed structure, even for zwitterionic or non-ionic surfactants.

10.4.6 Giant vesicles and critical phenomena with phospholipids

An astonishing observation was made by N. Gershfeld from studies on Langmuir–Blodgett films and monolayers of sodium double-chained C_{12} phospholipids. He found that the 'monolayers' that ought to exist at the air–water interface according to conventional ideas of surface chemistry were in fact 'bilayers'. Leaving that aside for the moment, these surface 'bilayers' appeared to be in equilibrium with vesicles of the usual size, about 200 nm diameter.

What is more interesting is that he extracted the lipid fraction from the plasma membranes of various single-celled organisms. These lived at a range of different temperatures. At precisely the temperature at which the organism lived, the small 200 nm vesicles transformed into giant vesicles of the size of a biological cell. The existence of a critical temperature for formation of cell membranes from lipids alone is interesting enough. The implications go far beyond this. For this physical capacity of membrane-forming lipids to form biological cell-sized membranes integral to the existence of a cell would probably be exploited by such organisms in responding to environmental changes. It is comprehensible that a change in temperature or salinity can induce a change in genetic expression of the mix of lipids integral to a cell membrane sufficient to accommodate that challenge. If borne out, the existence of critical temperature for giant vesicle formation has very important consequences which bear on the old debate on Lamarkism vs. Darwinism. For if cells can adjust their plasma membrane from the cell interior lipid reservoir by *physical* processes in response to environmental change, that is of interest [21–23].

The existence of 'bilayers', not monolayers, at an air–water interface has often been seen with neutron scattering and ignored. Such states of aggregation are probably 'mesh' phases, bilayers with holes of the right curvature to accommodate transmembrane proteins in real cells. Explicit confirmation for these states with CTAB vesicles has been given by Ref. 24, and more recently for 'blastula' aggregates that form from mixed cationic and anionic systems [25].

References

[1] J. N. Israelachvili, D. J. Mitchell and B. W. Ninham, *J. Chem. Soc. Faraday Trans. II* **72** (1976), 1525–1568.
[2] C. Tanford, *The Hydrophobic Effect*. 2nd edn. New York: John Wiley (1980).
[3] D. J. Mitchell and B. W. Ninham, *J. Chem. Soc. Faraday Trans. II* **77** (1981), 601–629.
[4] D. J. Mitchell, B. W. Ninham and D. F. Evans, *J. Coll. Interface Sci.* **101** (1984), 292–295
[5] A. Wulf, *J. Phys. Chem.* **82** (1978), 804–811.
[6] B. Jönson, B. Lindman, K. Holmberg and B. Kronberg, *Surfactants and Polymers in Aqueous Solution*. Chichester: John Wiley (1998).

[7] D. F. Evans, M. Allen, B. W. Ninham and A. Fouda, *J. Solution Chem.* **13** (1984), 68–84.
[8] B. W. Ninham and D. F. Evans, *Faraday Discuss. Chem. Soc.* **81** (1986), 1–17.
[9] J. E. Brady, D. F. Evans, G. G Warr, F. Grieser and B. W. Ninham, *J. Phys. Chem.* **90** (1986), 1853–1859.
[10] R. M. Pashley, P. M. McGuiggan, B. W. Ninham, D. F. Evans and J. Brady, *J. Phys. Chem.* **90** (1986), 1637–1642.
[11] B. W. Ninham, Y. Talmon and D. F. Evans, *Science* **221** (1983), 1047–1048.
[12] B. W. Ninham, D. F. Evans and G. J. Wei, *J. Phys. Chem.* **87** (1983), 5020–5025.
[13] B. W. Ninham, S. Hashimoto and J. K. Thomas, *J. Coll. Interface Sci.* **95** (1983), 594–596.
[14] A. Renoncourt, N. Vlachy, P. Bauduin, M. Drechsler, D. Touraud, J.-M. Verbavatz, M. Dubois, W. Kunz and B. W. Ninham, *Langmuir* **23** (2007), 2376–2381.
[15] E. Z. Radlinska, B. W. Ninham, J. P. Dalbiez and Th. N. Zemb, *Coll. Surfaces* **46** (1990), 213–230.
[16] B. W. Ninham, B. Kachar and D. F. Evans, *J. Coll. Interface Sci.* **100** (1984), 287–301.
[17] D. D. Miller, J. R. Bellare, D. F. Evans, Y. Talmon and B. W. Ninham, *J. Phys. Chem.* **91** (1987), 674–685.
[18] B. Kachar, D. F. Evans and B. W. Ninham, *J. Coll. Interface Sci.* **99** (1984), 593–596.
[19] D. F. Evans and B. W. Ninham, *J. Phys. Chem.* **90** (1986), 226–234.
[20] M. E. Karaman, B. W. Ninham and R. M. Pashley, *J. Phys. Chem.* **98** (1994), 11512–11518.
[21] N. L. Gershfeld, *Biochim. Biophys. Acta* **988** (1989), 335–350.
[22] N. L. Gershfeld, *Biochemistry* **28** (1989), 4229–4232.
[23] A. J. Jin, M. Edidin, R. Nossal and N. L. Gershfeld, *Biochemistry* **38** (1999), 13275–13278.
[24] J. Gustafsson, G. Orädd, M. Nydén, P. Hansson and M. Almgren, *Langmuir* **14** (1998), 4987–4996.
[25] N. Vlachy, A. Renoncourt, M. Drechsler, J.-M.Verbavatz, D. Touraud and W. Kunz, *J. Coll. Interface Sci.* **320** (2008), 360–363.

11

Bicontinuous phases and other structures: forces at work in biological systems

11.1 Cubic phases

11.1.1 Introduction to cubic phases

With any curved bilayer or topologically closed region such as single- or multi-walled vesicles, there is an inevitable conflict due to internal packing constraints (cf. Fig. 9.8 and 10.2).

In the energetically optimal configuration, lipids packed into the outer bilayer have normal curvature. The hydrocarbon chains are stretched. By contrast, the inner surfactants have reverse curvature. The chains are compressed. So although the interior of a bilayer can be considered fluid-like, above the Krafft (gel) temperature of the lipid it is asymmetric. As we have seen, and re-summarize, this necessary frustration has several consequences. The first is that when the radius becomes small enough, the interior layers of a multilamellar vesicle structure must collapse into aggregates different in structure to that of a bilayer. The interior can be micelles. Or it can be bicontinuous. This gives rise to supra-aggregation as a normal and expected state of self assembly. Or else the interior can be empty and surfactant-free. The second consequence is that a single-walled bilayer or vesicle has to have a radius (typically 100 nm) sufficiently large to accommodate this additional internal packing condition.

There is another way that the frustration can be relieved.

The bilayer-forming lipids or surfactants have a surfactant parameter around $v_H/(a_P l_H) \sim 1$, or <1. The average of the principal curvatures at the interface is zero for a flat bilayer. The frustration can then be accommodated if the self-assembled structure retains zero average curvature, but is rearranged so that the product of the principal curvatures, the Gaussian curvature, is zero [1].

This gives rise to an everywhere saddle-shaped aggregate that is bicontinuous. Three typical such shapes are those in Fig. 11.1.

(a)

(b)

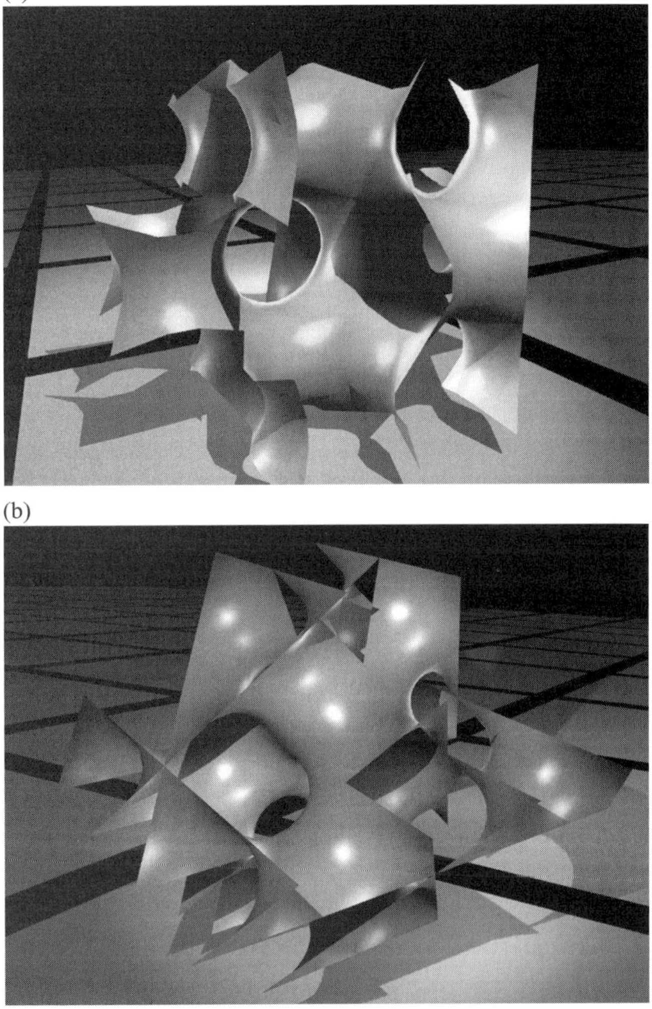

Fig. 11.1. Basic cubic phases. Lipid bilayers are wrapped on the cubic surface, which represents the mid-plane of the bilayer. The graphics are for the three fundamental P(a), D(b) and G(c) surfaces. Reproduced with permission, courtesy of S. Ramsden and S. T. Hyde.

A library of new structures in this rapidly developing field can be found on the internet [2].

These are called cubic phases because of their cubic symmetry. The surfaces themselves are called infinitely periodic minimal surfaces, IPMS. They are other non-Euclidean relatives from a huge developing area of geometry and topology that is opening up whole new concepts on structure. They are described by hyperbolic,

(c)

Fig. 11.1. (*cont.*)

non-Euclidean geometries [1]. In these geometries, the Euclidean postulate that two parallel lines can never meet is relaxed. The average curvature is zero everywhere but the Gaussian (product of) curvatures vary continuously over the surface. They were discovered with lipids first by Kåre Larsson 40 years ago. The discovery that lipids could form such apparently peculiar structures met with much opposition for three decades – an alternative interpretation of X-ray analysis of such lipid phases (difficult to resolve with the then technology) held that the lipid polymorphism had the form of lipid cylinders stacked in alternating orthogonal layers. That may well be possible sometimes for single-chained surfactants faced with strong repulsive hydration forces between them. But the interpretation defies packing constraints for double-chained lipids. The dispute on lipid polymorphism that had to rely for its description on a strange non-Euclidean geometry was not surprising in retrospect. Hyperbolic geometries were discovered by the mathematician Gauss two centuries ago. Even Gauss, the greatest mathematician of his century, found them so strange that he did not publish his results for 30 years, prompted to do so only after being informed that the son of his friend Lobachovski was about to do so. Subsequently his student Riemann laid the foundations for such counterintuitive geometries.

With lipid cubic phases, typical water channels have dimensions of the order of 13–20 Å depending on how one counts water bound to headgroups, with the thickness of the curved bilayer hydrocarbon regions twice the chain length, say 30 Å.

They can transform from one form to another with little energy, extravagant ease and via a process called the Bonnet transformation [1]. A further exploration of the area is the elegant essay in Refs. 1 and 3. See also the website of S. T. Hyde for recent developments [4].

Such saddle shapes always occur when there is conflict between surface area and conservation of volume constraints. A thin lamellar formed by a potato chip or popadom cooked in a microwave oven is a classic macroscopic exemplar. The surface area shrinks, and the internal volume remains much the same, a conflict resulting in a saddle-shaped chip. The same kind of frustration can be seen in operation with the drying of a smooth mud surface, and in a myriad regular patterns formed at the bark surface of a variety of trees in new growth phase. Apart from termites' nests (see Fig. 9.5) and mitochondria, cubic and bicontinuous phases seem to be used at any opportunity. The circulation of blood in the feet of ducks and penguins goes via a bicontinuous arrangement that protects against freezing.

11.1.2 Global packing and cubic phases

Besides the internal packing constraints on lipids that lead to their formation, bicontinuous cubic phases will also occur due to global packing conditions. An example is that of a surfactant that forms cylindrical micelles ($1/3 < v_H/(a_P l_H) \sim 1/2$). As $v_H/(a_P l_H)$ increases due to, for example, repulsive hydration forces that increase with reduced water volume fraction, the surfactants in the cylinders reach a different kind of frustration. When the cylinders are close packed in parallel arrays, global packing forces them to change to lamellar bilayers. This occurs at packing volume fractions of $V_s = 0.75$ in a square array of parallel cylinders, 0.91 in close-packed hexagonal arrays. If they are aligned in alternating stacks perpendicular to each other, the close packing conditions are lower, e.g. $V_s = 0.84$ in simple cubic symmetry. They can transform to cubic phases to relieve both the global packing frustration and the stretching of chains in this configuration.

11.1.3 Mesh phases: two-dimensional analogues of cubic phases

Monolayers vs. other microstructures at interfaces

Recall that the conventional view on microstructures formed by lipids and surfactants at air–water interfaces with increasing concentration is this: it supposes that a two-dimensional gas of surfactants will form a liquid-like close-packed monolayer. After the monolayer is filled, surfactants form micelles and other microstructures in the bulk water. Atmospheric gas is about 1 M in concentration at the interface, around 5 mM in bulk water. It associates with surfactant hydrophobic tails. This complication is almost always ignored. If the surfactants form micelles in bulk, they will probably segregate to form hemi-micelles at an interface, and multiple layers

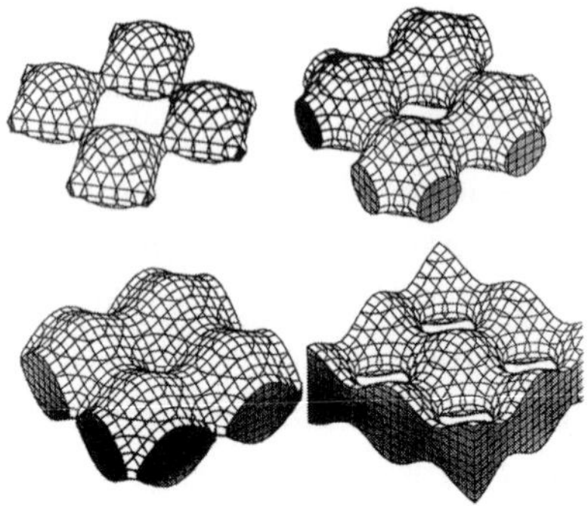

Fig. 11.2. Some square 'mesh' surfaces of differing (non-zero) mean curvature. Vesicles with holes to relieve the transbilayer tension form with constant average curvature. Reprinted from Ref. 1, p. 157. Copyright 1997, with permission from Elsevier.

with lipids. Surface structures are also likely to include so-called mesh phases (cf. Fig 11.2).

Despite vast amounts of work very little is really known about surface aggregate structures, a matter complicated by inevitable impurities, the dissolved gas problem and experimental artefacts of scattering and ellipsometric studies. Mesh phases with catenoidal-shaped holes of zero average curvature to accommodate trans-membrane proteins seem to occur with mixed naturally occurring biomembrane lipids at surfaces and in bulk phases. Such proteins are often catenoidal in shape [1]. Phase separation occurs within the two-dimensional bilayer to give required holes that fit the proteins. It would be surprising, given the vast range of bulk bicontinuous structures now known, if such two-dimensional analogues did not exist. Their existence has some important implications, discussed below.

An enormous range of self-assembled structures that characterize lipid and lipid–protein polymorphism became accessible once it was realized that their description involved the language of shape captured by hyperbolic geometries. The essential role of curvature in driving many biological processes also became clear. The reader is referred to Ref. 1 for this literature up to the year 2000 and the associated website [4].

11.2 Lung surfactants

An example of bicontinuity in surfactant organization is in lung surfactants. The term refers to the foamy layer of the alveoli of the lungs. It is composed in most part of phosphatidylcholine, about 90%, and two proteins, one, about 9%, being very hydrophobic, the remainder being hydrophilic. (In practically all other cells of the body there is a mixture of phospholipids.) Lung surfactant in mammals forms late

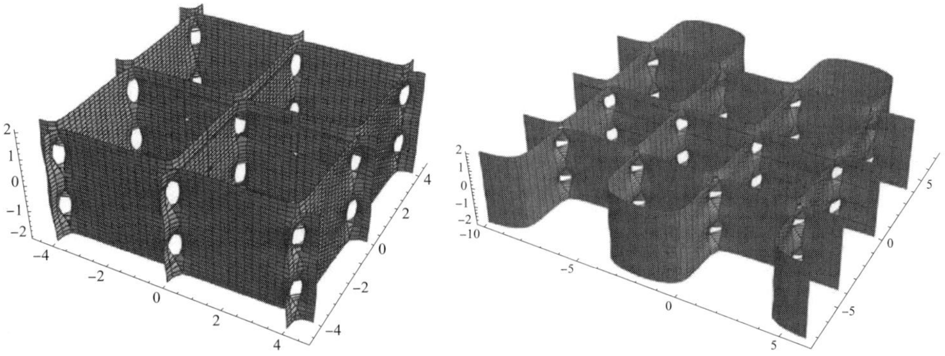

Fig. 11.3. Simplified schematic of lung surfactant structure obtained from cryo-TEM. The holes are occupied by hexamers of hydrophobic proteins that pin the bilayer walls down. From Ref. 5. Reprinted by permission of the publisher (Taylor & Francis Ltd, http://www.tandf.co/uk/journals).

in pregnancy, so that premature babies are at risk of breathing difficulties (sudden infant death syndrome). The lung surfactant allows breathing. This involves an enormous expansion and contraction of the surface area of the lungs at essentially zero work, every few seconds for a lifetime, say 70 years. Simultaneously with each expansion and contraction, oxygen uptake takes place and CO_2 and water are expelled. How this works is clearly a matter of some importance. Indeed, Descartes might have better put his proposition 'I think; therefore I am' as 'I breathe; therefore I am', as the question of his thinking depends on his ability to breathe.

Until a decade ago the structure of this lung surfactant lining was envisaged as that of tubular myelin, a cylinder-like arrangement of multiple bilayers of phospholipids and water. Such structures occur, but are an artefact of electron microscopy.

Difficulties with that then-accepted picture were that if it were so, the large interfacial tension of the system that occurs on expansion would not allow breathing and/or oxygen and CO_2 transport. Some other structure must be involved. Indeed it is. The actual structure is shown in Fig. 11.3.

It is a bicontinuous, tetragonal arrangement with bilayer planes of phosphatidylcholine bilayers. These are pinned down at the edges with the proteins that form a hexameric unit, assisted by the hydrophilic structure, rather like children's Lego blocks. The size of the units is about 40 nm. This is a beautiful example of physical supraself assembly. The structure opens up and expels the water and CO_2, and the lipid bilayers unravel to expose a hydrophobic monolayer that takes up oxygen, and then folds back down again.

More work has been done, particularly by M. Larsson, to elucidate the picture. Awareness of this remarkable structure and its capacity to reorganize has allowed the exploitation of specific ion effects for clinical applications. These change headgroup interactions to change curvature and therefore change structure to facilitate delivery.

The probable reason that phosphatidylcholine is essentially the sole lipid component of lung surfactants is that, among all lipids, its headgroup is the most strongly resistant to free radical damage. Free radicals occur in sub-nanobubbles that form in the air–water–CO_2 transfer process. Awareness of the structure of the lung surfactant allows a better understanding of viral recognition of proteins in healthy and cigarette-smoke-damaged lungs. An interesting example of the possible interplay between molecular forces and biological processes comes from the observation that workers in salt mines rarely have respiratory diseases. The workers are exposed to and breathe in salt aerosols. The colloidal aerosol particles in the alveolar region dissolve and release nanobubbles. The nanobubbles do not fuse as the concentration of salt is > 0.15 M (see Chapter 8). Proteins in the aqueous region which are recognized by the viruses adsorb at the bubble surfaces in a different orientation to that without the aerosol particles and bubbles. They then present to the viruses in unrecognizable form. That mechanism operates for vaccines that involve adjuvants in order to work. Adsorption of antibodies on the colloidal particle surface takes place in an orientation favourable for antigen recognition and binding.

11.3 Hydrophobin and cubic phases in fungi

Hydrophobins are extremely hydrophobic proteins similar to those integral to the lung surfactant structure. They appear to be unique to mycelial fungi. They play essential roles in the emergence of areal structures (mushrooms), gas exchange in fruiting bodies and lichens, and in pathogenesis.

It seems at least probable that such rapid expansion reminiscent of lung surfactant behaviour takes place via network structures, perhaps visualized as collapsed multiple-layered fishnets. On expansion they contain water similarly to the compartment in lung surfactants [6].

It may be that the similarity of hydrophobin's properties to the same role of hydrophobic proteins in lung surfactants explains the extreme toxicity of some mushrooms. Similar structures are probably very widespread in nature, especially in gels, and because of their often low supporting surfactant content have simply not been noticed (jellyfish are about 99% water).

11.4 Cubosomes and chloroplasts

Bicontinuous structures of cubic symmetry occur in chloroplasts of dark-adapted plant cells. They have been explored extensively [7,8].

The advantage that accrues to bicontinuity in transporting and directing reactants, here Mg^{2+} vs. H^+, in photosynthesis seems suggestive.

It is only recently that it has been realized that 'cubosomes' occur almost universally in the mitochondria of cells under various conditions. They were found first by K. Larsson two decades ago and literally thousands of such structures have been found and their symmetry classified by T. Landh [1]. Their role must be to direct and containerize cell biochemical traffic. They can transform from one form to another with extravagant ease depending on physicochemical conditions. This occurs by known topological rearrangements that are understood. Cubosomes form in bacteria and other single-celled organisms. Many other mammalian cells form cubosomes from multilamellar organelles. The transformations have been shown to occur, apparently universally, in cells deprived of nutrients in the beautiful work of Y. Deng et al. [9]. One such micrograph showing clearly the cubic symmetry in a mitochondrion of a starved amoeba is shown in Fig 11.4.

Cubosomes are different from the cubic phases formed by lipids and surfactants.

The cubic phases formed by lipids have water-filled pore dimensions of typically 1.3 nm. Their formation is understood in terms of theories of surfactant self-assembly as explained above. For the cubosomes, pore dimensions are typically 130 nm. This (almost macroscopic) size suggests that they are formed by hydrodynamic processes. They are formed from complex mixtures of polymers, lipids and proteins.

To see how they might form we consider a two-dimensional analogue. At an air–water or oil–water interface, or in a vertical column with opposing diffusion gradients, such as salt, gravity, temperature gradients, regular columnar self-organization takes place. This has been explained quantitatively by hydrodynamics. The driving force comes from the opposing diffusion gradients. The self organization occurs over a wide range of length scales. These are called Gibbs–Marangoni effects. It seems that the self-organized *three*-dimensional cubosomes of biology form in the same way as do two-dimensional structures, in response to similar opposing diffusion gradients related to nutrition.

The three-dimensional hydrodynamic problem seems not to have been solved. For further work see Ref. 3.

Double diffusion seems to be involved in templating of microfossils (Section 13.6).

11.5 Cubic membranes and DNA

One other issue deserves mention: the cubic membranes in vivo adsorb DNA very strongly. The ramifications of this have hardly been considered, let alone explored.

(a)

(b)

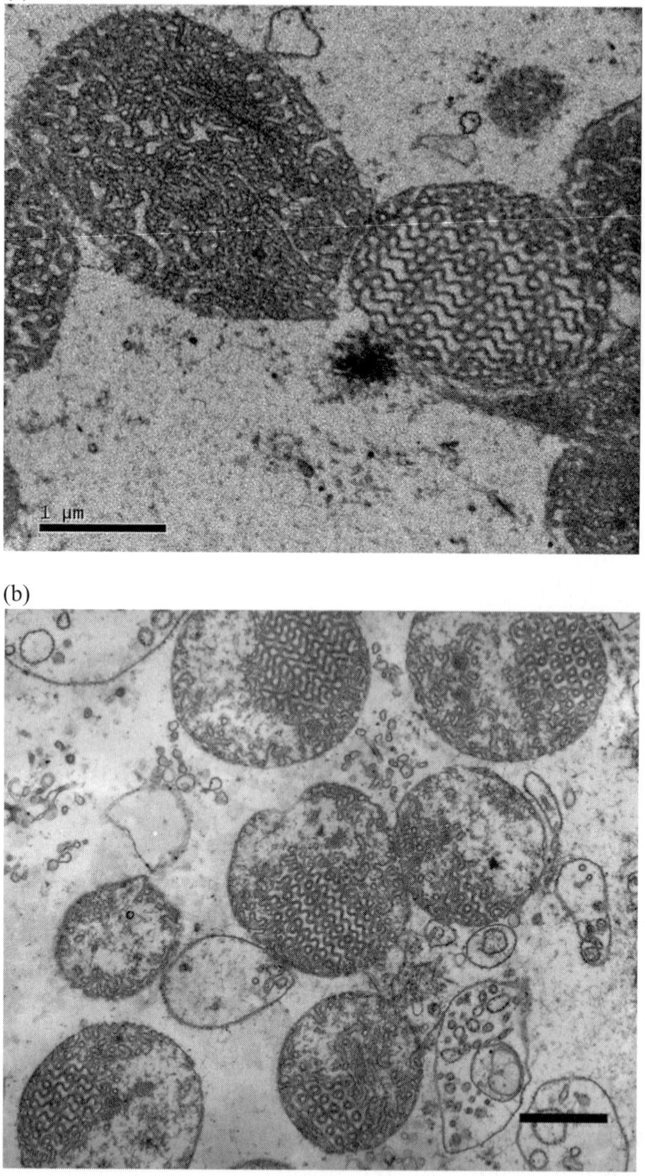

Fig 11.4. TEM micrographs of cubic-phase cubosomes in mitochondria of nutrient-deprived amoeba. They occur in all nutrient-deprived tissues, and in light-deprived protolamellar bodies of plants. Bar is 1 micrometre. The patterned figure is a cross section across a particular IPMS. Reproduced with permission, courtesy of Yuru Deng.

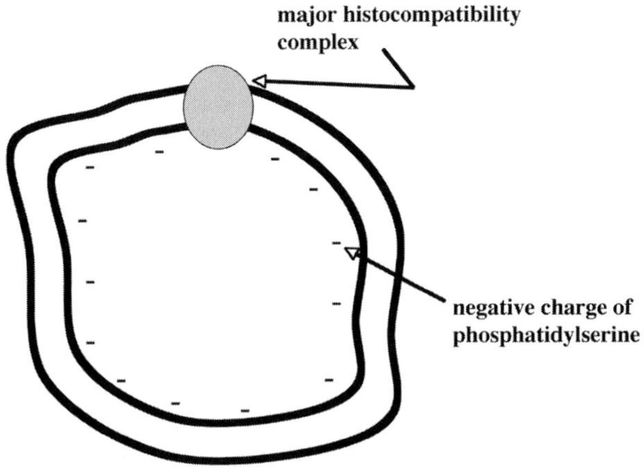

Fig. 11.5. Cartoon representing background for immunosuppression of a T cell induced by cationic surfactants.

But they are probably considerable, just as for the catalysis of alkanes in zeolites discussed in Chapter 5 (see Ref. 10).

11.6 Immunosuppression induced by cationic surfactants: an example of physical chemistry in biology

In Section 10.3 we have explained the antiseptic action of cationic surfactants that takes place at the cmc in physiological media. It is a good illustration of curvature change induced by mixing surfactants. This is an effect of physical chemistry alone. While it has long been known that excess use of such products for bacterial throat infections confirmed the old adage – cure the symptoms, prolong the disease – it was not known why. The prediction and demonstration that such widely used cationic surfactants at low dosages, below the cmc, are also excellent immunosuppressants came as a surprise. Indeed they are more effective (and reversible) immunosuppressants than a drug of choice in controlling organ transplant rejection, cyclosporin A. The effect is an example of physical chemistry principles operating in biology. It has not been exploited.

Immunosuppression occurs when the cells of the immune system can no longer recognize or respond to the challenge posed by a foreign antigen. The situation is outlined in Fig. 11.5.

Cationic surfactants below the cmc adsorb into the membrane of T cells, but not in sufficient quantity to destroy them. The adsorption depends on chain length and follows a Langmuir isotherm. That is, the amount of surfactant adsorbed at a given concentration is directly related to the hydrophobic free energy of transfer of the hydrocarbon tail to the oil-like region of the T-cell membrane. This has been measured for suspensions of T cells of the immune system. Thereafter what happens is that:

1 The surfactants flip-flop over to the inner side of the T-cell membrane.
2 The inner side of the T-cell membrane is negatively charged due to an excess of the anionic surfactant phosphatidylserine (PS). PS is present on the inner side of the cell membrane with little in the outer membrane layer. The adsorbed cationic surfactants neutralize the electrostatic charge of the PS. The surface potential decreases. This means that the electrical double layer is switched off. More importantly perimembrane water structure changes due to association of the anionic PS lipid and the cationic surfactant. The adsorbed calcium concentration at the inner membrane surface then drops to zero.
3 This bound calcium is crucial to the tertiary structure of the (transmembrane) major histocompatibility complex. The complex can no longer do its job of recognizing antigens.

A whole chain of sequential communication processes to a chain of other cells necessary for defence are then not triggered off as they should be.

This work, not all of which was published, was done both in vitro on mouse thyroid allograft rejection, and in vitro on T-cell mixed lymphocyte culture experiments [11,12]. It has been amply confirmed on human models since.

One result that emerged from this study was the fact the immunosuppression effect increased with increasing chain length. The effect kicked in precisely when the measured uptake of surfactant equalled (and neutralized) the estimated negative charge of membrane-associated phosphatidylserine in a suspension of T cells. This seemed to confirm the mechanism above.

But a curious result remained. The immunosuppressant effect disappeared when the surfactant hydrophobicity increased even further (diC_{16} and diC_{18}). This puzzle was resolved when it was realized that the flip-flop rate across membranes increased exponentially with chain length. With these longer-chained surfactants, the flip-flop rate is at least three months. So in tests of induced immunosuppression using an in vivo mouse allograft rejection experiment that lasted a month, or with an in vitro mixed lymphocyte culture test lasting 5 days, the effect does not show up.

There are some indirect medical consequences of the use of cationic surfactants.

The immunosuppression effects of cationic surfactants appear to be connected to several well-known observations. Allergic reactions due to over-use of eyewashes for contact lenses that contain cationics as antibacterial agents can be connected to the sensitive local immune lymphatic system near to and protective of eyes.

Muscle relaxants injected into patients before major operations prior to general anaesthetic administration used to be a concentrated artificial curare compound. These are pure cationic polymers and induce muscle relaxation. It was known anecdotally by anaesthetists that patients with skin allergies or sensitive skin, or women overexposed to cationics via washing machine products, cosmetics etc., came down 10/1 more than men with post-operative anaphylactic shock, an immune system response. So they moved away from using these curare compounds in cases they suspected were allergic.

The women had been sensitized by long exposure to cationics.

Immunocompromised patients with AIDS have strong skin reactions to cationic solutions.

Short-chained cationic surfactants were used for some time as spermicides. Above the cmc they disrupt cells (sperm membranes), and are then adsorbed in the body. It may not be coincidental that a world-wide outbreak of cervical cancer known to be related to immunosuppression coincided with the product's introduction and declined on its removal.

The *possible* environmental consequences of these matters deserve consideration. An alarmist view would be that since cationic surfactants are widely used, from oven cleaners to dish and clothes washers, to toothpastes and skin lotions and in agricultural sprays, that since they are compatible with skin lipids, and that since the turnover rate in the body is about 5 months, the entire developed Western world is probably immunosuppressed! A more pragmatic view is that without these surfactants, the entire Western world would be dead due to infectious disease anyway.

11.7 Bacterial resistance

Cationics are widely used as effective sterilizing agents, above their cmc. A more serious consequence of the inducement of changes in cell membrane properties via cationic surfactants leading to immunosuppression follows. There is a possibility of adaptation of bacteria to assault by these surfactants. Mesh membranes, with the appropriate curvature to accommodate the adsorbed single-chained surfactants, could emerge via mutations. Such mutated super-bacteria would actually *need* the surfactants that previously destroyed them, in order to survive. This would be an example of Le Chatelier's Principle a fortiori! Whether this is so, or not, remains to be seen. We are faced with a major problem of increased bacterial resistance to antibiotics. The use of surfactants as selective tools to induce and study membrane effects and metabolism in mixed populations of bacteria is a matter that has hardly been considered.

An example of such physicochemical effects on bacterial growth is that in Ref. 13.

Growth rates of the bacteria *Staphyloccus aureus* and *Pseudomonas aeuginosa* grown in different electrolytes follow a Hofmeister series. This is presumably due to specific ion binding to a particular rate-limiting enzyme, again a physical not a biochemical effect. There is every reason to assume that especially cationic surfactants can also be used to affect growth rates differentially via membrane-induced effects.

11.8 General anaesthesia: the possible role of lipid membrane phase transitions in conduction of nervous impulse in general anaesthesia

We have identified the action of local anaesthesia via lidocaine. A more sophisticated account is given in Ref. 1.

The mechanism of general anaesthetic action seems to be quite different. But again the effect induced involves reversible physical chemistry. Awareness of the existence of the wider class of states of curvature for bilayers and mesh phases allows a reasonable guess as to what goes on with general anaesthesia. The ideas were developed by K. Larsson. See Chapter 5 in Ref. 1, and many other papers.

Propagation of the action potential along nerve cell membranes, characterized by the original Hodgkin–Huxley equations, involves ion transport mediated by transmembrane proteins.

Nowhere in the notions of this standard 'theory' are lipids involved except to provide a passive material background support.

But the lipids have a more serious role to play. A general anaesthetic such as isoflurane (F_3C-CHCl-O-CHF_2) or halothane (F_3C-CHBrCl) is a simple, inert and very hydrophobic molecule. Any action it has can involve only physical chemistry. Its strongly hydrophobic nature demands that if it is supplied as gas in excess, then it must be taken up in the hydrocarbon core of a bilayer or membrane. The swelling of the internal volume of the hydrocarbon layer then has to induce a change in Gaussian curvature of the bilayer to that reminiscent of a mesh phase.

If conduction of the action potential involves a dynamic coupling with a change in state of the lipid, a bilayer–mesh phase-like transition is easy to envisage. Then the bilayer core swelling caused by hydrofluorane uptake would impose so strong a constraint on required curvature changes that the signal could no longer propagate.

Since fluorocarbons and hydrocarbons are incompatible, after some time the hydrofluorane would eventually leak out and the nervous system would be restored to normal.

The argument has some force in that bubbling hydrofluorane through a white suspension of the lamellar phase of phospholipids induces a phase transition to a clear viscous cubic phase. This is immediately obvious by inspection with polarized light with crossed polars. The system reverts to a lamellar phase after a few minutes when the gas flow is turned off. This experiment was done by one of us (BWN) with K. Larsson

The mechanism of action of xenon, another anaesthetic, with apparently few side effects, and a very hydrophobic molecule, has been much debated; similarly for other inert molecules. It was claimed by Linus Pauling that it forms clathrates that bind to the membrane surface, or transmembrane proteins. No one seems to have done the experiment of observing phase changes with phospholipids or brain lipids to settle the issue.

Whether this inert molecule adsorbs in the membrane or at its surface, it must also change membrane organization.

11.9 Metastasis and anaesthetics: other consequences of mesh phase transitions

The induction of metastasis by anaesthetics has been convincingly explained by K. Larsson and is connected to the same phase-change mechanism (see Chapter 5 in Ref. 1). Relations between anaesthetic agents, cancer and immunosuppression and membrane phase changes are also outlined there, drawn from many works of Larsson.

Other consequences of the existence of polymorphism in plant cell membranes on plant ripening, cell fusion and many other biological processes are also discussed. Many other inert molecules such as ethylene and nitrous oxide behave in the same way by inducing membrane phase changes. The matter is of such importance that we simply flag it and refer the reader to explore the reference above. It is, again, purely a matter of physical chemistry. The mechanism is also consistent with permanent brain damage occurring with 'petrol sniffing' – unlike fluorocarbons, which are immiscible with hydrocarbons, short-chain alkanes are almost impossible to extract from lipid membranes.

In transmission of calcium across synaptic junctions, the vesicles are known to be states of supraself assembly, micro-cubic bicontinuous phases contained inside few-walled vesicles. These required states of self assembly would also be affected by uptake of a curvature-changing anaesthetic just as for the membrane-mimetic supraself-assembled states already described.

Another example of physical chemistry affecting whole animal behaviour is the specific ion effect of lithium acetate or chloride used to control manic depression. There can be no real chemistry proper here. The salt is used at the rate of one or two grams per day. Since the blood of a normal human contains 5 litres of saline at 0.15 M, about 0.7 moles of sodium, the amount of lithium ion delivered is about

1/10–1/20 of the available sodium. That amount is not unreasonable to effect ion exchange by Hofmeister effects, so altering the structure of appropriate brain proteins and interfering with transmission processes.

Possible consequences if transmission of nervous impulses involves mesh and cubic phases:

The lipids in the brains of mammals have a very different mix to those in ordinary cells. The phase behaviour of these phosphatidylcholine, ethanolamine, serine etc. and the short-range hydration forces between them have been extensively studied. But lipids in the brain have a major component of predominantly polyunsaturated ω–3, ω–6 lipids and some $\omega-9$s. Because of this polyunsaturation they can be expected to have a considerable degree of flexibility in packing and reorientation within a 'bilayer'. In fact they are prime candidates for the formation of cubic and mesh phases. Since they oxidize fast on exposure to the atmosphere no work has been done on their phase behaviour until recently and nothing has been published. If it turns out that they do form cubic (and mesh) phases with or without cholesterol this points to a coupling of lipids and proteins in transmission of the nervous impulse. It is of interest that the 'vesicles' involved in transferring Ca^{2+} and acetylcholine across the synaptic junction are not multilamellae as used to be thought. They have recently been shown to be supra-aggregates, cubic phases surrounded by a few protective lamellae. There is a problem in neurophysiology which has been a puzzle for some time. Single-cell recordings placed at different points in the brain can track a conduction path, a 'thought' process, between two sites which extends across many synapses a distance of say 10 cm, over a period of say 0.01–0.1 s. Each time in the connected pathway calcium is transferred across successive synapses. The question is: how does the calcium get back to restore the resting potential? If the transport across the synaptic junction involves cubic phase supra-aggregates and lipid mesh phases are involved then it is possible to speculate as follows: as the conduction wave passes along the membrane then the same kind of aggregates might 'bleb off' along the membrane surface too. They would then act as reservoirs of calcium to restore the resting potential soon after passage of the impulse.

11.10 Inter-aggregate transitions

Regardless of the simplifications in the biological speculations above, there are a few unifying ideas that emerge. Transitions between cubic, hexagonal, vesicle, mesh and lamellar phases can be induced by changes of physicochemical parameters, such as inter-aggregate interactions, mixed surfactants and volume fractions. These all affect the curvature and are reflected in the effective surfactant parameter. Global packing requirements and bilayer asymmetry give rise automatically to states of supraself assembly that have not generally been recognized – careful phase diagrams that identify two phase regions in lipid–water systems are often in fact supraself-assembled states of aggregation, our prototype example being a bicontinuous structure protected by bilayers that rearrange and fuse with a target membrane to deliver their internal content.

Phase segregation of mixed surfactant and lipid membrane occurs on approach of an aggregate due to the intermolecular forces between them. That then facilitates fusion, impossible otherwise due to hydration forces between the membranes.

Many of the self-assembled structures that occur and that we can make are quite delicately poised and can transform from one form to another with extravagant ease.

One such example of such phase changes may be the rhodopsin dark-adapted eye response. The photosensitive pigment in mammalian eyes, rhodopsin, is embedded in a folded lamellar-like state within rods and cones. In the light-adapted state the protein conformational change triggered by adsorption of photons with subsequent biochemistry sets off a cascade of events in the communicating nerve cells. This is well understood. But in the dark-adapted state the eye can respond to as few as one or two photons. How could this be? It turns out that if the lipid mix is extracted [14] it is poised to self assemble in a region in the lipid–water phase diagram very close to where lamellar, cubic and hexagonal phases all adjoin. In the dark-adapted state with the admixture of a few extra lipids it may be that the system goes over to a mesh-like phase in a state of extreme (mechanical) tension. If one rhodopsin molecule receives a photon and undergoes a conformational change, this can trigger a wave of lipid phase transitions down the folded membrane that induces conformation changes in further rhodopsin molecules successively. That notion is not in conflict with what we expect for ordinary nervous conduction processes.

There is a substantial amount of work on Hofmeister effects with rhodopsin which change phase behaviour [15].

11.11 Drug delivery and bicontinuity

Drug delivery via cubosomes provides the necessary bicontinuity for slow release. These are now made by sonicating mixtures of pluronic polymers with surfactant with the drug in a witches' brew. A particularly ingenious application of cubic phases is that by Kåre Larsson. A problem with periodontal disease, as with all antibiotics, is that one wishes to target a particular site only – here the infected gums. He exploited a phase transition of polyoxyethylene surfactants or some mix thereof that passes from ordinary multilamellar liposomes to a (stiff) cubic phase at the body temperature, 37 °C. The fluid suspension of multilamellar-phase vesicles containing the antibiotic is injected into the infected region. The transition to cubic phase occurs, anchors in the site and local delivery is ensured without the necessity to dose the whole patient.

Another ingenious example of targeted drug delivery with brain cancer is due to B. Perly. An effective drug is called ellipticene, shaped like a rugby football [16]. It anchors into the cavity of a particular catenoidal brain membrane receptor. Presumably this induces a phase transition of the membrane to inhibit cancer cell fusion and metastatic activity as discussed above. Perly anchored the drug to a surfactant, to the other end of which he attached a genetically engineered group that recognized the target protein. The highly specific delivery so engineered worked very well.

The mechanisms by which specific drugs used by psychiatrists work in particular areas of the brain must involve membrane transitions that affect axon transmission.

11.12 Membrane fusion and unfolding

The physical mechanism by which bilayer lipid membranes fuse poses some apparently arcane difficulties. Direct force measurements show that it is virtually impossible to drive them together due to the very strong hydration forces between them. That is as it should be if a cell is to retain its integrity over its expected lifetime. The incompatibility of egg and sperm membranes of different mammals, with different lipid mixtures, probably has something to do with speciation. Cell fusion for molecular biology applications is easily effected by cationic surfactants. This involves some catastrophic membrane breakdown by changing curvature and re-annealing. But the physical rearrangements of lipids necessary for fusion and processes such as endocytosis and exocytosis have not been known.

Some insight into how membrane lipids rearrange to effect fusion can be found from the following considerations.

11.13 The tetradecane-DDAB microemulsion system: an exemplar for sponge and mesh phases

Recall from Chapter 10 our claim that the complicated microstructure of the DDAB-alkane or olefin microemulsion – including the cubic phases – could be systematically predicted by a simple geometric model. This is discussed more in Chapter 12. The energetics of oil uptake in the surfactant tails provides an intramolecular force that opposes headgroup interactions. The systematic changes of curvature with increasing alkane chain length imposed from the oil side of the oil–surfactant–water interface all make sense. These can be exploited in formulation with mixed surfactants and oils. But when the oil for the system is tetradecane, a different self-organization takes over. Unlike the lower-chain-length alkanes, tetradecane ($C_{14}H_{20}$) is indifferent to adsorption in the surfactant tails (diC_{12}). The alkane tetradecane is hardly adsorbed into the surfactant tail region at all – the monolayer interface must be flat. While in many respects it mimics the behaviour of the other alkanes in having a spontaneous emulsion region, the single-phase region does not extend to the oil corner. This means that it is not a simple reverse curvature phase. (On admixing a small amount of highly penetrating oil like hexane which partitions into the tails of DDAB it will so revert.)

This was a puzzle for some time. It was partially resolved in terms of a model involving disordered connected swollen lamellae. This is important as it shows how such a system can transit topologically from a system of bicontinuous monolayers in a microemulsion, to one of bicontinuous bilayers as exists in the cubic phases. (Holes must form at first in the monolayer to give mesh phases that can then rearrange to form sponge-like structures.)

The interface changes within the single-phase region, from a monolayer to a bilayer in a complex topological reorganization. This is a technically complicated matter. It is important as it is undoubtedly a process used in biology to effect rearrangements of self-assembled structures. The biological ramifications have not been published yet. They involve the same transformation, in reverse, giving a collapse (in the biological case) from a sponge bilayer to an (ordered) multi-bilayer stack. The ordering is induced by the enhanced stiffness of the multilayer. That explains the folding and unfolding of cubic membrancs in vivo and the numbers fit published data well [17].

The original idea was presented in Ref. 18.

11.14 The anti-parallel, extended or splayed-chain conformation of amphiphilic lipids

Some novel interesting ideas on membrane fusion have been developed by R. Corkery [19].

Lipids with amphiphilic character and two or more hydrocarbon chains can exist in a molecular conformation such that the molten, gel or crystalline hydrophobic chains point away from a central polar headgroup. This is the anti-parallel, extended or splayed-chain conformation of lipids. In general the concept refers to assemblies of several to many individual molecules, but the definition has a mono-molecular basis. The splayed-chain concept was first rejected by Langmuir in 1917, and has rarely been considered as a valid conformational state for multi-chain lipids since. A few studies emerged over the last decade that suggest that the splayed-chain state is a possibility in lipidic systems. A collection of previously published evidence provides a historical and conceptual framework and specific evidence for the existence of the splayed-chain state. A role for the splayed-chain state is illustrated for various biological and non-biological processes. These include the behaviour of Langmuir–Blodgett films, swelling and liquid crystalline behaviour of amphiphiles, biomembrane fusion, exocytosis, endocytosis, various membrane–protein interactions, artherosclerosis, apoptosis and skin barrier function.

11.15 Specific ion partitioning in two-phase systems: a contribution to ion pumps?

There is an interesting consequence of specific ion binding that occurs in a two-phase self-assembled system of finite volume. An example is this: the interior of a red blood cell contains a high density of haemoglobin, almost close-packed globular proteins that bind oxygen cooperatively. The red cell is, if we like, a 'phase' of proteins of finite volume separated from the exterior physiological electrolyte medium by the cell membrane. The protein phase has a huge surface area available for specific ion adsorption. The cells themselves are quite densely packed in the blood, and have available to them a finite volume also. The ions of the electrolyte partition between inside and outside the cell. There is a much higher concentration of potassium inside the red cell than outside, and vice versa for sodium ions. A similar asymmetry exists for chloride ions. Classical theories of electrolytes based on electrostatic forces could not account for the phenomenon. The source of this asymmetry was then attributed to active biochemical 'ion pumps' that work against the concentration gradients. Although this is accepted as a fundamental tenet of modern biology, the existence of such pumps has been the subject of occasional dispute for many years. An entree to a contentious, indeed scandalous, literature can be found in Ref. 20.

However, if Hofmeister effects are recognized, there is a source of asymmetric ion partitioning within the finite volume system that has not been taken into account. Just as oxygen binds to haemoglobin so potassium may do so preferentially. Such a source of partitioning exists. It may or may not be significant. Proof-of-concept of this neglected source of ion partitioning has been recently published [21].

The authors used dioctanoylphosphatidylcholine (diC_8PC), a lipid which has a cloud point at about 46 °C in pure water. After phase separation, the upper layer comprises a diluted micellar solution of diC_8PC, while the bottom phase contains about 99% of the surfactant and is highly viscous [22]. The volumes of each are known by inspection. Systems are made up with a fixed overall concentration of different electrolytes, NaCl, NaF, NaBr or $NaNO_3$. The specific partitioning of ions between upper and lower phases is extraordinary and reflects the balance between specific anion binding to the condensed phase, and finite-volume effects. In effect the system shows an equilibrium 'ion pump' with no pump and no membrane [21].

Such an effect must be operating in red cells. How much of a contribution to an ion pump this mechanism makes is not yet known. It is testable with a model two-phase haemoglobin system.

Similar partitioning of ions occurs in some fungi, where it is used to explosive effect in dispersing ascospores. There is clearly no ion pump here [23].

There are many problems of salt partitioning unresolved by classical physical chemistry that ignore Hofmeister effects. They occur, for example, in understanding the mechanism of excretion of salt via the kidneys of dugongs that feed exclusively on sea grass, and in exclusion of sea salt by jellyfish and by the specific glands of seabirds, and in compartmentalization of salt-tolerant plants.

References

[1] S. T. Hyde, S. Andersson, K. Larson, Z. Blum, T. Landh, S. Lidin and B. W. Ninham, *The Language of Shape. The role of curvature in condensed matter physics chemistry and biology*. Amsterdam: Elsevier (1997).
[2] http://anusf.anu.edu.au/Vizlab/viz_showcase/stephen_hyde/embed/imgs
[3] Z. A. Almsherqi, S. D. Kohlwein and Y. Deng, *J. Cell Biol.* **173** (2006), 839–44.
[4] http://www.rsphysse.anu.edu.au/~sth110/sth_papers.html
[5] M. Larsson, K. Larsson, S. Andersson, J. Kaqkahr, T. Nylander, P. Wollmer and B. W. Ninham, *J. Disp. Sci. Tech.* **20** (1999), 1–12.
[6] J. G. H. Wessels, *Mycologist* **14** (2000), 153–158.
[7] B. E. S. Gunning, *Protoplasma* **215** (2001), 4–15.
[8] B. E. S. Gunning and M. W. Steer, *Plant Cell Biology: structure and function*. Boston: Jones & Bartlett (1996).
[9] Y. Deng, S. D. Kohlwein and C. A. Mannella, *Protoplasma* **219** (2002), 160–167.
[10] Z. A. Almsherqi, S. T. Hyde, M. Ramachandran and Y. Deng, *J. R. Soc. Interface* **5** (2008), 1023–1029.
[11] R. B. Ashman and B. W. Ninham, *Mol. Immun.* **22** (1985), 609–612.
[12] R. B. Ashman, R. V. Blanden and B. W. Ninham, *Immunol. Today* **7A** (1986), 278–283.
[13] P. Lo Nostro, B. W. Ninham, A. Lo Nostro, G. Pesavento, L. Fratoni and P. Baglioni, *Phys. Biol.* **2** (2005), 1–7.
[14] M. Mahalingam, K. Martínez-Mayorga, M. F. Brown and R. Vogel, *Proc. Natl. Acad. Sci. USA* **105** (2008), 17795–17800.
[15] R.Vogel, G. B. Fan, F. Siebert and M. Sheves, *Biochemistry* **40** (2001), 13342–13352.
[16] P. Sizun, C. Auclair, E. Lescot, C. Paoletti, B. Perly and S. Fermandijan, *Biopolymers* **27** (1988), 1085–1096.
[17] M. Olla, A. Semmler, M. Monduzzi and S. T. Hyde, *J. Phys. Chem. B* **108** (2004), 12833–12841.

[18] S. T. Hyde, *Coll. Surf. A* **129–130** (1997), 207–225.
[19] R. W. Corkery, *Coll. Surf. B* **26** (2002), 3–20.
[20] L. Edelmann, *Cell. Mol. Biol.* **51** (2005), 725–729.
[21] M. Lagi, P. Lo Nostro, E. Fratini, B. W. Ninham and P. Baglioni, *J. Phys. Chem. B* **111** (2007), 589–597.
[22] P. Lo Nostro, S. Murgia, M. Lagi, E. Fratini, G. Karlsson, M. Almgren, M. Monduzzi, B. W. Ninham and P. Baglioni, *J. Phys. Chem. B* **112** (2008), 12625–12634.
[23] F. Trail, I. Gaffoor and S. Vogel, *Fung. Genet. Biol.* **42** (2005), 528–533.

12

Emulsions and microemulsions

12.1 Emulsions

When a third component, 'oil', is added to a mixture of surfactants and water, the system can form an emulsion. The 'oil' can be any predominantly hydrophobic solution or solid paraffin particles that phase-separate from 'water' without surfactant. The 'surfactant' can include any amphiphilic materials like proteins or long-chain alcohols. The hydrophilic part can be water or another liquid, and can contain salts, sugars, whatever. So the term emulsion, including foods, cosmetics, lubricants, drug delivery, etc., embraces if we like an entire major phylum in the Chemical Kingdom.

Here the 'surfactant' adsorbs at the water–oil interface to segregate 'oil' from 'water' into macroscopic pools. The surfactant adsorbed can be a monolayer, or it can consist of multiple layers. Because of their size, emulsions usually scatter light and appear opaque. The pools of water in oil, or of oil in water, often contain within them smaller oil–water–surfactant microstructures. Sometimes they are bicontinuous. Emulsions are mostly thermodynamically unstable. That is, the droplets will coalesce with time. It used to be thought that the formation of emulsions always required work, such as stirring or sonication, but this is not so. Some, as we shall see, form spontaneously and are thermodynamically stable systems that exist as an apparent single 'phase' with complicated microstructure.

With ionic surfactants, electrostatic double-layer forces act to oppose coalescence of droplets. Depletion forces (see Chapter 4) due to micelles in water are probably even more effective in stabilizing emulsions. Hydrophobes, impurities that associate with surfactant tails, can induce micellization at low surfactant concentrations and enhance this process.

The removal of atmospheric gas can switch off hydrophobic interactions, as discussed in Chapter 8, and enhance emulsion stability, a matter that has been little exploited [1].

Long-chained insoluble surfactants in a monolayer around 'oil' drops inhibit fusion. A mechanism by which emulsion drops fuse has been explored by Kabalnov and Wennerström [2]. Probably this is the same mechanism involved in gas bubble–bubble fusion.

Emulsions often have more complicated microstructure than just drops of oil in water or vice versa, as they are usually envisaged. The structure of even simple emulsions is not known in general.

However, it seems that the same overriding principles hold as for geometric packing of aggregates in surfactant self-assembly in water.

This has been demonstrated in some very beautiful experiments by Lissant. He made emulsions with kerosene, water and a non-ionic surfactant with a kitchen mixer. These emulsions could form as oil-in-water systems with as little as 0.5% water. Simple geometry gave the observed structures (Fig. 12.1) [3,4].

12.2 Microemulsions

Microemulsions are transparent solutions of mixtures of oil, water and surfactant with a rich variety of microstructures. With most industry formulations a so-called cosurfactant is also required. This is typically a short-chained soluble alcohol, the role of which will be explained below. Microemulsions are thermodynamically stable; that is they are, for example, reversible to temperature cycles. They are transparent because the liquid microstructures within have dimensions so small that they do not scatter light. Sometimes they comprise droplets of oil in water (called normal phase) or of water in oil (reverse phase). The oil–water interface of a droplet is separated by a surfactant layer. The microstructure is commonly more reminiscent of bicontinuous connected cylinders of oil in water or of water in oil. There are a huge diversity of other microstructures available. They can be one, two, three and more phase systems. The ternary phase diagram for the system DDAB-dodecane-water is shown in the Appendix of Chapter 9, Fig. 9.31. The distinction between emulsions and microemulsions and other self-assembled aggregates is somewhat artificial.

It is important to note that these are all dynamic aggregates. The lifetime of a surfactant in the layer between oil and water varies by at least 10 orders of magnitude depending on chain length, as for surfactant and lipid–water systems.

Microemulsions were discovered by Shulman in the 1940s. However, Mother Gardiner's recipe for washing woolen clothes in Australia predates this by 100 years – it comprises methylated spirits (methanol), grated bar soap and a little eucalyptus oil. The most sophisticated modern washing detergents do not surpass it in effectiveness.

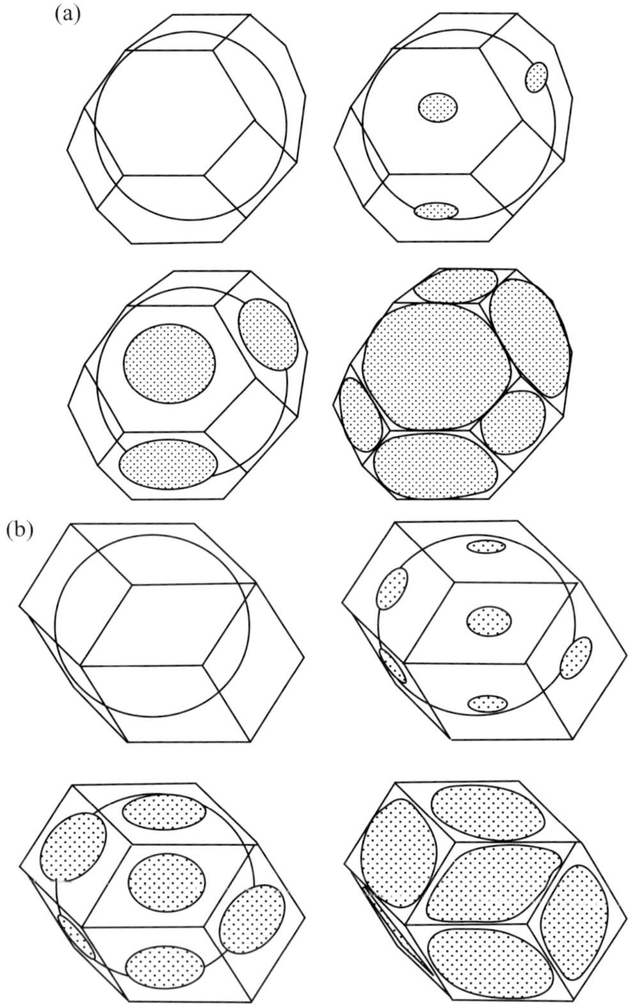

Fig. 12.1. Structure of low-external-phase emulsions. Spheres of oil coated by surfactant in water merge with increasing oil to give connected channels of water. The external phase is as low as 0.5%. The same happens with low water and surfactant content microemulsions. Reprinted from Ref. 3. Copyright 1966, with permission from Elsevier.

In fact even this is predated by about 200 million years by the Australian sawfly larvae. The caterpillars of this primitive wasp eat leaves of eucalyptus trees. They encapsulate within a diverticulum toxic terpenes in a microemulsion, using lipids from the leaves. The same use of microemulsions is made by koala bears. Both use the microemulsions in defence, the one against birds, the ill-tempered bears by spitting them at each other.

There is in the literature a distinction made between 'reverse micelles' formed from the surfactant AOT (see Appendix Chapter 9) and microemulsions. This particular double-chained surfactant forms droplets of water in oil. The size of the drops varies depending on volume fractions of oil, water and surfactant, the water activity being adjustable depending on drop size. They are used for synthesis of nanoparticles. But the distinction between reverse micelles and microemulsions is quite artificial.

Likewise it is generally claimed that the cause of microemulsion formation is the existence of the ultralow interfacial tension between a microemulsion and oil. This shows up in an extremely flat interface in a test tube containing the two phases. This is not true and puts the cart before the horse. The ultralow interfacial tension comes about because of the existence and microstructure of the microemulsion [5].

12.3 Three-component ionic microemulsions

Most of the work and interest in microemulsions and other ternary or quaternary mixtures has been devoted to zwitterionic lipids such as phospholipids, and triglycerides and oils and protein mixtures in connection with food, and single-chained anionics and non-ionic surfactants with cosurfactants and salt. A truly vast amount of work went into mapping the phase behaviour of complex systems and trying to unravel microstructure. By and large these attempts failed completely.

A particular model system predicted by theory and chosen deliberately to elucidate microstructure cut through this difficulty, and the problem was resolved. We devote most of this chapter to this system.

For a long time it was believed that the formation of a microemulsion with ionic surfactants required the addition of a 'cosurfactant', typically a short-chained alcohol such as butanol. This is simply not so.

A general theory of self assembly of micelles, vesicles and microemulsions identified the appropriate conditions [5].

The theoretical prediction is that microemulsions can form without cosurfactant. One condition necessary is that to solubilize enough oil a low enough surfactant curvature is required, in other words $v_H/(a_P l_H)_{eff} \sim 1$. This means essentially a double-chained surfactant. With a single-chained (micelle-forming) surfactant, $v_H/(a_P l_H) < 1/3$. (A cosurfactant like an alcohol which is also amphiphilic can partition into the interfacial region, and associate with the surfactant to give what is the effectively a double-chained surfactant.) The condition $v_H/(a_P l_H)_{eff} \sim 1$ is necessary, but not sufficient. More is required. The oil has to be compatible with the (liquid-like) surfactant tails. Experimental and theoretical studies of alkane uptake in thin films of phospholipid bilayers provide a guide. The rule that emerges is that when the chain length of the alkane is less than that of the surfactant, the alkane penetrates strongly and swells the bilayer. When the alkane chain length exceeds that of the surfactant, no or little penetration occurs. So one expects as a working rule that to set curvature at the oil–water interface, an oil like hexane will adsorb strongly and give large reverse curvature, while a longer-chained oil like dodecane will adsorb less and give a less-curved surface. In other words, the nature of the 'oil' sets intramolecular forces and the effective surfactant parameter at an interface from the oil side.

With those notions as a guide, it became possible to make three-component ionic microemulsions. The model system that elucidated microstructure and allowed its prediction was chosen as the surfactant didodecyl dimethyl ammonium bromide (DDAB), illustrated in the Appendix of Chapter 9. Unlike most microemulsions studied previously this surfactant is weakly soluble in both oil and water. Therefore it has to sit at the oil–water interface. Further the diC_{12} chain length of the surfactant spans the strongly penetrating (C_6) to weakly penetrating (C_{14}) alkanes. This system allowed microstructure to be worked out by studying phase behaviour as a function of alkane chain length [4].

The phase diagrams turned out to be very rich in microstructure (see Fig. 9.31) [6].

Within the ternary phase diagram that demarks phase boundaries as a function of volume fractions of components, there are two lamellar phases, three cubic phases, hexagonal phases of reverse curvature, a large single-phase microemulsion region, a spontaneous emulsion region and other regions containing one-, two- and three-phase systems. The oil penetration depends on chain length. With strongly penetrating oils like hexane or cyclohexane, these microemulsions can form with as little as 2% water [7].

12.4 Bicontinuity and spontaneous emulsions

In the very large single-phase region these microemulsions are predominantly bicontinuous. The conductivity varies over 10 orders of magnitude along a water dilution line, and undergoes a critical percolation transition to essentially zero, the conductivity of oil, along a water dilution line of constant surfactant/oil ratio. The viscosities varied similarly by 5 orders of magnitude along the same dilution line (Fig. 12.2).

For each oil there occurs a region adjoining this line where a spontaneous white emulsion forms. The conductivity goes to essentially zero, that of oil, across this boundary. These emulsions are thermodynamically stable, being reversible to temperature cycles. It became clear that once curvature is set by oil uptake in the surfactant region, the microstructure is determined by global packing conditions alone [4,7,8].

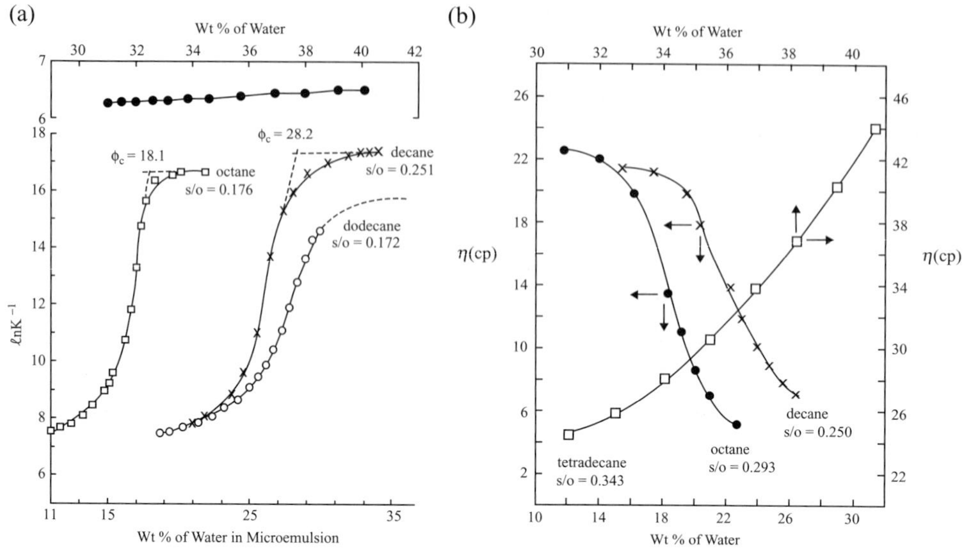

Fig. 12.2. Plots of conductivities (a) and viscosities (b) for DDAB/water/alkane systems along a dilution line in a single microemulsion phase reflect changing microstructure within a single phase. (a) The alkanes are octane (□), decane (×), dodecane (○) and tetradecane (•); (b) octane (•), decane (×) and tetradecane (□); s/o is the surfactant-to-oil weight ratio. Reprinted from Ref. 8 with permission. Copyright 1984 American Chemical Society.

The microstructure is essentially that of random connected cylinders. These cylinders have constant reverse curvature, effective surfactant parameter (surfactant + 'bound' oil) >1. These surround water conduits. The diameter and connectivity of these channels depend on volume fractions alone. The headgroup area, from neutron scattering, varies little over the entire phase diagram. Since the surfactant is virtually insoluble in both water and oil, the only variable left is global packing. Bicontinuity was evident from measurements of conductivity [4,7,8] and diffusion NMR. With increasing water the connected 'cylinders' lose the number of their connections progressively, until, along a line of constant water to surfactant ratio, they disconnect to form non-conducting reverse spheres of water in oil.

The entire single-phase boundaries in the ternary phase diagram and microstructure within could be calculated by imposing constant mean curvature – at the interface, different for each oil – depending on penetration. It was then just a matter of global geometric packing constraints again [9–15].

There are also other phases that form hexagonal (ordered cylinders), swollen and compressed lamellar, and at least three different contiguous cubic phases besides the stable emulsion phase [16].

12.5 Percolation exponents

What really clinched the matter of microstructure beyond doubt was later work still that showed that the model predicted critical exponents for percolation precisely! The system was indeed a perfect model porous medium [17–20].

12.6 Interfacial tensions at the oil–microemulsion interface

A very curious and interesting result emerged in careful measurements of the interfacial tension between a two-phase octane-DDAB-water microemulsion and octane system. The microemulsion comprised long connected cylinders of water surrounded by surfactant in oil. The measured interfacial tensions were much too high by orders of magnitude. They could in no way be accounted for in terms of conventional ideas regarding van der Waals interactions of cylinders of water across oil. But they could be accounted for in terms of the elusive giant dipole fluctuation forces discussed in Chapters 5 and 6 [21].

12.7 Single-chained surfactants and non-ionics

Much work was done over 30 years on water-soluble ionic single-chained surfactants with soluble cosurfactants that form microemulsions. Extensive, unconscionably boring mindless studies by slave graduate students of phase diagrams used either five-component systems like sodium dodecyl sulfate (SDS)-butanol-toluene-brine-water, or non-ionic surfactant-water-oil-cosurfactant systems. Little progress was made. It turns out that a 'cosurfactant' like butanol associates with the surfactant at the interface and the combined entity is effectively a double-chained surfactant with $v_H/(a_P l_H)_{eff} \sim 1$. This is required in order to encapsulate sufficient oil in this case. But the process of formation involves two stages: first alcohol partitions into the hydrocarbon tail region of micelles and swells them. After that the previously excluded toluene is taken up by the swollen micelles and partitions the alcohol to the micellar surface. The microstructure is not that of swollen droplets as claimed, but again mostly a bicontinuous structure determined by packing.

The same complication was a huge impediment to unravelling microstructure of microemulsions with non-ionic surfactants. Cosurfactants, necessary to provide sufficient interfacial curvature, partition between 'oil' and aqueous phases, and the interface depends on volume fractions of each, and no rhyme or reason seemed to attend their microstructures.

For references to some of the literature on these systems and the problems as they were, see Refs. 24 and 25.

12.8 Specific ion effects and 'impurities' change microstructure

We have repeatedly seen the universal occurrence of Hofmeister specific ion effects. They are equally available for exploitation in microemulsion design by changing curvature through headgroup interactions. Alkane uptake by the surfactant depends on alkane chain length. This changes the curvature from the oil side of the oil–water interface. We can turn on opposing intramolecular forces from the water side of the interface by changing the counterion.

The effects are quite dramatic. With Br^- as counterion there is, as we have seen, a very broad single-phase microemulsion region. With Cl^- and I^- as counterions the one-phase region collapses to a tiny part of the ternary diagram. The ion binding is determined by the specificity of dispersion forces, local water structure and compatibility of hydration between the quaternary ammonium group and the halide counterion. The difference in these forces changes interfacial curvature and therefore the microstructure dramatically [26].

12.9 Competitive anion binding

An even more startling effect is seen if one ion-exchanges Br^- for SO_4^{2-} as counterion with the same DDAX cationic microemulsion system. With sulfate as counterion the system has normal, not reversed, curvature. It forms oil-in-water droplets rather than a water-in-oil structure. The headgroup repulsions are larger than with bromide and cause oil adsorbed in the surfactant tails to be expelled. So that now $v_H/(a_P l_H)_{eff} < 1$. The system is quite delicately poised. Thus if one titrates into the oil-in-water sulfate microemulsion system a very small amount of NaBr ($\sim 10^{-2}$ M), the microemulsion flips back to the reversed curvature structure! This is apparently absurd as the system contains about 0.5 M sulfate counterions. The sulfate is divalent and is present at near molar concentrations, in large excess over the monovalent Br^-. Yet the Br^- ion displaces the sulfate, 'binds' strongly and changes the interfacial curvature.

What this tells us is that electrostatic forces are irrelevant. Headgroup forces are determined by specific dispersion forces and hydration [27].

12.10 Cationic binding to cationic surfaces

An even more extraordinary effect occurs if DDAB microemulsions are made up with successively increasing amounts of low-concentration electrolytes containing alkali bromide salts rather than water. The added bromide ions are the same as the counterion of the DDAB surfactant, which is present in typically >1 M concentration without the added electrolyte. The additional counterions should have little effect. At an added electrolyte concentration of $>\sim 0.125$ M the microemulsions no longer form. The system collapses to a polydisperse emulsion. Further, NMR studies show that the *positively* charged cations bind to the *positively* charged DDAB surfactant surface. The same kinds of qualititative changes in self-organization effects also happen with low-concentration buffers. Such effects are probably due to partitioning of the added electrolytes into the pools of water formed from the junctions of the intersecting connected cylinders. The point of these remarks is to note that microstructure can be changed dramatically by apparently very small changes in physicochemical conditions. Such large changes in self organization make sense if one is aware of the underlying microstructure and of the molecular forces operating [28].

12.11 Impurities and mixtures

An example of a similar change in microstructure occurs when a small amount of vitamin K (1%) is added to a cubic phase of a monolein–water system. The vitamin K is oil-soluble. It induces a change in structure from cubic to lamellar and hexagonal phases. The amount of additive is so small that it cannot be attributed to changes in average surfactant parameter or local curvature. The cubic phase has constant average curvature but varying Gaussian curvature. The reorganization in microstructure is probably due to the separation of and association of the vitamin K into the connected junction regions of the cubic phase.

These two examples demonstrate an additional, delicately poised flexibility for self organization in bicontinuous microstructures. Whether the average curvature is zero as in cubic phases, or constant as in microemulsions, the Gaussian (product of curvatures) varies continuously. So the pools of water can have different local water structure (and reactivity) to that in the cylinder-like connections that form them.

Similar effects are seen with small amounts of 'impurity' lipids in membrane mimetic systems. Phase segregation of lipids (rafts) within the membrane bilayer causes variations in local curvature along the membrane surface and causes phase changes [29].

12.12 Specificity of oils, *cis* and *trans* oils, alcohols and cholesterol

The oil (DDAB)–water microemulsion model system is rich in complicated microstructure. It can be understood in terms of simple ideas, local curvature preset at the interface plus global packing constraints. The DDAB model surfactant has in common with biological lipids the feature of double chains, and relative insolubility of the surfactant and lipids in both oils and water. Double chains are necessary for lipids to form membranes. Insolubility is necessary if the supporting lipid structure of a biomembrane is to exist for sufficient time. Further, the chain length has to be sufficiently long to be compatible with a range of additives or physicochemical conditions that it has to deal with. This is necessary for changes in curvature required for transport and endo- and exocytosis. But the lipid chain length cannot be too long, as the Krafft temperature would then be too low and the chains frozen at physiological temperature.

Local curvature and global packing seem sufficient to capture the essence of self organization. We have shown that there are a variety of ways we have available to manipulate curvature. Among these surfactant chain length, mixed surfactants, cosurfactants, mixed alkanes, temperature, Krafft temperatures, alkane chain length, counterion, ion binding, electrolytes and buffers all can be used to prescribe microstructure.

Another subtle variable of some interest is oil saturation.

If the DDAB microemulsions are made with strongly penetrating oils like *n*-hexane and cyclohexane the phase behaviour and microstructure of the one-phase microemulsions at very low water content are somewhat different. Just as for the emulsions of Lissant [3] (Section 12.1) the phase can exist at water contents as low as 0.5%. The conductivity at first increases up to 4%, showing abrupt changes on the way that correspond precisely to the changes in structure (bcc, fcc packing) elucidated by Lissant. It then decreases by 10 orders of magnitude with additional water as outlined above.

If we use, instead of hexane or cyclohexane, the oils 1-hexene or 2-*trans*- or 2-*cis*-octene, more subtlety emerges. The double bonds towards the ends of the chains are more 'hydrophilic' and penetrate the surfactant region more strongly. What is of interest from a biological perspective is that the adsorption of the *cis* and *trans* isomeric oils into the surfactant tails and their effects on changing curvature are very different. Presumably they also change membrane permeability. This is relevant to debates on *trans* vs. *cis* fats in diet [7].

Long-chain alcohols that are insoluble in water, such as decanol, behave much like the alkanes [6,11]. Short-chain alcohols such as methanol, ethanol or pentanol that are soluble in water dramatically affect headgroup hydration and forces between surfactant layers. Again this has some intersection with problems of biological interest in so far as the choline end group of the most common lipid phosphatidylcholine and the tetramethyl ammonium group of our surfactant are very similar [30,31].

Omega−3, 6 and 9 lipids, along with cholesterol, are major components of membranes of the brain. The mixtures of these easily oxidized lipids with double bonds in positions 3, 6 and 9 give considerable flexibility to these lipids (see Fig. 9.25). Presumably this allows greater flexibility in rearrangements and two-dimensional phase behaviour of bilayers with ion conduction and transport with the nervous impulse.

12.13 Supraself assembly and other 'phases'

12.13.1 The DDAB system

The model DDAB system forms a microemulsion that spans a large range of component ratios. The water content can range from 0.5% to about 70%. Within that phase the microstructure varies enormously, but predictably as we have seen. The microstructure comes about at essentially fixed headgroup curvature set by opposing interfacial forces. It is due just to global packing constraints. One open question on microstructure here, and for other self-assembled systems, concerns the hexagonal phase (parallel cylinders of water coated by a surfactant layer in oil).

Microstructures within ordered phases such as multilamellae, hexagonal and cubic phases are revealed by their optical textures with polarized light (cf. Fig. 9.31). It can be unambiguously determined by techniques such as SAXS and SANS. The most convincing assignment of structure from scattering is to the ordered hexagonal phase. This gives sharp reflections. Surprisingly, it is possible that these ordered cylinders of water in oil may be more complicated. It seems that in the DDAB system, and perhaps for all hexagonal 'phases', there can another layer of complexity. Random 'struts' of water channels can form within the phase that connect the ordered cylinders. Their number would increase with increasing water content within the phase. These would be missed by scattering, electron microscopy and other probes. If so, this would explain how hexagonal phases transit to cubic phases. The point is that the definition of what we mean by 'phase' in self-assembled systems is not always clear.

Here are some examples:

1 Take a freely flowing DDAB microemulsion. Its structure is that of random connected surfactant-coated water cylinders in oil. If it is stirred in a flask with a magnetic stirrer the phase becomes stiffer and stiffer, so viscous that eventually the stirrer comes to a halt. The phase does not change with time. If the stirrer is then reversed, the system unwinds back to its original state. If we like, it has a memory.

2 In a monolein–water–glycerol system, over a large region in the phase diagram standard characterization probes show the existence of two phases, hexagonal and lamellar. In fact it seems to have a microstructure involving lamellae that separate interior and external hexagonal phases, as we have seen before.

3 The structure of the spontaneous emulsion phases in DDAB has not been explored systematically. It seems probable that the emulsion droplets in oil are multilamellae that surround microphases of hexagonal or microemulsions in the interior. The notion of such

supra-aggregated states as local surfactant curvature changes emerges again from global packing.

12.13.2 Supra-aggregation a general phenomenon

We have emphasised in Section 10.4 that some apparently peculiar states of self assembly with vesicles formed by didodecyl dimethyl ammonium hydroxides and acetates are indicative of the more general phenomenon of supraself assembly. Depending on concentration the vesicles can contain within their interiors a dense packing of micelles or of smaller vesicles, or even cylindrical or cubic microstructures. It is as if the interior of a multilamellar 'onion' phase of bilayers collapses into micelles. The spontaneous emulsions that occur with the DDAB microemulsion system exhibit the same behaviour.

Such complex microstructures are a consequence of global packing constraints, of the necessary asymmetry of any curved bilayer, of the necessary different physicochemical conditions (water structure) that exists in the interior and exterior of any topologically closed container; and of the 'ion binding' that affects intramolecular and intermolecular forces. The recognition that such structures exist calls into question what we mean by the term 'phase' or equilibrium borrowed from macroscopic Gibbsian thermodynamics.

At least, with bounding bilayers of typical membrane lipids, or other double-chained surfactants, supraself-assembled states can exist for months, the lifetime for biological cells [32].

12.13.3 The copper AOT–isooctane–water microemulsion

We now give further evidence from another system that such states of self assembly are to be expected.

The double-chained surfactant (illustrated in the Appendix in Chapter 9) sodium AOT–isooctane–water microemulsion system has been a standard model system for a long time. Much work has been done using these microemulsions. They were considered to be just simple reverse-curvature spherical water-in-oil micelles. Work has been focussed on such systems as vehicles to template nanoparticles. The motivation was to use these reverse spherical droplets as nanoreactors with water structure and size prescribed by volume fractions. These nanoreactors can then be used to precipitate metal nanoparticles using a reducing agent. These can then be used to fabricate surface monolayers with, e.g., mooted interesting optical properties for devices. This procedure works very effectively for all kinds of metal ions as counterions to the AOT surfactant. Or else the system has been mooted as an artificial photosynthesis vehicle, a forlorn hope given the success of plants in this field.

More interesting from our point of view is the phase diagram for the copper AOT system. Copper ions have a capacity to take up integral numbers of water of hydration: 1, 2, . . . 8 water molecules. Consequently the surfactant parameter varies, quite abruptly, with water content. It is unlike the DDAB–alkane–water system where curvature, and hence microstructure, is set by oil penetration into the surfactant chains. Here the oil, iosooctane, is compatible with the chains (see the Appendix in Chapter 9 for AOT's molecular structure) and the curvature is set by the area per molecule, which changes with hydrated water content. As is evident from a very complicated phase diagram, phase behaviour changes discontinuously, abruptly at water/surfactant mole ratios 1, 2, . . . 8 [33–38]. Each triangle in the ternary diagram itself contains several phases. The entire complicated phase diagram and microstructure can be predicted from packing considerations alone.

For details we refer to the original and later work of Pileni [33–38]. In brief, one can divide the system into reverse micelles enclosing water which contain 'oil' (isooctane) in the AOT chains, an 'effective micelle', with excess oil (oil penetration is here completely energetically indifferent and takes up the volume available in the splayed double chains).

On addition of water – at constant $v_H/(l_H a_p)$ – when these 'effective' hard spheres reach close packing, a fraction must elongate and become ellipsoids at first, and then grow to cylinders. The system separates into spheres and cylinders. When the cylinders and their 'bound' oil can no longer pack (75% for square array, and 91% in close-packed array) some must be removed into a multilamellar phase of closed liposomes. The two-phase system proceeds until all are liposomes. But here we have a difference. The interior surfactant molecules of the liposome cannot pack. These interior layers then spit out excess water and revert to the hexagonal exterior phase with which they are in equilibrium; *but* with the same global packing constraints! Remarkably the spacing of the surrounding lamellar-phase liposomes is predicted by these simple packing constraints. So we now have a micro-supraself-assembled hexagonal phase surrounded by lamellar phase immersed in a hexagonal phase. The zero interfacial tension boundary dividing these two 'phases' is not a boundary at all. As we proceed to add more water the spherical 'liposomes' containing this microphase can themselves no longer pack. Geometric constraints alone dictate that the interior collapses further. The internal hexagonal or micellar or cubic microphase within collapses further to become interdigitated micelles, expelling all excess oil. An extraordinarily complicated phase diagram and supraself-assembled aggregates turns out to be a fairly simple affair, due to requirements of mass balance alone. The microstructure of the system is reasonably convincing in that the 'templated' precipitated copper nanoparticles, spheres and cylinders are exactly what is predicted from the microstructure! This opens up new insights.

By adding additional specific salts to the water, one changes $v_H/(l_H a_p)$ due to specific ion effects and can template, more or less to order, a rich variety of nanoparticles and alloys.

12.14 Polymerization of microemulsions

The features of microstructure described above appear to be fairly universal. Work on the same DDAB model system with fluorocarbon oils or styrene or methacrylate gives the same structures [39–41]. The microstructures again follow, after fixing interface curvature by balancing oil penetration against headgroup repulsion, from the requirement of global packing alone. It is difficult to gainsay this overriding constraint. The complication of multicomponent systems, where surfactant and/or cosurfactant partition between oil and water depending on volume fractions, has made it difficult to unravel these systems so easily. One example is the system sodium AOT–water–oil–methacrylate. Without the (water-soluble) methacrylate this system forms reverse micelles. Polymerization of the methacrylate led to a new technology to form highly monodisperse spherical nanoparticle latices. Neutron scattering was used to fit the microstructure to spherical microstructures prior to polymerization. But the scattering could be fitted only by invoking a Hamaker constant, an interaction force between the droplets at least 10-times larger than any conceivable van der Waals force. In fact, prior to polymerization the system is disordered connected cylinders, as for the DDAB system, with methacrylate as cosurfactant. On addition of initiator the methacrylate migrates to junctions of the connected cylinders that form the bicontinuous microemulsion. Without the methacrylate cosurfactant, the cylinders then break off to form disconnected polymerized methacrylate nanospheres. The problem is one of enormous industrial importance and the mechanism of polymerization posed a puzzle. The bicontinuity of the original microemulsion could have been determined by the use of a conductivity meter rather than expensive neutron scattering, the software for which imposed an invalid microstructure.

With the DDAB system and styrene or methacrylate as oil a good deal of work had been done to polymerize the oil. The idea was to set microstructure by volume fractions as above. Then polymerization of the oil produces a porous medium with known porosity and pore size. This has not worked so far in bulk systems. However, it does work perfectly well to make membranes for ultrafiltration or reverse osmosis.

12.15 Non-swelling lamellar phases

A remarkable system was studied by Corkery [42]. He made a series of ionic double- and triple-chained stearate surfactants. These had long been studied at the turn of the last century and forgotten. (The aluminium salts are used in napalm.) The lamellar phase of this surfactant cannot swell with addition of water because of an unusual arrangement of the triplet of chains. A consequence is that the lamellar phase swells enormously with oil. It takes up to 98% oil. With environmental oil spills on beaches or in the ocean conventional methods employ surfactants to disperse the spill. Here though the surfactant causes aggregation of the oil. It is not an academic matter, and the process works very effectively with salt water and heavy oils spread on beach sand. If the oil is dosed with colloidal magnetite the pollutant oil is easily removed with a magnet. The oil surfactant and magnetite can be reused.

The extraordinary behaviour of these really very ordinary surfactants which come about because of a simple rearrangement of the chains may well be a key factor in biological cell fusion and recognition.

12.16 Gels

The literature on gels is vast, and we do not attempt to address it, except to refer the reader to that literature. Gel structure is accessible from and directly parallels that of surfactant self-assembly, although this seems not to be recognized.

12.17 Some remarks on ion-binding models

We have had no recourse so far to a long-standing phenomenological characterization of micelle formation. This is called the ion-binding model. The same simple model is used in one form or another to characterize and quantify ion binding to proteins, polymers and membranes. From measurements interpreted by this model one deduces pK_as, so the matter is really quite fundamental. It seems to work well. But nowhere in the theoretical ideas on self assembly we have described has the ion-binding model appeared.

We wish to establish a connection with this phenomenology.

The ion-binding model describes micellization by the equation:

$$x_M = x_1^N \cdot x_2^Q$$

Here x_M is the mole fraction of micelles consisting of N surfactant molecules, x_1 the mole fraction of monomers, x_2 the mole fraction of counterions and Q the number of 'bound' counterions. The equilibrium constant $K = \exp[N\Delta G/(RT)]$ where ΔG is the hydrophobic free energy change of the transfer of a mole of hydrocarbon chain from water to the oil-like interior of the micelle and R is the gas constant, T the temperature. The hydrophobic free energy of transfer can be obtained from measurements of partition coefficients of oil and water. The equation seemed to fit well for typical anionic and cationic surfactants, with a fraction of bound counterion, Q, about 0.8. The 'binding' was measured, usually by NMR. The apparent success of the phenomenological model led to a plethora of work with the same phenomenology to measure calcium, Mg^{2+}, etc., binding to proteins, and proteins at interfaces. There was a problem here. The ion-binding model seemed to conflict with the less specific, more general statistical mechanical-nucleation model of Section 10.1 in which the term ${X_2}^Q$ does not appear. It is replaced by a term that represents the surface free energy of a monomer at the curved micellar interface.

The phenomenological ion-binding model for micelles appeared to conflict with models based on curvature energy, from which our surfactant parameter derives. A theory of surface curvature free energy based on the double-layer electrostatic Poisson–Boltzmann equation reconciled the two approaches. But the classical ion-binding model emerged as a special case that held only at the limit of 'strong' binding. 'Binding' comes out in terms of a complicated physisorption excess of counterions about the micelle. There is no actual binding. The ion-binding model is rigorously correct in this theory, but only at the asymptotic limit that all counterions are bound! For standard cationic and anionic surfactants like SDS and CTAB the theory worked up to 130 °C, and cmcs and aggregation numbers came out right, and as a function of salt concentration; and it was confirmed by neutron scattering for a range of other surfactants by J. Hayter, who also simplified the theory. What was appealing in this model was that the balance of forces at the interface necessary to find the optimal micelle that forms could be calculated analytically. There was no need to worry about hydrocarbon chain forces or hydration or other unknown forces. They were all seemingly taken into account automatically [43–46]. The same theory could be extended to include the phenomenological Manning ion-condensation model for polyelectrolytes [47].

12.18 When and why ion binding breaks down: Hofmeister effects

The theory was a major advance. It provided a theoretical basis for measurements of ion binding. But it worked and was predictive only for the most common counterions and salts such as Na^+ or Br^-. It failed to account for 'peculiar' ions. These are those that we have already seen form vesicles with double-chained surfactants: hydroxide, acetate or carboxylates. No binding occurs. At least any NMR or other measurement assuming the ion-binding model for interpretation gave values of 0–50% for the fraction of 'bound' counterions. Since the ion-binding model is valid only when nearly all ions are 'bound', the numbers were meaningless. Here was something very odd indeed. For these surfactants cmcs were twice as large and aggregation numbers half as big as those for 'normal' surfactants. The matter seemed at first to be rationalized by recognizing that forces between bilayers with OAc^- as counterion in Na^+ OAc^- were 50-times larger than those for Br^- as counterion with Na^+ Br^-. This would be consistent with no binding (high surface charge, greater forces between bilayers) for OAc^- ions and strong binding (low surface charge, smaller forces) for Br^- ions [48,49].

12.19 Inconsistency of the ion-binding theory with direct force measurements

12.19.1 The DLVO theory and ion binding

Now we have a big problem. To summarize again: the ion-binding model, widely used to characterize binding of ions like Ca^{2+} to proteins, was equivalent to an extension of electrostatic double-layer theory. Ion binding, as already remarked, emerges as related to a physisorption excess of counterion. There is *no* actual chemical binding and the theory works rigorously only when counterions are strongly ($>\sim 80\%$) physisorbed or 'bound'. It works for Br^- as counterion to a cationic surfactant like CTABr. It works for other alkali halides as counterions. But the force measurements between bilayers of the same kind of surfactant adsorbed onto mica could only be fitted to the DLVO theory, and as a function of salt concentration, by assuming 80% 'bound' counterion. This cannot be correct. The surface charge that describes the interaction must be the bare charge of the interacting surfactant bilayers. There is a clear contradiction here. On the other hand, for OH^-, OAc^- and our 'peculiar' ions that do not satisfy the ion-binding expectations from electrostatic theory, the force measurements as a function of added salt, K^+OAc^-, fit perfectly to double-layer theory. There is no fitting parameter of bound surface charge. That is as it should be if double-layer theory is correct. Double-layer theory works for what we regarded as 'peculiar ions'. It does not work for 'normal' counterions. The conclusion is inescapable. The theories have to work for all ions. Hence the theory of micelles is incorrect – it is based on the double layer for a spherical surface, or double layer theory is incorrect for a planar surface, or both. This serious problem had to wait until recently for its resolution. As will be clear by now, both the theories of self assembly based on ion-binding notions and the DLVO theory at the heart of colloid science were flawed and incorrect.

We have belaboured this somewhat technical point at some length.

The reason is that the inclusion of specific-ion Hofmeister effects, driven by dispersion forces missing in conventional theory, has to be admitted. The effects cannot be due to 'water structure' alone.

The inclusion of dispersion forces to give consistency in theories has been attempted in a large number of papers, with varying degrees of success, in accounting for Hofmeister phenomena (see Chapter 7). Confidence in predictability has been inhibited by lack of dynamic polarizability data for ions. That deficiency has recently been remedied and explains secondary hydration and reversal of Hofmeister series [50–53].

The result is that interpretation of binding to self-assembled aggregates, to proteins and polymers, of ionic self energies, of pK_as, the problem of pH, of surface and membrane potentials, and of force measurements remains to be revisited and reinterpreted.

References

[1] R. M. Pashley, M. J. Francis and M. Rzechowicz, *Curr. Op. Coll. Interface Sci.* **13** (2008) 236–244.
[2] A. S. Kabalnov and H. Wennerström, *Langmuir* **12** (1996), 276–292.
[3] K. J. Lissant, *J. Coll. Interface Sci.* **22** (1966), 462–468.
[4] D. F. Evans, D. J. Mitchell and B. W. Ninham, *J. Phys. Chem.* **90** (1986), 2817–2825.
[5] D. J. Mitchell and B. W. Ninham, *J. Chem. Soc. Faraday Trans. II* (1981), 601–629.
[6] K. Fontell, A. Ceglie, B. Lindman and B. W. Ninham, *Acta Chem. Scand. A* **40** (1986), 247–256.
[7] B. W. Ninham, S. J. Chen and D. F. Evans, *J. Phys. Chem.* **88** (1984), 5855–5857.
[8] S. J. Chen, D. F. Evans and B. W. Ninham, *J. Phys. Chem.* **88** (1984), 1631–1634.
[9] T. N. Zemb, S. T. Hyde, P. J. Derian, I. S. Barnes and B. W. Ninham, *J. Phys. Chem.* **91** (1987), 3814–3820.
[10] B. W. Ninham, I. S. Barnes, S. T. Hyde, P. J. Derian and T. N. Zemb, *Europhys. Lett.* **4** (1987), 561–568.
[11] I. S. Barnes, S. T. Hyde, B. W. Ninham, P. J. Derian, M. Drifford and T. N. Zemb, *J. Phys. Chem.* **92** (1988), 2286–2293.
[12] S. T. Hyde, B. W. Ninham and T. N. Zemb, *J. Phys. Chem.* **93** (1989), 1464–1471.
[13] T. N. Zemb, I. S. Barnes, P. J. Derian and B. W. Ninham, *Coll. Polymer Sci.* **81** (1990), 20–29.
[14] I. S. Barnes, P. J. Derian, S. T. Hyde, B. W. Ninham and T. N. Zemb, *J. Phys. France* **51** (1990), 2605–2628.
[15] T. N. Zemb, *Coll. Surf. A* **129–130** (1997). 435–454.
[16] P. Barois, S. T. Hyde, B. W. Ninham and T. Dowling, *Langmuir* **6** (1990), 1136–1140.
[17] M. A. Knackstedt and B. W. Ninham, *Phys. Rev. E* **50** (1994), 2839–2843.
[18] M. A. Knackstedt and B. W. Ninham, *AIChE J.* **41** (1995), 1295–1305.
[19] M. Monduzzi, M. A. Knackstedt and B. W. Ninham, *J. Phys. Chem.* **99** (1995), 17772–17777.
[20] M. A. Knackstedt, M. Monduzzi and B. W. Ninham, *Phys. Rev. Lett.* **75** (1995), 653–656.
[21] M. Allen, D. F. Evans, B. W. Ninham and D. J. Mitchell, *J. Phys. Chem.* **91** (1987), 2320–2324.
[22] M. Olla, A. Semmler, M. Monduzzi and S. T. Hyde, *J. Phys. Chem. B* **108** (2004), 12833–12841.

[23] S. T. Hyde, *Coll. Surf. A* **129–130** (1997), 207–225.
[24] B. Lindman and B. W. Ninham, Self assembly: a biassed random walk through the literature. In *Proceedings from Ettore Majorana School of Microemulsion in Progress in Microemulsions*, ed. S. Martellucci and A. N. Chester. Plenum Press (1989), 85–111.
[25] B. W. Ninham, In search of microstructure. In *Structure and Reactivity of Reverse Micelles*, ed. M. P. Pileni. Elsevier (1989), 3–13.
[26] V. Chen, D. F. Evans and B. W. Ninham, *J. Phys. Chem.* **91** (1987), 1823–182.
[27] M. Nydén and O. Söderman, *Langmuir* **11** (1995), 1537–1545.
[28] S. Murgia, F. Portesani, B. W. Ninham and M. Monduzzi, *Chem. Eur. J.* **12** (2006), 7889–7898.
[29] F. Caboi, T. Nylander, V. Razumas, Z. Talaikyté, M. Monduzzi and K. Larsson, *Langmuir* **13** (1997), 5476–5483.
[30] R. M. Pashley, P. M. McGuiggan, B. W. Ninham, D. F. Evans and J. Brady, *J. Phys. Chem.* **90** (1986), 5841–5845.
[31] M. C. Pinna, P. Bauduin, D. Touraud, M. Monduzzi, B. W. Ninham and W. Kunz, *J. Phys. Chem. B* **109** (2005), 16511–16514.
[32] B. W. Ninham and D. F. Evans, *Faraday Disc. Chem. Soc.* **81** (1986), 1–17.
[33] P. André, B. W. Ninham and M. P. Pileni, *New J. Chem.* **25** (2001), 563–571.
[34] M. P. Pileni, B. W. Ninham, T. Gulik-Krzywicki, J. Tanori, I. Lisiecki and A. Filankembo, *Adv. Mater.* **11** (1999), 1358–1362.
[35] I. Lisiecki, P. André, A. Filankembo, C. Petit, J. Tanori, T. Gulik-Krzywicki, B. W. Ninham and M. P. Pileni, *J. Phys. Chem. B* **103** (1999), 9168–9175.
[36] I. Lisiecki, P. André, A. Filankembo, C. Petit, J. Tanori, T. Gulik-Krzywicki, B. W. Ninham and M. P. Pileni, *J. Phys. Chem. B* **103** (1999), 9176–9189.
[37] A. Filankembo, P. André, I. Lisiecki, C. Petit, T. Gulik-Krzywicki, B. W. Ninham and M. P. Pileni, *Coll. Surf. A* **174** (2000), 221–232.
[38] P. André, A. Filankembo, I. Lisiecki, C. Petit, T. Gulik-Krzywicki, B. W. Ninham and M. P. Pileni, *Adv. Mater.* **12** (2000), 119–123.
[39] M. Olla, M. Monduzzi and L. Ambrosone, *Coll. Surf. A* **160** (1999), 23–36.
[40] C. C. Co, R. de Vries and E. W. Kaler, *Macromolecules* **34** (2001), 3224–3232.
[41] R. de Vries, C. C. Co and E. W. Kaler, *Macromolecules* **34** (2001), 3233–3244.
[42] R. W. Corkery, *Coll. Surf. B* **26** (2002), 3–20.
[43] B. W. Ninham and D. F. Evans, *J. Phys. Chem.* **87** (1983), 5025–5032.
[44] B. W. Ninham and D. J. Mitchell, *J. Phys. Chem.* **87** (1983), 2996–2998.
[45] D. F. Evans, M. Allen, B. W. Ninham and A. Fouda, *J. Sol. Chem.* **13** (1984), 87–101.
[46] D. J. Mitchell, B. W. Ninham and D. F. Evans, *J. Phys. Chem.* **88** (1984), 6344–6348.
[47] U. Mohanty, B. W. Ninham and I. Oppenheim, *Proc. Natl. Acad. Sci. USA* **93** (1996), 4342–4344.
[48] R. M. Pashley, P. M. McGuiggan, B. W. Ninham, D. F. Evans and J. Brady, *J. Phys. Chem.* **90** (1986), 1637–1642.
[49] J. E. Brady, D. F. Evans, G. Warr, F. Grieser and B. W. Ninham, *J. Phys. Chem.* **90** (1986), 1853–1859.
[50] D. F. Parsons and B. W. Ninham, *J. Phys. Chem. A* **113** (2009), 1141–1150.
[51] D. F. Parsons and B. W. Ninham, *Langmuir* (2009), DOI: 10.1021/1a902533x.
[52] A. Salis, D. F. Parsons, M. Boström, L. Medda, B. Barse, B. W. Ninham and M. Monduzzi, *Langmuir* (2009), DOI: 101021/1a902721a.
[53] D. F. Parsons, M. Boström, T. J. Maceina, A. Salis and B. W. Ninham, *Langmuir* (2009), DOI: 10.1021/1a903061h.

13

Forces at work: a miscellany of issues

We have nearly finished now. To recapitulate, the philosophical question we began with was posed by D'Arcy Thompson in his book *On Growth and Form* [1]. How can we link structure and function, geometry and forces? How do we develop a Language of Shape? As D'Arcy Thompson tells us, it was axiomatic to the founders of the cell theory of biology and of physiology that progress in their disciplines would depend on advances in physical chemistry and colloid and surface chemistry. Half a century later, the molecular biology revolution took off. The expectation was that the more developed physical sciences ought to have had something to contribute to the new biology. And indeed the contributions through the bringing to bear of a suite of new experimental techniques that developed, X-ray and neutron scattering, electrophoresis, nuclear magnetic resonance, electron microscopy, ultracentrifuges and so on, are undeniable.

Although biologists and earth scientists and chemical engineers used the tools of the physicists and chemists, somehow the conceptual linkages were unclear. This was a puzzle.

But despite our collective frustrations we have, maybe, made some progress. At least some dim images of what might be a bigger picture are beginning to emerge.

13.1 Al Khemie

This can be seen perhaps if we step back and take a long view of colloid and surface science, of Al Khemie and the chemistry of the Egyptians. This, a main foundation of the then economy, was directed to the preservation of mummies for an afterlife. The astronomy and formulations of the Egyptian priests formed the basis of Greek science. The Egyptian motivation is an argument of sorts for the benefits of directed, applied research. See one of Bertrand Russell's favourite authors, the polymath John William Draper [2]. Russell and Whitehead faced a problem similar to ours, when they attempted in *Principia Mathematica* to rigorize mathematics and remove its inconsistencies. In the end they failed, and Russell gave up mathematics [3]. George Smith of the British Museum went down to Nineveh after the stupendous discoveries by Layard in Mesopotamia. Smith found the stories of Genesis and of Noah of the Old Testament in the rubble of the library of the palace of King Ashurbanipal. The stories on clay tablets, in cuneiform, date back to the earlier Sumerian civilization ca. 4000 BC. Smith made his discoveries around the same time that Hofmeister was doing his experiments on specific ion effects in the 1870s. One such 'volume' was rediscovered 30 years ago by David Tabor. It was stored away in the Ashmolean Museum at Oxford. Tabor had it translated for Zettlemoyer's 60th-birthday Festschrift, writing a paper entitled 'Babylonian lecanomancy: an ancient text on the spreading of oil on water' [4]. (Interestingly again, it was just at that time that Tabor, with his student Israelachvili, was pioneering the first direct measurements of molecular forces.) In what is probably the earliest known 'scientific paper', the Sumerian manuscript dealt with the spreading of oil on water, linked, as we know, to molecular forces. The subject has come back into vogue as petroleum engineers, as a matter of urgency, and environmental chemists have become aware that the matter needs to be taken up with a vengeance. It is a crucial matter in enhanced oil recovery where specific ion adsorption affects the wetting properties of (microporous) rock surfaces, a key to optimizing oil recovery by reservoir flooding. The Sumerian priests, however, were motivated by the need for divination by lecanomancy.

A more recent important formulation for applications in war is the recipe for Greek Fire (due to Marcus Graecus, tenth century, quoted by John Julius Norwich [5]): 'Take pure sulphur, tartar, sarcocolla (Persian gum), pitch, dissolved nitre, petroleum (obtainable from surface deposits in Mesopotamia and the Caucasus) and pure resin; boil these together, then saturate tow with the result and set fire to it. The conflagration will spread, and can be extinguished only by wine, vinegar or sand.' This recipe was closely guarded and crucial to the survival of 1000-year Roman (Byzantine) and Greek civilizations.

Microemulsions made from ammonium nitrate and napalm are essential modern-day explosives for the mining industry and for war. The Byzantines used colloid science regularly, and to great effect, in a different connection. They were much devoted to neutralizing opposition by blinding their enemies or families. In the most famous such extravaganza Gibbon, in volume 3 of *The Decline and Fall of the Roman Empire* [6], recounts how 100 000 captured Bulgarians were all blinded except for every 100th man, each left with a single eye to lead them all back home. The method is simplicity itself: one puts the head of the subject under a towel over a basin of hot vinegar for half an hour. This is a specific ion effect. The transparent cornea made up of ordered proteins is disordered and denatured by the ion exchange of the natural counterions of the bathing tear medium by acetate ions. The process can probably be used more positively in understanding cataracts.

Another not so useful formulation: ‘Boil and bubble, toil and trouble? . . . ’, from witches formulating spells. Formulation uses toad bladders, lizards tongues, salt, vinegar, but full recipe lost. Quoted by William Shakespeare, *Collected Works, Macbeth*.

Benjamin Franklin at the time of the French revolution and the American war of independence was reputed to have worked out the size of a molecule by spreading oil on Clapham pond in London in a famous experiment. In fact he got bored and went to visit some ladies at a local pub. An Englishman near the turn of the nineteenth century did the experiment properly, and reported what Franklin would have got had he stayed.

Useful formulations:

Housewife to husband – recipe for removing stains: Try vinegar, you idiot!

Cleaning and the denaturing of proteins by innumerable household recipes, sugar soap, bleaches, sodium bicarbonate, bactericides and so on are so familiar that we do not think of or analyse them in terms of specific molecular forces, mainly because these forces were not understood. The dewatering and drying of codfish with lye (sodium hydroxide) was the basis of the wealth of the Hanseatic League for centuries and more than any other factor led to the rise of the Nordic civilizations.

Mother Gardiner’s recipe for washing woollens: take grated bar soap and water. Add methylated spirits and a small quantity of eucalyptus oil to produce a foaming liquid. Washes woollens better than any present product (Australia, nineteenth century, still available in supermarkets).

The controlled burning and distilling of eucalyptus trees in Australia was a major industry for a century, the charcoal being used for gold extraction, and the eucalyptus oils used for mineral extraction by froth flotation, adsorption thereof rendering the desired ore particles hydrophobic.

‘Essential oils’ extracted principally from Australian and New Guinean trees are currently in much demand, typically for cosmetics, skin diseases and control of head lice in ‘alternative medicine’. More or less magical properties are attributed to such cures. The industry takes no account of the fact that the first point of call for mixtures of such, usually toxic, short-chained hydrophobic molecules has to be cell membranes, and that they are not metabolized.

An example that illustrates one of the surface chemical techniques to which nature puts such oils is the Australian sawfly larva. This ancient parthenogenic wasp saws open a leaf and lays its eggs in a perfect hexagonal array. The insects hatch into ugly black caterpillars that voraciously consume the entire foliage of the tree. Mission accomplished, they proceed head to tail, in convoy, down the tree trunk and on to the next tree. When finished they go back into the ground in hexagonal array again to pupate and emerge later as adults. How they process the toxic oils is seen when the migrating caterpillar columns are attacked by birds. The caterpillars rise up as one covering themselves with a clear microemulsion gel formed of water, lipids and toxic oils that have been segregated for precisely this additional defensive purpose.

13.2 Wishing reason upon the ocean!

Jan Morris had this to say on trying to make sense of the complex jumble that formed the British Empire [7]: ‘If to the Queen herself all the myriad peoples of the Empire did seem one, to the outsider their unity seemed less than apparent. Part of the purpose of the Jubilee Jamboree was to give the Empire a new sense of cohesion: but it was like wishing reason upon the ocean, so enormous was the span of that association, and so unimaginable its contrasts and contradictions.’

Wishing reason on the ocean seems an apt enough form of words that encapsulates the difficulties that have faced colloid science and surface chemistry in trying to make sense of the real world.

These examples represent real world chemistry. Typically such recipes lack any predictive principles. Another phenomenon, apparently extremely simple, lies in the ion-specific bubble–bubble interaction experiments of Chapter 8. They are apparently incomprehensible in terms of established theories, beyond the statement and listing of facts. We can expect some sense will be made of these when the compounding and confounding faults of classical theories are removed. A better quantitative understanding of the interplay between specific ion effects, local hydration and dissolved gas is emerging.

A better analogy to the situation we faced, and face still, is encapsulated in the Rochamadour phenomenon. Rochamadour is a place along the route to the shrine of St. James of Compostela. It has been visited for centuries by millions of pilgrims. This ancient and famous site is reached by a steep stone stair path from the banks of the Dordogne. Along the path are the stations of the Cross. It is said that Henry II of England had to do penance in the twelfth century for the assassination of Thomas à Beckett, by following the stations of the Cross on his knees. What is germane to our story is that the stone steps on the route are full of fossils. Yet it took 700 years before the emerging science of geology was able to contemplate the notion that the fossils were the remains of real unimaginably old life-forms, not just artefacts.

13.3 Drawing threads together

But there are hints of an emerging unity.

The answer became clearer and is neatly summarized in an aphorism of Stephen J. Gould, the famous writer on evolution in his book Eight Little Piggies. He said:

I have long believed that *conceptual locks* are more important barriers to progress in science than *factual lacks*.

In this book we have been re-examining the foundations in a search for some conceptual locks.

We found some, but not all, in :

1 Molecular forces: specific ion (Hofmeister) effects were a mystery and unexplained.
2 Self assembly: the language of shape in nature is often non-Euclidean (bicontinuous structures).
3 Long-range 'hydrophobic' forces turn out to be driven by many things, and depend on dissolved gas. They are cooperative, and interplay with salt specificity and 'water structure'.
4 Theories had been forgotten about dissolved atmospheric gas. This really changes everything. The entire scientific world forgot about this, except for some odd Russians who used the bubble–bubble inhibition in salt to float coal. The omission of the effects of gas is inexplicable.

13.4 Some consequences of conceptual locks

The theories were often pretty much impotent. Hofmeister specific ion effects in molecular forces, explored extensively a century ago, are ubiquitous and were not encompassed by theory. They are as important in the scheme of things as Mendel's work was in genetics. They affect matters from pH, buffers, pK_as, ion binding to membrane and proteins, to mechanisms and energetics of enzyme action (driven by effects associated with dissolved gas). We could not explain what appears to be the simplest of all experiments, bubble–bubble interactions and their remarkable specific ion- and concentration-dependence.

Apart from the missing dissolved gas effects, the 150-year-old theories of physical chemistry had made a fundamental mistake at the very beginning. It was a big, not a trivial, mistake that has underpinned everything else. The very foundations of colloid science embodied in the DLVO theory started with a separation of quantum mechanical and electrostatic forces which we have seen is invalid. Concepts such as the Born energy and Debye–Hückel theory of electrolytes, on wetting and interfacial tensions and electrochemical potentials, do not give a quantitative accounting of the specific ion effects. These are dictated by quantum fluctuation forces that were simply not included in the classical theoretical framework.

The matter goes beyond lack of predictability and acceptance of the classical theories as approximations that can be handled with adjustable phenomenological parameters. This is so for three reasons that we rehearse again because it is so fundamental a matter.

The first is that the classical theories that do not account for specific ion effects are still used to *interpret* measurements of entities as fundamental as pH, or ion binding or to set pH with buffers, or membrane and zeta potentials, or forces between surfaces. This is usually forgotten. If the underlying theory is flawed then so the meaning of such measurements is in doubt. Interpretations that are sometimes justifiable at very low concentrations become invalid at the real-world concentrations of most interest.

A second reason is this. If a theory is wrong that may not seem to be too important from a pragmatic engineering viewpoint. But the theory that 'works' for activity coefficients or pH with fitting parameters for one electrolyte cannot *predict* the properties of mixed electrolytes at high concentrations and temperatures. That is a very important issue industrially and in the earth sciences.

A third reason is that the entire intuition we have has derived from classical theories. It is deeply embedded in our collective psyche. The rigidity that attends the consequent dogmatic framework inhibits flexibility in the design and choice of experiments that may reveal surprising insights.

13.5 The tyranny of theory when theory meets reality: some examples

To illustrate we recall again some examples of how fixed ideas can indeed inhibit matters. Emulsions and single-walled vesicles of biological lipids had always been thought to be unstable. We have seen that this is not so. The spontaneous emulsions formed in the DDAB microemulsion system, the stability of droplets of oil on removal of dissolved gas and spontaneous vesicles formed by changing counterion point to something new. The idea that surface states of surfactant aggregation at an air– or oil–water interface in equilibrium with bulk microstructured phases had a limited repertoire like monolayers is too restricted. The idea that self-assembled microstructures in the bulk solution were limited to micelles, cylinders and lamellae was far too restricted. The fact that there are as many surface states of self assembly as there are for bulk microstructures has implications for membrane fusion. The fact that there are naturally occurring equilibrium states of supraself assembly changes the idea of a phase.

If we do not understand such matters we know at least how to change from one form to another by using mixed surfactants and forces, via specific ion effects, to arrive at a desired microstructured formulation.

We can arrive at a separation technology for clustered hydrophobic molecules engineered by molecular biology in bacteria by removing dissolved gas, so turning off the hydrophobic interaction. We can add specific ion effects to disperse proteins by imposing repulsive forces, and then gently order the precipitation of a protein crystal for X-ray structural analysis by titrating in a small amount of different counterion that binds more strongly.

We can do the same for polymer adsorption, for specific ion adsorption to change wettability for surfaces, and for compaction of ceramic colloids for nuclear waste disposal and other applications.

If we know that extension of the primitive model of electrolytes to include dispersion forces gives a good account of electrolyte activities, we can use that same model to give a predictive account of activities and osmotic coefficients of mixed electrolytes. If we know that the Debye length in mixed electrolytes with very small amounts of 'pixie dust' injected by multiply charged polymers is very different to what we had expected from the classical linear theory then we can better tackle better problems like bacterial adhesion.

If we understand that a measurement of pH or 'fixing' pH by buffers is not at all what we think, then we are in better shape to design experiments. Similarly for membrane potentials and ion transport across membranes.

Throughout this book we have alluded to problems in biochemistry and biology where there had appeared to be no intersection with physical chemistry. But with the bringing to bear of new insights, we have seen that we can sometimes come to a seamless intersection. At least we can hazard where to start. The story is not complete, but now we can use the modified intuition and theories with more confidence both in biology and in areas from enhanced oil recovery to minerals beneficiation. The work of P. Cremer and co-workers on Hofmeister effects in systems close to biology is an exemplar [8].

Our story goes a little further. It turns out that what we might call a giant blind spot was not restricted to colloid science. It occurred equally in the physics of molecular interactions, in quantum electrodynamics. There also classical results of over 50 years standing seem to be incorrect, and qualitatively so. One of these, on the resonance excited state–ground state interactions, is important in matters such as photosynthesis, in unravelling electron transfer from light capture, and even in mooted atom-based quantum computing.

13.6 Known unknowns

13.6.1 Microfossils and biomorphs

A recent, continuing debate concerns what we might call an inverse Rochamadour effect.

The discovery of 'microfossils' has attracted much attention. These tiny structures found in ancient rocks in Western Australia and elsewhere to all intents and purposes mimic precisely minute life forms. Their age predates the origins of life by a billion years. They are orders of magnitude smaller than known fossils of single-celled organisms. That poses a difficulty. Having been subject to enormous temperatures and pressure they contain no carbon, which would have been substituted by other elements. If indeed they were microfossils that would call for a complete revamp of our ideas on the history of the earth. However, a large body of work by García-Ruiz and Hyde and their students shows how to make the 'fossils'.

In a typical experiment, a layer of solution of 0.5 M barium nitrate is placed on top of a liquid column, and allowed to sit. Double diffusion gradients are presumably set up – the salt is heavier than the column liquid, and there are opposing density and salt-concentration gradients. Sampling of different levels of the column after a day or two reveals precise copies of all kinds of 'microfossils'. Many more different structures can be made easily with different salts and conditions [9–11].

They are generally states of self assembly that involve hierarchies of colloids, e.g. thin plates that coalesce under the influence of molecular forces into spiral super-structures in regular repeatable ways.

A great deal of work is developing on biomorph assembly and the templating of materials proper for medical applications. One of the holy grails is the synthesis from solution of amorphous apatite for bone replacements. None of this work systematically exploits effects due to dissolved atmospheric gas, CO_2 or Hofmeister effects in altering structures. Nor does it exploit effects that can be induced by adding sugar isomers or other solutes.

Much exciting work is developing on elucidating systematic hierarchies of new structures theoretically and experimentally in Refs. 10 and 12. This continues and develops much further the pioneering work of Sten Andersson, von Schnering, Nesper and Kåre Larsson on the application of minimal surfaces to structure in inorganic and lipid chemistry outlined in the book *The Language of Shape* [13].

13.6.2 Frescoes and nanoparticles

A process of extraordinary world heritage importance concerns the non-invasive restoration of frescoes and paper at the University of Florence that dates back to the pioneering work of Enzo Ferroni [14] after the Arno floods that damaged most of the art masterpieces collected in several museums, including the Uffizi collection. This work can be tracked through the website and papers of Luigi Dei, Piero Baglioni and colleagues [15–17], and is illustrated very well in *Nature* [18]. The best-known examples are the restorations of the frescoes of Piero della Francesca in Arezzo, and those of Fra' Lippi in Prato. The technique uses special microemulsions for cleaning and nanoparticles of calcium hydroxide for rebuilding the frescoes [19]. The nanoparticles are precipitated from bulk solution. The sizes of the nanoparticles can be changed by exploiting Hofmeister effects through the addition of salts. (This takes us back to Berthelot's original observations on soda lime that we started with.) More interesting is what happens with addition of sugars to the mix, and with variation of temperature. Different isomers produce different results. If increasing amounts of sugars are added in the range from 0.1 to 1 M the size and polydispersity of the nanoparticles can be tuned more or less at will. Natural gums containing special sugars were used by Mayans, evidently to control nanoparticle size for their frescoes [20]. The same Hofmeister and water structure effects on nanoparticle synthesis occur in reverse micelles where the water is all 'water of hydration'. It is clearly involved in the still-mysterious synthesis of pseudopolyrotaxanes [21].

Effects of sugars on nanoparticle growth and form might well occur in mammalian milk in the first few days after birth. The milk, colostrum, contains a high concentration of lactose, which is absent later on. One can reasonably hypothesize that the lactose–physiological medium mix has an additional job to metabolism – soluble calcium and phosphate might form nanoparticles used to seed bone growth. Nanoparticles of iron oxide of precisely the right size and magnetic properties are synthesized in bird brains to assist navigation, and again the templating synthesis probably requires the right isomeric sugar mixtures.

13.7 Water structure

We come now to a contentious issue that can induce in normally mild-mannered scientists a state of apoplexy. The term 'water structure' is as ill defined as the words 'hydrophobic' or 'hydrophilic'. There is probably a consensus that 'water structure' is short-range. It refers to an ordering of water molecules that can span only about three layers from a surface, although the definition of the word surface is often problematic. With ions this 'hydration' extends less. On this most would agree, be they experimental theoreticians engaged in computer simulation or theoreticians working from an analytic point of view who compare their results with those of experimentalists working in real science. There is a limit to the size of a simulation experiment that restricts it to small systems, and there is a problem in comparing notes as it were. This is because in the world of real science there is generally dissolved gas.

So a comparison of a simulation of interactions between hydrophobic surfaces, or of adsorption of ions at interfaces, with a molecular model for water and for ionic interactions ignores dissolved gas at interfaces and in bulk solution. The comparisons are for two different systems. The presence of dissolved gas is not a factor at high electrolyte >1 M, but strongly influences interactions at lower concentrations. The present consensus on 'water structure' might then reasonably be viewed by sceptics with a jaundiced eye.

By contrast the sceptics have adopted a heresy almost Manichean in its implications. Taking a holistic view, they imagine, and adduce evidence, that water structure can extend for literally thousands of angstroms. The arguments recur and have been presented forcefully in Ref. 22. What is envisaged here is not the capillary forces that account for the transport of water in trees and biomimetic clues for the engineering of wettability in technological applications such as microfluidics [23]. The arguments have been put most eloquently in a superb essay on the state of water in living systems: from the liquid to the jellyfish [24]. A particular problem remarked on by Henry is the problem of how a jellyfish with at least 98% water can get on with its job if the consensus view on water structure holds. It is a fair question that is not easy to dismiss. It is rendered more cogent by the knowledge that jellyfish can apparently handle fairly extreme changes in salt concentration with equanimity.

The consensus considers the advocates of very long-range water structure as mad. But they have no means to model such 'very long-range' water effects even if they were to take the matter seriously. The heretics assert what to them is self-evident truth.

If we looked for evidence that might bolster the Manicheans, we can find some clues in issues that have occurred in several phenomena that this book has been concerned with. The first is the observation that the tensile strength of liquids that contain atmospheric gas (or indeed any other weakly soluble solutes) is 1000-times less than what would be calculated from molecular potentials. The solutes behave as defects in solids. We can recall the polywater saga, discredited as due to trace impurities on glass contaminated by handling. Real biological water is always so contaminated. We can recall the real polywater phenomena with cold fusion of silica over very large distances discussed in Chapter 4. The experiments of Bunkin discussed there on the absence of laser-pulse-induced cavitation in gas free water and the different long-range structure that shows specific ion dependence are of much relevance. Then we have the bubble–bubble fusion problem at distances of hundreds of nanometers and its ion-pair dependence; and finally the measured very long-range hydrophobic interaction forces of various sorts that cannot be ascribed to polymer bridging. These forces and bubble–bubble interactions and emulsion instability, we have argued, are due to an interplay between dissolved gas and salt molecule ion-pair defects that can percolate and manifest themselves over very large distances. We have seen that removal of atmospheric gas from a suspension of hydrophobic but still charged colloids switches off hydrophobic interactions. From that (experimental) evidence, which is certainly real, the Manicheans may have a point. The consensus may be correct for clean ideal systems, and the heretics possibly correct for dirty systems that comprise the real world. We remark too that we have little understanding of dilute gels.

With thanks to Morris Kline, from whom we borrowed this quotation [3], the situation we have surveyed can be summarized in the words of Greek philosopher Xenophanes:

> *The gods have not revealed all things*
> From the beginning.
> But men seek and so find out better in time.
> Let us suppose these things are like the truth.
> But surely no man knows or ever will know
> The truth about the gods and all I speak of,
> For even if he happens to tell the perfect truth,
> He does not know it,
> But appearance is fashioned over everything.

Or else the lesson is embodied in the Jesuits' saying that *God writes straight in crooked lines.*

And with that we end our story.

References

[1] D. W. Thompson, *On Growth and Form*. Cambridge: Cambridge University Press (1917).
[2] J. W. Draper, *History of the Intellectual Development of Europe*. New York: Harper Brothers (1864).
[3] M. Kline, *Mathematics: the loss of certainty*. Oxford: Oxford University Press (1980).
[4] D. Tabor, *J. Coll. Interface Sci.* **75** (1980), 240–245.
[5] J. J. Norwich, *Byzantium, The Early Centuries*. London: Penguin (1990).
[6] E. Gibbon, *The History of the Decline and Fall of the Roman Empire*. London: Macmillan (1910).
[7] J. Morris, *Farewell the Trumpets. An imperial retreat*. London: Penguin (1978).
[8] X. Chen, T. Yang, S. Kataoka and P. S. Cremer, *J. Am. Chem. Soc.* **129** (2007), 12272–12279.
[9] J. M. García-Ruiz, E. Melero-García and S. T. Hyde, *Science* **323** (2009), 362–365.
[10] http://wwwrsphysse.anu.edu.au/~sth110/sth_papers.html

[11] J. M. García-Ruiz, S. T. Hyde, A. M. Carnerup, A. G. Christy, M. J. van Kranendonk and N. J. Welham, *Science* **302** (2003), 1194–1197.
[12] S. T. Hyde, M. O'Keeffe and D. M. Proserpio, *Angew. Chem. Int. Ed.* **47** (2008), 7996–8000.
[13] S. T. Hyde, S. Andersson, K. Larsson, Z. Blum, T. Landh, S. Lidin and B. W. Ninham, *The Language of Shape. The role of curvature in condensed matter physics, chemistry and biology*. Amsterdam: Elsevier (1997).
[14] E. Ferroni, Environment and conservation of cultural heritage. In *Ecological Physical Chemistry*, ed. C. Rossi and E. Tiezzi. Amsterdam: Elsevier (1991).
[15] http://www.csgi.unifi.it
[16] P. Baglioni, R. Giorgi and L. Dei, *Comptes Rendus Chimie* **12** (2009), 61–69.
[17] P. Baglioni and R. Giorgi, *Soft Matter* **2** (2006), 293–303.
[18] P. Ball, *Nature News* (12 Jul 2001), DOI: 10.1038/news010712-8.
[19] M. Ambrosi, L. Dei, R. Giorgi, C. Neto and P. Baglioni, *Langmuir* **17** (2001), 4251–4255.
[20] R. Giorgi, D. Chelazzi, R. Carrasco, M. Colon, A. Desprat and P. Baglioni, The Maya site of Calakmul: 'in situ' preservation of wall paintings and limestone by using nanotechnologies. In *The Object in Context: Crossing Conservation Boundaries, Munich IIC – International Institute for Conservation Congress 2006: Proceedings*, ed. D. Saunders, J. H. Townsend and S. Woodcock. (2006), 162–169.
[21] P. Lo Nostro, J. R. Lopes, B. W. Ninham and P. Baglioni, *J. Phys. Chem. B* **106** (2002), 2166–2174
[22] G. H. Pollack, *Cells, Gels and the Engines of Cells*. Seattle: Ebner and Sons (2000).
[23] M. M. Kohonen, *Langmuir* **22** (2006), 3148–3153.
[24] M. Henry, *Cell. Mol. Biol.* **51** (2005), 677–702.

Index